DATE DUE

MAY 7 82			

DEMCO 38-296

Modeling and Simulation in Science, Engineering, & Technology

Series Editor
Nicola Bellomo
Politecnico di Torino, Italy

Advisory Editorial Board

Mathematics of Climate Modeling

Valentin P. Dymnikov
Aleksander N. Filatov

Birkhäuser
Boston • Basel • Berlin

QC981 .D94 1997
Dymnikov, V. P.
Mathematics of climate
modeling

117334 Moscow, Russia

Aleksander N. Filatov
Hydrometeorological Center of Russia
Bolshoj Predtechenskij Pereulok
123242 Moscow, Russia

Printed on acid-free paper
© Birkhäuser Boston 1997

ISBN 0-8176-3915-2
ISBN 3-7643-3915-2
Formatted by the authors in LATEX.
Printed and bound by Maple Vail, York, PA.

9 8 7 6 5 4 3 2 1

CONTENTS

Preface

The present monograph is dedicated to a new branch of the theory of climate, which is titled by the authors, "Mathematical Theory of Climate." The foundation of this branch is the investigation of climate models by the methods of the qualitative theory of differential equations. In the Russian edition the book was named "Fundamentals of the Mathematical Theory of Climate." Respecting the recommendations of Wayne Yuhasz (we are truly grateful to him for this advice), we named the English edition of the book "Mathematics of Climate Modelling." This title appears to be more appropriate, since the constructive results of the theory are at present preliminary and have not been fully tested with experiments in climate modelling. This branch of science is yet developing and its practical results will be obtained only in the near future.

Nevertheless, we want to keep the terminology which we have used in the introduction to the Russian edition of the book, since the authors hope that this term will be accepted by the scientific community for identification of a given branch of climate theory.

On preparing the English edition, new ideas were established connecting some significant new research results obtained by the author.

We are deeply grateful to G. Marchuk for continual encouragement of this scientific enterprise and fruitful discussions, to our young colleagues A. Gorelov, E. Kazantsev, A. Gritsun, and A. Ilyin, who worked with us on the problem of the mathematical theory of climate, and to Dr. V. Shutyaev. We thank M. Rhein for preparing the preliminary translation of the book, and V. Kazantseva, who did the main part of the work on the translation and preparation of the manuscript for printing.

V. Dymnikov, A. Filatov

Introduction

Every new terminology as well as every new theory must stand the test of time. Because the term "mathematical" as applied to the theory of climate is not yet completely accepted, we must explain to the reader what we associate with it.

In [29,31] the mathematical theory of climate is defined as a branch of the theory of climate treating of the laws of the behavior of the climatic system trajectories by the methods of qualitative theory of differential equations. It may be that such a definition gives the reader not familiar with these methods poor information, for this reason we shall reformulate the above definition in the following way:

The mathematical theory of climate is a branch of the theory of climate, which investigates the behavior of the climate models solutions on the arbitrarily large time-scales by the use of a collection of mathematical methods.

As is well-known, the general theory of climate deals with the evolution of climatic system, i.e. the joint evolution of the atmosphere, the ocean, the criosphere and the biosphere.

The climatic system can be regarded from different points of view and studied by a variety of techniques. For example, it can be studied by the statistical method using the multi-year observed data or by investigation of the physical processes responsible for the climate formation or, finally, by the modelling of climatic system, producing the mathematical model of climatic system on the base of the thermohydrodynamics laws. It is just this aspect which is of much interest to us and we shall discuss it further.

The question how much one or another model of climate is correct, i.e. how much the model is suitable for the climate changes prediction, presents the important problem of the theory of climate. We shall discuss this problem below.

It should be pointed out that in the last decade the development of the general circulation coupled atmosphere-ocean models is, undoubtedly, one of the main directions of the climate modelling. For the considered climate models we shall assume that the following hypotheses are true:

1. *Evolution of the climatic system is determined.* This means that for the vector of the climatic system parameters there exists a mapping $f : X(t) \to X(t + \tau)$, $t \geq 0, \tau \geq 0$ allowing us to determine the state of the system at instant $t + \tau$, on condition that the state at instant t is known.

2. *The mapping f* is the solution of some "ideal" hydrodynamical model of the nature and the observed data are a realization of the solution of the "ideal" model at discrete time.

It follows from these hypotheses that the evolution of climatic system can be described with sufficient for practice range of accuracy by the system of thermohydrodynamic equations and that the models of climatic system must have a global solution.

This means that for the solution of the system of equations governing the given climate model the uniqueness and existence theorems must be valid for all t. These theorems are very important to the mathematical theory of climate. We notice that the above hypotheses do not forbid to consider a stochastic description of the climate in the following sense.

Initial data and forcing (and possibly other parameters of the system) can not be determined, but may be presented by their probability distributions. In this case we study the statistical solutions of climate models. However, if the mentioned parameters are known precisely, the climate models are said to be deterministic, although they may have a complex (chaotic) behavior.

Let us take one more hypothesis, which is true practically for almost all climate models.

3. *Climate models belong to the class of nonlinear dissipative systems.* A system is dissipative one, if it possesses an absorbing set. The absorbing set is a subset of the phase space characterized by the following property: any orbit of a system will get into this set sometime and remains there forever.

The absorbing set is an infinite-dimensional set. Inside the absorbing set trajectories of dissipative system move to an attracting set named the attractor. The attractor is already a finite-dimensional set.

Thus, the dissipative system goes into some finite-dimensional subset of the phase space after a certain (maybe large) time interval and further evolution of the system is actually realized on this subset.

In many cases the dissipative systems have the inertial manifolds. The inertial manifold is the finite-dimensional smooth invariant manifold exponentially attracting all the solutions of the system. The inertial manifolds contain the global attractor.

Thus, we shall proceed from the fact that a climate system is an infinite-dimensional system. This brings up the question: what functionals of the model solutions describe the real climate and which of them must be compared with observed data.

Since the dynamics of climatic system develops (in time) entirely either on the attractor or on the inertial manifold, the climatical characteristics, as characteristics of the final motions, will be one way or another associated with the characteristics of attractors, inertial manifolds and external forcings.

It is possible that these characteristics will help us to display the usefulness of the studied model for the prediction of climate changes, which is one of the crucial problems of the mathematical theory of climate.

It should be noted that we shall deal with followings models:

1) deterministic models (possibly with a complex chaotic behavior of solutions);

2) stochastic models having either initial data or forcing (or both) given as some probability distributions.

The mathematical theory of climate goes back to the classic work of Lorenz [95]. By now many works have been dedicated to the investigation of the Lorentz attractor structure, however, the number of them does not decrease. The results of some works will be considered in the last chapter of the monograph.

Let us describe the main characteristics of attractors (and of the inertial manifolds). They are: 1) dimension, 2) structure, 3) stability, 4) probable motions on attractors and inertial manifolds, 5) probability measure on the attractor, and ergodicity. We dwell on these characteristics briefly.

Dimension shows how much the dynamics of the given infinite-dimensional climatic system is finite-dimensional. Because of this, it is important to obtain the more precise estimates of the Hausdorff dimension of the attractor.

Attractor structure describes the lattice, i.e. the geometry of the attractor.

Stability of attractors and inertial manifolds just as structural stability, so Lyapunov stability determines the sensitivity of attractors and inertial manifolds to the variations of the system parameters. If parameters of the system (forcing, viscosity coefficient, and etc.) vary slowly, we can observe, in some sense, continuous dependence of variations of an attractor on variations of the above parameters. However, if parameters of a system reach some critical values, such a continuous dependence can be broken and bifurcation will take place in the system. Then the original attractor disappears, and, instead of it, a new one will arise, it has different dimension, properties and structure.

The Lyapunov stability of the attractor is closely connected with the possibility of construction of such spectral-difference approximations of the climate model, which will conserve, in some sense, a closeness between the attractor of the original model and the attractor of finite-difference approximation of this model.

Analysis of possible motions on the attractor is connected with the possibility of the consideration of the attractor as a phase space of dynamical system arising on the attractor due to the extension of the solutions on it for all $t \in R$, $R = (-\infty, +\infty)$. This phase space will be a compact metric space.

Probability measures on the attractor appear in deterministic case and when the probability distributions of the forcing or the initial conditions are given. If the initial conditions are not determined and some probability distribution of them is given, in process of evolution on the attractor the distribution arises, which is characterized by the invariant probability measure. Practically, there occurs "the forgetting" of the initial distributions. This leads to the necessity of the detailed consideration of the invariant probability measures on the attractor.

It is worthy to note that *in deterministic case* on the attractor we can always construct, at least, one invariant measure on the attractor by using the compactness of the attractor and by applying the Krylov–Bogolyubov theory. But there is a question: which of the constructed measures is essential, i.e. which of them is consistent with observed distributions. The problem is not so simple as it seems to be at first glance. The matter is that we do not possess the constructive algorithms to construct the above measures. By virtue of the fact that there is the invariant probability measure on the attractor, the recurrent theorems of Poincaré and the first Birkhoff theorem must be fulfilled, and if the system is ergodic, the second theorem of Birkhoff must hold also. It is just these of attractor characteristics we shall take into account when speaking about the "climate" of the system.

If we remember the traditional definition of the climate as the ensemble of states which the system passes through in a sufficiently large time interval, then one can say that the most adequate to it characteristic is the essential invariant measure concentrated on the attractor or, in the case of time-independent forcing, it is the statistical stationary solution associated with the essential invariant measure by the existence and uniqieness theorems.

We are coming now to the problems of approximation of climate models and their numerical realization. For this purpose it is convenient to distinguish the climate models of three levels.

The first level model is a set of the partial differential equations describing the climatic system with respect to the initial and boundary conditions. All the aforesaid about the climate models can be referred to the first level models. Models of this class are developed at the present time for the study of different problems concerning the climate. It is important to note that a first level model represents an infinite-dimensional system, because its solution by the Fourier-series expansion in some set of functions reduces to the solution of an infinite system of nonlinear ordinary differential equations.

However, it is hardly probable that the solution of these systems of equations can be obtained analytically. Because of this, the methods of numerical integration are naturally applied here. In this case we replace the spatial and time derivatives by their finite-difference approximations or look for a solution by applying the Galerkin approximations. In both cases we replace the infinite-dimensional system by some its finite-dimensional approximation.

The second level model is any finite-dimensional approximation of the first level model. For a performance of numerical computations on the second level model we need a set of program codes, which will realize this model at the concrete computer.

The third level model is a system of programs realizing the second level model at the computer of one or another class.

Let us consider a connection between the first level model and the second level model. The main problem here is to answer the following questions: 1) if we observe some phenomenon in the second level model, can we hope that the analogous phenomenon will be observed in the first level model, 2) alternatively, if there is some phenomenon in the first level model, can we be assured that the same phenomenon will be observeded in the second level model.

It is important to know, which properties of the first level model are approximated by the solutions of the second level models on the arbitrarily large time-scales. The crucial moment by the analysis of the closeness of the solutions generated by the models of the first and second levels is the fact that the constants in a priori estimates of the closeness of solutions are the functions of the time-interval T, on which, as a rule, the considered solutions go to the infinity for T tending to the infinity. Since investigating the climate we deal with large time-scales (in the limit as $T \to \infty$), the concept of the closeness of two climate models as a concept of the closeness of the solutions (in sense of smallness of some difference of solutions norm) loses its sense. We must formulate new methods of approximation for the climate models and on the base of these methods consider a problem of the predictability of the climatic characteristics.

It is obviously that the traditional climatologists will positively accept the mathematical theory of climate only in the case when there will be the constructive results obtained which will be useful for other branches of climate theory also.

By convention the general theory of climate can be divided into five branches.

1. Statistical theory of climate. The aim of this branch is the description of climate by modern statistical methods on the base of the measurements data (or results of four-dimensional analysis of data).

2. Physical theory of climate which subject is the study of the physical processes responsible for the climate formation.

3. Hydrodynamical theory of climate. The prime objects of this branch are the study of linear and nonlinear wave processes taking place in the climatic system and the study of their stability.

4. Mathematical theory of climate. The aim of this theory is the analysis of the climate models solutions behavior on the arbitrarily large time-scales.

5. Numerical modelling of climatic system. The purpose of this branch is a development of models, which are able not only to describe the current climatic system, but, in some sense, to predict the climate changes by the changes of external forcing.

It follows from this classification that mathematical theory of climate is the connecting link between the first three branches and the last one, which provides the answer to the fundamental question: what means the phrase "is predictable in some sense".

We will briefly describe the content of some chapters.

Chapter 1 contains the introduction to the general theory of the nonlinear dissipative dynamical systems. A good deal of attention is given here to the theory of invariant measure concentrated on the attractor. It is the author's opinion that the theory of the essential invariant measure concentrated on the attractor is one of the most important branches of mathematical theory of climate.

Such a detailed description of the theory of dissipative systems is given not for the convenient reading of this book only (especially for the scientists on geophysical hydrodynamics and the theory of climate, whom the book is dedicated to), but also because the authors hope that the book can be accepted as a fundamental for the special course of lectures on the mathematical theory of climate in the uneversities preparing students in geophysical hydrodynamics and the theory of climate.

In Chapter 2 the theory of the nonautonomous nonlinear dissipative systems is presented. Such a system arises, for example, when the external forcing obviously depends on the time and when the coefficients or the model parameters vary with time.

Autonomous systems are connected to the semigroup of nonlinear operators, nonautonomous systems are connected to the processes and the family of processes. Process is a natural generalization for the notion of the semigroup. On considering attractors of the nonautonomous dissipative systems we make accent not on the invariance of the attractor, but on such its properties as the attraction and the minimality.

For the nonautonomous systems the concepts of the kernel and section of kernel are used. The kernel and the section of the kernel are natural generalizations for the concepts of the invariant set for the nonautonomous nonlinear dissipative systems. Our primary interest in this chapter is in the methods of the averaging in Bogoluybov sense.

By these methods the some relation between autonomous and non-autonomous system can be established. For the sufficiently general nonlinear systems the theorem on the averaging on the finite interval is proved. The closeness of the attractors of the initial and averaged systems is established.

Chapter 3 contains a detailed analysis of the barotropic model of the dynamics of the atmosphere. We believe that such a level of description of the climatic system is simplest but at the same time it has all characteristic properties of the first level model (nonlinear dynamics, infinite dimension of phase space, description of characteristic atmosphere processes etc.). For this model it is proved the existence of the global attractor and the estimate is given of its Hausdorff dimension with respect to the orography and Rayleigh friction in the boundary layer. Here the estimate of the dimension of the attractor of the corresponding Galerkin approximations is given and its dependence on the order of approximation is pointed out. Further, statistical solutions of model under consideration are studied.

The existence of the invariant probability measure on the attractor is proved. Since the attractor is a compact set, so the existence of invariant measure on the attractor permits to assert that almost all the motions on the attractor are stable according to Poisson. If the dissipation operator of considered model has a power more than two, in that case, the existence of the inertial manifold can be proved. (Note, that recently in works [86, 126] the existence of the inertial manifold has been proved also for the case, when the power of the dissipative operator is equal to 2). In dependence on the power of the dissipative operator of the model, the estimate of the dimension of the attractor is obtained.

Chapter 4 is concerned with the general theorems on approximations for the infinite-dimensional dissipative systems (on the space and time). These theorems are very essential from the point of view of the keeping by the second level models of the metric and topological invariants of the first level models.

In Chapter 5 the results of the numerical investigation of the structure of attractors generated by the barotropic equations on the sphere are presented. The particular attention is given to the analysis of such a situation when the invariant measure is concentrated in the neighborhood of the stationary points. This situation permits us to introduce the concept of the quasistationary regimes of the atmosphere circulation. (The theory of regimes of the atmosphere circulation corresponding to different definitions of regimes is the object of Chapter 8.) In the chapter there is given the analysis of the applicability of the analytic estimates of the dimension of the attractor of atmospheric circulation barotropic models both for the different asymptotical cases and for the real atmospheric conditions.

In Chapter 6 we derive the results for the two-layer baroclinic model which is an intermediate stage between barotropic and general primitive baroclinic equations. Here the existence of the global attractor for the model under consideration is proved. The estimate of the attractor dimension is obtained. From this estimate we can see the contribution of the barotropic and baroclinic components in the attractor dimension. In the last section of the chapter there are given the results of the numerical investigations of the characteristics of the attractor of two-layer baroclinic model for the different values of forcing which corresponds to the different regimes of the atmospheric circulation.

Chapter 7 is dedicated to the problem of the study of the structure of the attractor of the "ideal" climate model by the series of observed data. This problem is the crucial one, since we do not have at our disposal an "ideal" climate model, but the identification of the climate models for the purposes of the climate changes prediction is the key task. The existing methods for the reliable estimate of the dimension of the attractors for their sufficiently large values require such long data series that for the class of models under study these methods are practically not applicable. Therefore in this chapter one studies alongside with the Grassberger–Procaccio method the connection between the attractor dimension of the systems with the chaotic behavior of the trajectory on the attractor and the number of statistically independent degrees of freedom. The numerical analysis is carried out for the above-mentioned two-layer baroclinic model of the atmosphere.

Chapter 8, as it was mentioned above, provides the new approach to the problem of classification of the atmospheric regimes. In the chapter there will be formalized the concept of the atmospheric circulation regime. Particular emphasis is placed here on description of the two-regime circulation corresponding to the two-modal density distribution function in the space of states. From our point of view such a situation may appear only on the attractors of systems possessing the low dimension. This problem also is discussed in Chapter 6. The results of the numerical experiments on the modelling of the regimes of the atmosphere circulation are presented.

Finally, in Chapter 9 the recent theory of the dissipative systems generated by the primitive equations of the thermohydrodynamics of the atmosphere and the ocean is described. The main results of this chapter are connected first of all with the existence and uniqueness theorems in general for the models of atmospheric and oceanic circulation. We note that the scientists of the Siberian school have the priority in this branch of the mathematical theory of the climate. The development of the general theory for the models of such a level is also one of the goals of the mathematical theory of climate.

Chapter 1

Dynamical Systems. Attractors, Invariant Measures

In the development of the general theory of climate it is convenient to abstract from the specific form of one or another model of climate and consider such their properties which are typical of certain sufficiently wide classes of the models. From our standpoint the climate model is a system of partial differential equations with the corresponding boundary and initial conditions.

As is well known, it is common practice to classify systems of differential equations as autonomous and nonautonomous ones. Nonautonomous systems will be considered in the next chapter. We now shall take a close look at autonomous systems. Among autonomous systems the two following classes are of interest to us:

1) systems whose solutions are determined for all t, where t is time, i.e. $t \in R$, $R = (-\infty, +\infty)$;

2) systems whose solutions are determined for all $t \geq \tau$, where $\tau \in R$ is a certain real number.

One usually takes $\tau = 0$ and considers systems with solutions determined for all $t \geq 0$. From the physical point of view the difference between these two systems is following.

In the first case, if we know the state of the system at any given instant of time $t = t_0$, then we can determine its state at another given instant of time just as in the past, i.e. if $t < t_0$, so in the future, i.e. if $t > t_0$. In the second case we can determine only the future states of the system, i.e. if $t > t_0$, when the state of the system at $t = t_0$ is known.

The conservative systems belong to the first class and *the dissipative* systems belong to the second class. Models of climate fall into the class of nonlinear dissipative systems. However, this does not necessarily mean that one must restrict oneself to the study of such systems only.

It is known that many of dissipative systems possess an attractor. As is shown in [88], the solutions on the attractor are already determined for all t. Because of this, both of cases are important for our purposes. One usually studies systems with unique solvability. Such systems possess a solution which is not only exists for all t or for all $t \geq 0$, but is also unique, i.e. it is uniquely determined by its initial state. The solutions of such systems possess the property of great importance – so-called *group property*. The group property of solutions of autonomous systems and the property of continuous dependence of solution on the initial state and time (that takes place, as a rule, in models under study) represent two main elements of the concept of *dynamical* system.

In other words, the climate models, which we intend to study, must generate a dynamical system in a phase space, if the solution is determined for all t, or a semidynamical system, if the solution is determined for all $t \geq 0$. (One uses often the terms "flow" and "semiflow" to indicate that the variable t varies continuously.)

Nonlinear dissipative systems generate a semidynamical system in the corresponding phase space. Nevertheless, the term "semidynamical system" (or "semiflow") is used in the theory of these systems more rarely than the term "semigroup", which is generated by the initial boundary value problem for one or another set of equations. All the results of this theory are formulated in terms of the semigroups of continuous operators generated by the corresponding problem and acting in the phase space of the system.

In this chapter we give some necessary information from the theory of dynamical and semidynamical systems. We note, that this is very important for an understanding of the results given in the book, since this chapter contains the terms, main concepts, definitions, propositions, which will be repeatedly used in the subsequent chapters. We are coming now to the consideration of autonomous systems. It will be shown that the solutions of such systems possess the group property. Let us consider the Cauchy problem for the evolution equation

$$\partial_t u = F(u), \, u \mid_{t=t_0} = u_0, \qquad (1.1)$$

where F is a nonlinear operator from the Banach space B_1 into the Banach space B_0 and it is assumed that $B_1 \subseteq B_0$. Let there exists a Banach space $B, B_1 \subseteq B \subseteq B_0$ such that problem (1.1) has a unique solution $u(t) \in B$ for all $t_0 \in R$ and for all $u_0 \in B$ which is determined for all t. (At the moment it is of little importance in what sense one understands the solution of the problem (1.1)). Introduce a notation for this solution:

$$u = f(t, t_0, u_0), \ t \in R, \ u_0 \in B.$$

Replacement $t - t_0 = \tau$ brings (1.1) to the form

$$\partial_\tau u = F(u), \quad u \mid_{\tau=0} = u_0. \tag{1.2}$$

The solution of this problem takes the form $u = f(\tau, 0, u_0)$. Since the right hand sides of equations (1.1) and (1.2) coincide and the solutions of these equations pass through one and the same point u_0, the solutions of these systems must coincide by uniqueness:

$$u = f(t, t_0, u_0) = f(\tau, 0, u_0) = f(t - t_0, 0, u_0) \equiv f(t - t_0, u_0), \tag{1.3}$$

i.e. the solution of (1.1) contains parameters t and t_0 in combination $t - t_0$. Let $u_1 = f(t_1 - t_0, u_0)$. Then the solution of (1.1) with initial condition $u \mid_{t=t_1} = u_1$ will be

$$u = f(t - t_1, u_1). \tag{1.4}$$

Since the solutions (1.3) and (1.4) coincide for $t = t_1$, they will coincide for all t, i.e.

$$f(t - t_1, u_1) = f(t - t_0, u_0).$$

Substituting the expression for u_1 into this equality, we find

$$f(t - t_1, f(t_1 - t_0, u_0)) = f(t - t_0, u_0).$$

Setting $t_1 - t_0 = s$, $t - t_1 = \sigma$, we obtain

$$f(\sigma, f(s, u_0)) = f(\sigma + s, u_0), \quad \forall \ \sigma, s \in R. \tag{1.5}$$

The equality (1.5) is the group property of solutions of the problem (1.1). If the solutions of (1.1) are determined only for $t \geq t_0$, then we rewrite (1.1) in the form of (1.2) and find the solution $u = f(\tau, u_0)$, for all $\tau \geq 0$ and for each $u_0 \in B$. In this case the solution is usually written in the following form

$$u(\tau) = f(\tau, u_0) = S(\tau) u_0,$$

where $S(\tau)$ is the nonlinear operator from B onto B. By virtue of the property (1.5), the family of operators $S(\tau) \tau \geq 0$ is a semigroup

$$S(\sigma) S(s) u_0 = S(\sigma + s) u_0, \quad \forall \ \sigma \geq 0, \ s \geq 0, \quad S(0) u_0 = u_0. \tag{1.6}$$

For this reason it is natural to study these systems in terms of semi-groups of nonlinear operators.

Before proceeding to the presentation of the theory of dynamical systems and the theory of nonlinear dissipative systems we want to emphasize the essential role of the phase space in which the semigroup generated by this nonlinear dissipative systems is acting.

By convention the theory of dynamical systems can be classified into three branches:

1) theory of smooth dynamical systems (differential dynamics);

2) topological theory of dynamical systems (topological dynamics);

3) metric theory of dynamical systems (ergodic theory, dynamical systems with invariant measure).

The theory of smooth dynamical systems assumes that the phase space of dynamical system is a smooth manifold. Since the global attractor of dissipative system, in general case, is not a manifold, it is unlikely that the results of this theory can be used for our purposes. (Although, for dissipative systems with inertial manifold the results of this theory can be applied, evidently, for an analysis of the behavior of the system solutions on inertial manifolds.)

In the topological theory of dynamical systems the phase space as a rule is a complete metric space and it is often compact. From our point of view, the results of this theory can be used wholly, because a global attractor is a compact invariant set and it can be considered as a phase space of the corresponding dynamical system. In this case we obtain a dynamical system, the phase space of which is a compact metric space.

The metric theory of dynamical systems assumes that on the phase space of dynamical system a measure is given (often it is a probability measure) which is invariant relative to the dynamical system on the phase space. The phase space itself is a sufficiently arbitrary space, it may be a metric space.

For certain classes of the climate models it is proved that on the attractor in the natural way the invariant probability measure arises, if some probability distribution of initial data is given. In this case the attractor can be considered as a compact phase space with a probability measure, which is invariant relative to the dynamical system on the attractor. For this reason the results of the metric theory of dynamical systems on the whole are adaptable to our case.

If we consider nonautonomous problem of the type:

$$\partial_t u = F(t, u), \quad u\mid_{t=\tau} = u_\tau, \ t \geq \tau, \ \tau \in R,$$

then its solution can be written as follows

$$u(t) = U\left(t, \tau\right)u_\tau,$$

where $U(t, \tau)$, $t \geq \tau$, $\tau \in R$ is a two-parametric family of operators acting in the Banach space E. This family is said to be the process associated to the above problem. Here the main question is to study the behaviour of $u(t)$ as $t - \tau$ tends to infinity.

The theory of the processes, and the theory of the family of processes and that of families of semiprocesses will be presented in Chapter 2.

1.1 Metric Spaces. Compactness

In relation to the above-said, it will be helpful if we recall some information about metric spaces, in particular, about the compact metric spaces (see, for example, [82]), before we begin to study a dynamical system.

A metric space is a pair (X, ρ) consisting of a set X (called space) and a distance $\rho(x, y)$ between each elements x and y of X, where $\rho(x, y)$ is a two-argument positive real-valued function satisfying the following conditions:

1) $\rho(x, y) = 0$ if and only if $x = y$;
2) $\rho(x, y) = \rho(y, x)$, $\forall \, x, y \in X$;
3) $\rho(x, z) \leq \rho(x, y) + \rho(y, z)$, $\forall \, x, y, z \in X$.

Any set $Y \subset X$ with the same metric ρ is a metric subspace (Y, ρ) of the space (X, ρ). Henceforth, the metric space will be denoted by X. An open ball in a metric space is a set of the form $K_r(a) = \{x : \rho(x, a) < r\}$, where a, the center of the ball, is a point of X and r, the radius, is a positive real number; a closed ball is a set of the form $\overline{K_r(a)} = \{x : \rho(x, a) \leq r\}$ and a sphere is a set of the form $S_r(a) = \{x : \rho(x, a) = r\}$. Diameter $d(A)$ of the set $A \subset X$ is defined as

$$d(A) = \sup_{x, y \in A} \rho(x, y).$$

The set $A \subset X$ is said to be *bounded* if its diameter $d(A) < \infty$.

A sequence of points x_n, for $n = 1, 2, \ldots$, of the space X is said to be *convergent*, if there exists a point $x \in X$ such that $\rho(x, x_n) \to 0$, as $n \to \infty$.

The point $x \in X$ is called *a limit point* of the set A, if there exists a sequence of points x_n, for $n = 1, 2, \ldots$, $(x_n \in A, n = 1, 2, \ldots)$, which converges to the point x. (The limit point x itself may or may not belong to the set A.)

Each point of the set A is its limit point. The set $\bar{A}$ of all limit points of the set A is said to be *a closure* $\bar{A}$ of the set A. The set F is said to be *closed*, if it contains all its limit points. A complement $G = X \setminus F$ of the closed set F is called *an open* set.

Hence, an open set may be defined in the following way. The set $G \subset X$ is said to be an open set, if for each its point $x \in G$ there exists an open ball $K_r(x)$ contained in the set G, i.e. $K_r(x) \in G$. Any open set $G \subset X$ containing x is called a neighborhood of the point $x \in X$.

The sequence of points x_n of the metric space X is said to be *fundamental* if for each $\varepsilon > 0$ there exists $N(\varepsilon)$ such that $\rho(x_k, x_m) < \varepsilon$ for all $k > N(\varepsilon)$, $m > N(\varepsilon)$. Any convergent sequence is fundamental. The opposite is not true.

The metric space X is said to be *complete*, if any fundamental sequence converges to a point $x_0 \in X$. Any closed set F of the complete metric space X is also a complete metric space with the same metric.

It is known, that every metric space can always be embedded into a complete metric space. This procedure is called a completion of a metric space. If the metric space X is to be complete it is necessary and sufficient that any sequence of the closed balls embedded one into another, with radii tending to 0, has a nonempty intersection in this space. The set $F \subset X$ is said to be *everywhere dense* in the metric space X if its closure coincides with X, i.e. $\bar{A} = X$. To put it differently: the set $A \subset X$ is called everywhere dense in X, if each ball $K_r(x)$, $x \in X$ contains the points of the set A.

The set $A \subset X$ is *nowhere dense*, if each open ball $K \subset X$ contains another open ball K' having no point in common with the set A ; in other words: A is nowhere dense if $\overline{X \setminus \bar{A}} = X$. (A set A contained in a metric space is nowhere dense if the closure $\bar{A}$ of the set A does not contain any open set). It is known that the complete metric space X cannot be represented in the form of a countable union of nowhere dense sets.

A metric space is said to be *separable* if it contains a countable everywhere dense set. A subspace of a separable space is separable itself. A collection of the open sets G of the metric space X is called *a base* if any open set in X may be represented as an union of a number (finite or infinite) of the sets of G.

If the number of elements of the base is finite or countable, then we have a countable base and the space with the countable base, respectively. The metric space X possesses a countable base if and only if it is separable.

We shall give now the very essential for later use concepts of *precompact* and *compact* sets of a metric space. The set M of the complete metric space X is said to be *precompact* (*relatively compact*), if from any infinite sequence of points $\{x_n\}$ of the set M for $n = 1, 2, \ldots$, there can be chosen a subsequence $\{x_{n_k}\}$ convergent to a point x_0 of the space X, i.e.

$$\lim_{n_k \to \infty} x_{n_k} = x_0, \quad (\rho(x_{n_k}, x_0) \to 0, \quad n_k \to \infty).$$

The limit point x_0 need not be a point of the set M. If the more strong condition is valid, namely: if from any infinite sequence of points of the set M there can be chosen a subsequence convergent to a point of the set M itself, then M is called *a compact* set.

It follows from these definitions that if M is a precompact set, then its closure $\bar{M}$ is a compact set. In the finite-dimensional case the precompactness of the set M coincides with its boundedness, i.e. every bounded set is precompact and conversely. Further, the compactness of the set M in the finite-dimensional case means that M is bounded and closed.

In the infinite-dimensional case the compactness means that the set M is not only bounded and closed, but that M is, in the some sense, a "thin" set similar to the Gilbert "cube".

One can give another (equivalent) definition of the compactness of a set. A system of the sets $\{E_\alpha\}$, $E_\alpha \subset X$ is said to be *a covering* of the set $A \subset X$ if $A \subseteq \cup_\alpha E_\alpha$. If E_α are open sets, then the covering is said to be open. The set $M \subset X$ is compact if from any open covering of the set M there can be chosen *a finite subcovering*. The metric space X is *compact* if from any infinite sequence of points of this space there can be chosen a convergent subsequence. In other words, in the compact metric space each its infinite subset possesses at least one limit point.

One can give the definition of the compactness of the metric spaces which is equally valid for the topological spaces. Namely, a metric space is said to be compact if any its open covering (i.e systems of open sets $\{G_\alpha\}$ such that $\cup_\alpha G_\alpha \supseteq X$) contains a finite subcovering (i.e. from $\{G_\alpha\}$ there can be chosen a finite number of open sets which cover X). A compact metric space is also called *a compactum*.

A system of subsets $\{A_\alpha\}$ of the metric space X is called a system with the finite intersection property if any finite intersection $\cap_{j=1}^n A_j$ of elements of the system is not empty.

If the metric space X is to be compact, *it is necessary and sufficient* that it satisfies the condition: any system with the finite intersection property of its *closed* subsets has a nonempty intersection. Every closed subset of a compact metric space is a compact metric space itself with the same metric, i.e. a closed set of a compactum is a compactum.

The set M_ε lying in the metric space X is said to be ε-net of the space X if for each point $x \in X$ there exists a point $\hat{x}$ of M_ε such that $\rho(x, \hat{x}) < \varepsilon$, i.e. ε-neighborhood of each point of the space X contains at least one point of the set M_ε. The metric space X is *totally bounded* if for each ε there exists *a finite* ε-net in this space.

If the metric space X is totally bounded, then it is *separable*, i.e. it contains a countable everywhere dense set. The metric space X is compact if and only if it is *complete* and *totally bounded*.

The metric space X is said to be *locally compact* if each its point has a neighborhood which closure is a compactum, i.e. a compact metric space. It follows that a compact metric space is locally compact. The Euclidian n-dimensional space (a real line, a plane etc.) is locally compact, but it is not a compact metric space.

Notice, that a phase space for the partial differential equations and also for the nonlinear dissipative systems is, as a rule, not locally compact. A continuous function on a compact space is bounded and takes its largest and smallest values. Let us give now a definition of the distances between sets of a metric spaces. A number

$$\rho(x, A) = \inf_{a \in A} \rho(x, a)$$

is a distance from the point x to the set A. By $O_\varepsilon(A)$ we shall denote ε-neighborhood of the set A, i.e. $O_\varepsilon(A) = \{x : \rho(x, A) < \varepsilon\}$.

Introduce a concept of departure of the set A from the set B, which will be frequently used in the sequel:

$$\text{dist}_X(A, B) = \sup_{a \in A} \rho(a, B) = \sup_{a \in A} \inf_{b \in B} \rho(a, b).$$

The operation "dist" has the following properties:

1) $\text{dist}_X(A, B) \neq \text{dist}_X(B, A)$;
2) $\text{dist}_X(A, B) = 0$ if and only if $A \subseteq B$;
3) if $\text{dist}_X(A, B) < \varepsilon$, then $A \subset O_\varepsilon(B)$.

Consider a family of the sets A_λ, where $\lambda \in \Lambda$ is some parameter (Λ is a metric space). A family of the sets A_λ of a metric space X is said to be *upper semicontinuous* at a point λ_0, if

$$\text{dist}_X(A_\lambda, A_{\lambda_0}) \to 0, \quad \lambda \to \lambda_0.$$

This family of sets is said to be *lower semicontinuous* at a point λ_0 if

$$\text{dist}_X(A_{\lambda_0} A_\lambda) \to 0, \quad \lambda \to \lambda_0.$$

If both upper and lower semicontinuities at λ_0 take place, then the family A_λ is continuous at this point.

Introduce now the Hausdorff distance between the sets A and B :

$$\text{dist}_X^H(A, B) = \max\{\text{dist}_X(A, B), \text{dist}_X(B, A)\}.$$

The Hausdorff distance possesses all properties of a metric, i.e. it is symmetrical, it satisfies the triangle inequality and it is equal to zero if and only if $A = B$. The set of all compact sets of the metric space X becomes a metric space if the distance between the compact sets is given by Hausdorff metric. Finally, an conventionalal distance between two sets A and B of the metric space X is defined as

$$\rho(A, B) = \inf_{x \in A, y \in B} \rho(x, y).$$

1.2 Dynamical Systems. Main Properties

The theory of dynamical systems in general deals with the properties of autonomous systems of differential equations whose solutions are determined for all $t \in R$, $R = (-\infty, +\infty)$ [108, 119].

Definition 1.1. A dynamical system in a metric space X is a family of mappings $f(p,t) : X \times R \to X$, which to any point $p \in X$ for any real number $t \in R$ puts in correspondence some point $q \in X$, $q = f(p,t)$ and satisfies the following conditions.

1. For each point $p \in X$

$$f(p,0) = p. \tag{1.7}$$

2. For each point $p \in X$ and every real numbers $t \in R$ and $\tau \in R$

$$f(f(p,t), \tau) = f(p, t+\tau). \tag{1.8}$$

3. Mapping $f(p,t)$ is continuous in the pair of variables p and t, i.e. for each point $p_0 \in X$, every number $t_0 \in R$ and each $\varepsilon > 0$ there exists $\delta > 0$ such that

$$\rho(f(p,t), f(p_0,t_0)) < \varepsilon \tag{1.9}$$

if $|t - t_0| < \delta$, $\rho(p,p_0) < \delta$.

A space X we shall call *a phase* space of dynamical system $f(p,t)$ and the parameter t the time.

The equality (1.7) represents an *identity* condition, the equality (1.8) is a *group property* and the inequality (1.9) is a *continuity condition*. (A dynamical system often is called a flow.)

As for the continuity properties of dynamical system, one can obtain more strong proposition, namely to prove the integral continuity property and the uniform integral continuity property.

Theorem 1.1. (*Integral continuity property.*) *For each point $p_0 \in X$, each segment $[a,b] \subset R$ and each $\varepsilon > 0$ there exists $\delta > 0$ such that*

$$\max_{t \in [a,b]} \rho(f(p,t), f(p_0,t)) < \varepsilon \tag{1.10}$$

if $\rho(p,p_0) < \delta$.

Proof. Suppose that for some $p_* \in X$, some segment $[a_0, b_0]$ and some $\varepsilon_0 > 0$ there exists no $\delta > 0$ satisfying the conditions of the theorem. Then one can take a sequence $\delta_n \to 0$ as $n \to \infty$ such that for all $n = 1, 2, \ldots$ there exist a point p_n and a number $t_n \in [a_0, b_0]$ satisfying the conditions $\rho(p_n, p_*) < \delta_n$ and

$$\rho(f(p_n, t_n), f(p_*, t_n)) \geq \varepsilon_0. \tag{1.11}$$

Since $\delta_n \to 0$, it follows from the condition $\rho(p_n, p_*) < \delta_n$ that $p_n \to p_*$. Further, by the boundedness of the sequence t_n there can be chosen from it a convergent subsequence $\{t_{n_k}\} : t_{n_k} \to t_0 \in [a_0, b_0]$ as $n_k \to \infty$. Taking into account that $f(p, t)$ is continuous in the pair of variables (p, t), we find

$$f(p_n, t_{n_k}) \to f(p_*, t_0), \quad f(p_*, t_{n_k}) \to f(p_*, t_0).$$

Now, replacing t_n by t_{n_k} in (1.11) and passing to the limit as $n \to \infty$, we get $0 \geq \varepsilon_0 > 0$. We use here the continuity property of the function $\rho(p, q)$. Namely, if $p_n \to p_0$, $q_n \to q_0$, then $\rho(p_n, q_n) \to \rho(p_0, q_0)$, i.e. the possibility of passing to the limit under the sign of distance.

Theorem 1.2. (*Uniform integral continuity property.*) *For any compact set $K \subset X$, any segment $a \leq t \leq b$ and each $\varepsilon > 0$ there exists $\delta > 0$ such that*

$$\max_{t \in [a,b]} \rho\Big(f(p, t), f(p_0, t)\Big) < \varepsilon, \quad \forall\, p_0 \in K, \tag{1.12}$$

if $\rho(p, p_0) < \delta$, $\quad \forall\, p \in K$.

Proof. Suppose that for some $[a_0, b_0]$, K_0 and $\varepsilon_0 > 0$, where K_0 is a compact set, there exists no δ mentioned in condition of the theorem.

Then one can take a sequence $\delta_n \to 0$ as $n \to 0$ such that for every $n = 1, 2, \dots$ there exist $\hat{p}_n \in K_0$, $p_n \in O_{\delta_n}(\hat{p}_n)$ (i.e. $\rho(p_n, \hat{p}_n) < \delta_n$) and $t_n \in [a_0, b_0]$ such that

$$\rho\Big(f(p_n, t_n),\ f(\hat{p}_n, t_n)\Big) \geq \varepsilon_0. \tag{1.13}$$

Since K_0 and $[a_0, b_0]$ are compact sets, the sequences $\hat{p}_n$ and t_n may be thought of as convergent, i.e. $\hat{p}_n \to \hat{p}_0$, $\hat{p}_0 \in K_0$, $t_n \to t_0$, $t_0 \in [a_0, b_0]$ and $p_n \to \hat{p}_0$ as $n \to \infty$. Write the inequality

$$\rho\Big(f(p_n, t_n), f(\hat{p}_n, t_n)\Big) \leq \rho(f(p_n, t_n), f(\hat{p}_0, t_0))$$
$$+ \rho(f(\hat{p}_0, t_0), f(\hat{p}_n, t_n)).$$

It is evident that by sufficiently large n the right side of the inequality will be less than ε_0, i.e.

$$\rho\Big(f(p_n, t_n), f(\hat{p}_n, t_n)\Big) < \varepsilon_0, \quad \forall\, n \geq N,$$

but this contradicts (1.13).

We now proceed to the analysis of a dynamical system taken as a family of mappings. It will be shown that a set of transformations $f(p, t)$, when t passes through the set of real numbers, forms a group of transformations of the space X into itself.

Introduce a notation $G = \{f(p,t), t \in R\}$, $g_t = f(p,t)$. Show that G is a group with respect to the multiplication. To each t an element $g_t \in G$ is assigned.

Introduce an operation of multiplication of elements $g_t \in G$, setting

$$g_t \circ g_\tau = g_{t+\tau}, \tag{1.14}$$

that is

$$g_t \circ g_\tau = f(f(p,\tau),t) = f(p,t+\tau) = g_{t+\tau} \in G.$$

As an unit element in the set G an element g_0 will serve (i.e. identity transformation):

$$g_0 \circ g_t = g_{0+t} = g_{t+0} = g_t \circ g_0 = g_t. \tag{1.15}$$

As an inverse element for each $g_t \in G$ an element $g_t^{-1} = g_{-t}$ will serve. Really,

$$g_t \circ g_t^{-1} = g_t \circ g_{-t} = g_{t-t} = g_0. \tag{1.16}$$

It follows from (1.14), (1.15) and (1.16) that G is a commutative group with respect to the operation of multiplication. We formulate this in the following form.

Theorem 1.3. *A family of transformations* $G = \{f(p,t), t \in R\}$ *of the phase space* X *into itself determined by a dynamical system* $f(p,t)$ *forms a group with respect to operation of multiplication.*

Remark. Instead of the equality (1.14) defining the operation of the multiplication for the transformations g_t, we can introduce the operation of addition for the transformations g_t, setting

$$g_t + g_\tau = g_{t+\tau}, \quad \forall\, t, \tau \in R.$$

Then the set of transformations $\{f(p,t), t \in R\}$ will be an additive group of transformations (i.e. a group with respect to the operation of addition), which we denote by G_+. A set of real numbers R forms an additive group also. One can establish one-to-one correspondence between the elements of these two groups according to the rule: every number t in R is assigned to an element g_t in $G_+ : t \sim g_t$, $\forall\, t \in R$. In this case the relationship $t + \tau \sim g_{t+\tau}$ takes place for all t and τ in R.

Let us consider now the properties of an arbitrary element g_t of group G (or G_+), i.e. single mapping $g_t = f(p,t)$ for fixed t.

Theorem 1.4. *Each element* g_t *of the group* G *of transformations is a homeomorphism of the space* X *onto itself.*

Proof. Notice that if $q = f(p,t)$ then $p = f(q,-t)$ for each point $p \in X$, i.e.. the mapping $f(p,t)$ has the inverse one. This follows from the evident equality:

$$q = f(p,t) = f(f(q,-t),t) = f(q,0) = q.$$

Further, if $p_1 \neq p_2$ and $q_1 = f(p_1, t)$, $q_2 = f(p_2, t)$, then $q_1 \neq q_2$. In fact, let $q_1 = q_2$. Then we have

$$p_1 = f(q_1, -t) = f(q_2, -t) = f(f(p_2, t), -t) = f(p_2, 0) = p_2,$$

that is $p_1 = p_2$. So, the mapping $f(p, t)$ is *one-to-one* mapping for any fixed t. The continuity of direct and inverse mappings $q = f(p, t)$ and $p = f(q, -t)$ follows from the continuity of mapping $f(p, t)$ in the pair of arguments p and t. It is important to consider the properties of the mapping $f(p, t)$, when p belongs to a set $A \subset X$ and to consider the image of the set by the mapping and also the image of the union (intersection and so on) of sets. For that purpose we introduce now some definitions which will be used in the sequel.

Definition 1.2. We shall call a function $f(p, t)$, $t \in R$ for fixed $p \in X$ a motion of the point p or simply a motion.

Notice that the motion of each point $p \in X$ is continuous, because the function $f(p, t)$ is continuous in t.

If $p_n \to p_0$ as $n \to \infty$, then the sequence of motions $f(p_n, t)$ uniformly converges to the motion $f(p, t)$ on each interval $a \leq t \leq b$ by the integral continuity.

If $p_1 \neq p_2$, then the corresponding motions $f(p_1, t)$ and $f(p_2, t)$ are different, i.e. $f(p_1, t) \cap f(p_2, t) = \emptyset$ for all $t \in R$.

Definition 1.3. Let $A \subset X$ be a set in the phase space X. By $f(A, t)$ we shall denote a set of motions of all its points, i.e.

$$f(A, t) = \{f(p, t), p \in A\} = \bigcup_{p \in A} f(p, t).$$

The expression $f(A, t)$ we shall call a motion of the set A or the translation of the set A on t.

Theorem 1.5.
1. If $A \subset B$, then $f(A, t) \subset f(B, t)$.
2. If $p \in A$, then $f(p, t) \in f(A, t)$.
3. If $p \in f(A, t)$, then $f(p, -t) \in A$,
   $f(p, \tau) \in f(f(A, t), \tau) = f(A, t + \tau)$.
4. $f(A \cup B, t) = f(A, t) \cup f(B, t)$.
5. $f(A \cap B, t) = f(A, t) \cap f(B, t)$.

Equalities 4 and 5 hold for every number of the set A_α, that is:
6. $f(\cup_\alpha A_\alpha, t) = \cup_\alpha f(A_\alpha, t)$.
7. $f(\cap A_\alpha, t) = \cap_\alpha f(A_\alpha, t)$.

Proof. Properties 1, 2 and 3 are evident. Prove the equalities 6 and 7. Let $\cup_\alpha A = B$. Then

$$f(\cup_\alpha A_\alpha, t) = f(B, t)$$

$$= \bigcup_{p \in B} f(p,t) = \left(\bigcup_{p \in A_1} f(p,t)\right) \cup \left(\bigcup_{p \in A_2} f(p,t)\right) \cup \ldots$$

$$= f(A_1,t) \cup f(A_2,t) \cup \ldots = \cup_\alpha f(A_\alpha,t).$$

Now, the equality 6 is proved, we shall prove equality 7. Let $q \in f(\cap_\alpha A_\alpha, t)$. Then

$$f(q,-t) \in f(f(\cap_\alpha A_\alpha, t), -t) = \cap_\alpha A_\alpha \subseteq A_\alpha, \quad \forall\, \alpha,$$

that is $f(q,-t) \in A_\alpha, \forall\, \alpha$. Hence it follows that $q \in f(A_\alpha, t), \forall\, \alpha$ and, consequently, $q \in \cap_\alpha f(A_\alpha, t)$. Conversely, let $q \in \cap_\alpha f(A_\alpha, t)$. Then $q \in f(A_\alpha, t), \forall\, \alpha$. Therefore $f(q,-t) \in A_\alpha, \forall\, \alpha$, i.e. $f(q,-t) \in \cap_\alpha A_\alpha$. Consequently, $q \in f(\cap_\alpha A_\alpha, t)$. Notice that the property 3 has been repeatedly used here.

Theorem 1.6. *If a set $A \subset X$ is open (closed or compact), then for every $t \in R$ the set $f(A,t)$ is also open (closed or compact).*

Proof. A set $f(A,t)$ is open (closed or compact) as an image of the open (closed or compact) the set by the homeomorphism $f(p,t)$.

Introduce the following definitions.

Definition 1.4. The set of points $\{f(p,t), t \in R\}$ for fixed $p \in X$ is said to be a trajectory (an orbit or a complete trajectory) passing through a point p. The trajectory will be denoted by

$$\gamma(p) = f(p,R) = \{f(p,t), t \in R\}. \tag{1.17}$$

A set of points $\{f(p,t), t \in R_+\}$, $R_+ = [0,+\infty)$ is called a positive semitrajectory starting from a point p

$$\gamma^+(p) = f(p,R_+) = \{f(p,t), t \in R_+\}. \tag{1.18}$$

Similarly, negative semitrajectory emanating from a point p is a set of points $\{f(p,t), t \in R_-\}$, $R_- = (-\infty, 0]$. A notation:

$$\gamma^-(p) = f(p,R_-). \tag{1.19}$$

Let T be nondegenerate interval on R, i.e. either segment or half-segment (finite of infinite), or interval (finite or infinite).

Definition 1.5. A set of points $\{f(p,t), t \in T\}$, where T is non-degenerate interval on the real axis is called the arc of the trajectory $\gamma(p)$. In general, let $M \subset R$ be some set of real numbers. Then $f(p,M)$ is a set of points lying on the trajectory $\gamma(p)$, that is

$$f(p,M) = \{f(p,t), t \in M\} = \bigcup_{t \in M} f(p,t) \subset \gamma(p).$$

Theorem 1.7. *Let T_α be a system of the sets on R. Then*

1) $f(p, T_\alpha) \subset f(p, R)$ for T_α;

2) $f(p, \cup_\alpha T_\alpha) = \cup_\alpha f(p, T_\alpha)$;

3) $f(p, \cap_\alpha T_\alpha) \subseteq \cap_\alpha f(p, T_\alpha)$;

4) $f(f(p, T_1), T_2) = f(p, T_1 + T_2)$, $T_1 + T_2 = \cup_{t_1 \in T_1, t_2 \in T_2}(t_1 + t_2)$.

Proof. Let us prove the inclusion 3. Let $q \in f(p, \cap_\alpha T_\alpha)$. Then there exists $t_{\alpha_q} \in \cap_\alpha T_\alpha$ such that $q = f(p, t_{\alpha_q})$. In this case t_{α_q} is included in all sets T_α.

Because of this $q \in f(p, t_\alpha)$ for all α, that is $q \in \cap_\alpha f(p, T_\alpha)$. The equalities 1, 2 and 4 are proved analogously. Finally, we give some simple, but useful propositions.

Proposition 1.1. If $q \in \gamma(p)$, then $\gamma(q) = \gamma(p)$.

Proposition 1.2. If $p \neq q$, $q \notin \gamma(p)$, then $\gamma(p) \cap \gamma(q) = \emptyset$, i.e. trajectories corresponding to the different points do not intersect themselves.

Proposition 1.3. A motion of a point p uniquely determines the motions of all points of the trajectory $\gamma(p)$.

Proposition 1.4. Any finite arc of the trajectory of the form $f(p, [T_1, T_2])$, where T_1 and T_2 are numbers, is compact.

1.3 Invariant Sets

The invariant sets play an important role in the theory of dynamical systems. The invariant sets of dynamical system, in some sense, are similar to the stationary points of dynamical system. However, these sets can have a complicated structure. (As will be shown below, attractors represent the patterns of such sets). Now we shall give the strict definition of an invariant set.

Definition 1.6. A set $A \subset X$ is called *invariant* with respect to the given dynamical system $f(p, t)$ (or simply invariant), if $f(A, t) = A$ for all $t \in R$.

Proposition 1.5. The trajectory $\gamma(p)$ through p is an invariant set with respect to dynamical system.

In fact, taking into account the definition of trajectory $\gamma(p)$, we have

$$f(\gamma(p), t) = f(f(p, R), t) = f(p, R + t) = f(p, R) = \gamma(p).$$

Theorem 1.8. *A union of a number of invariant sets is an invariant set.*

Proof. Let A_α be invariant sets, that is $f(A_\alpha, t) = A_\alpha$. Let $A = \cup_\alpha A_\alpha$. Then

$$f(A, t) = f(\cup_\alpha A_\alpha, t) = \cup_\alpha f(A_\alpha, t) = \cup_\alpha A_\alpha = A.$$

Proposition 1.6. A union of a set of trajectories is an invariant set with respect to dynamical system.

This is a consequence of Proposition 1.5 and Theorem 1.8. We shall show, that the converse proposition is also true.

Proposition 1.7. Any invariant set is a collection of trajectories.

In fact, let A be an invariant set and p belongs to A ($p \in A$). Then $f(p,t) \in f(A,t) = A$ for all $t \in R$, i.e. $f(p,R) = \gamma(p) \subset A$. The above propositions we can represent in the form of the following theorem.

Theorem 1.9. *If a set is to be invariant, it is necessary and sufficient that it consists of trajectories.*

Theorem 1.10. *Nonempty intersection of invariant sets is invariant.*

Proof. Let A_α be invariant sets. Let $A = \cap_\alpha A_\alpha$. Then

$$f(A,t) = f(\cap_\alpha A_\alpha, t) = \cap_\alpha f(A_\alpha, t) = \cap_\alpha A_\alpha = A.$$

Theorem 1.11. *If A is invariant, then its complement $X \setminus A$ is also invariant.*

Proof. Let $q \in X \setminus A$ be any point of the set $X \setminus A$. Then $f(q,t) \in X \setminus A$ for all $t \in R$. In fact, if it is not the case, then we could have at some t^* $f(q,t^*) \in A$, but then $q \in f(A, -t^*) = A$, that is $q \in A$.

Thus, $f(q,t) \in X \setminus A$ for all $t \in R$, in other words, an complete trajectory $\gamma(q)$ belongs to $X \setminus A$, that is $X \setminus A$ consists of trajectories, therefore $X \setminus A$ is invariant.

Theorem 1.12. *A closure of an invariant set is invariant.*

Proof. Let A be invariant set and $p \in \bar{A}$. Then there exists a sequence of points p_n convergent to p and $p_n \in A$, $n = 1, 2, \ldots$. Then $f(p_n, t) \in A$, $t \in R$, by invariance of A. Further, $f(p_n, t) \to f(p, t)$ as $n \to \infty$, that is the point $f(p,t)$, for every $t \in R$, is a limit of sequence of the points $f(p_n, t)$ which belong to A. Consequently, $f(p,t) \in \bar{A}$, $t \in R$ and $\gamma(p) \subset \bar{A}$. Thus, the set $\bar{A}$ together with each its point contains the complete trajectory $\gamma(p)$ of the point, i.e. $\bar{A}$ is invariant.

Theorem 1.13. *If A is invariant set, then the following sets are also invariant:* a) int$A$, *i.e. the interior of the set;* b) ∂A, *i.e. the boundary of the set.*

Proof. Since int$A = X \setminus \overline{(X \setminus A)}$, the invariance of int$A$ follows from the above theorems. The invariance of ∂A follows from the equality $\partial A = \bar{A} \cap \overline{(X \setminus A)}$. Thus, a boundary of invariant set consists of complete trajectories.

Each dynamical system $f(p,t)$ uniquely defines a dynamical system on the arbitrary invariant subset $A \subset X$.

In conclusion we note that the set A is said to be positively (negatively) invariant if $f(A,t) \subseteq A$ for all $t \in R^+$ ($t \in R^-$).

A complementation of a positively invariant set is a negatively invariant set. A closure of any positively (negatively) invariant set is a positively (negatively) invariant set. A union of a system of positively (negatively) invariant sets is a positively (negatively) invariant set. An intersection of a system of positively (negatively) invariant sets is a positively (negatively) invariant set.

1.4   Classification of Motions

In this section we introduce the concepts of ω-limit and α-limit sets of dynamical system and give some classification of motions with respect to properties of these limit sets.

Definition 1.7. A point q is called ω-limit point of motion $f(p,t)$, if there exists a sequence $\{t_n\}$, $t_n \to +\infty$ $(n \to \infty)$ such that a sequence $\{f(p, t_n)\}$ converges to q, i.e.

$$\lim_{n \to \infty} f(p, t_n) = q, \qquad t_n \to \infty.$$

The point q is a limit point of the positive semitrajectory $f(p, R_+)$, therefore $q \in \overline{f(p, R_+)}$. Similarly, the point q is called α-limit point of motion, if there exists a sequence $\{t_n\}$, $t_n \to -\infty$ as $n \to \infty$ such that

$$\lim_{n \to \infty} f(p, t_n) = q, \qquad t_n \to -\infty.$$

The sets of all ω-limit and α-limit points of the motion $f(p,t)$ will be denoted by Ω_p and A_p. Note, that

$$\Omega_p \subset \overline{f(p, R_+)}, \quad A_p \subset \overline{f(p, R_-)}, \tag{1.20}$$

since a closure of semitrajectory contains all limit points of the semitrajectory. It can be shown that

$$\Omega_p = \bigcap_{\tau \geq 0} \overline{\bigcup_{t \geq \tau} f(p, t)} = \bigcap_{\tau \geq 0} \overline{F_\tau}, \quad F_\tau = \bigcup_{t \geq \tau} f(p, t). \tag{1.21}$$

Theorem 1.14. *If the point q is to be ω- (or α)-limit point of the motion $f(p,t)$, it is necessary and sufficient that for every $\varepsilon > 0$ and for every $T > 0$ there can be found a moment $t > T(t < T)$ such that*

$$f(p,t) \in O_\varepsilon(q). \tag{1.22}$$

Proof. Let $q \in \Omega_p$. Then there exists a sequence $t_n \to +\infty$ such that $f(p, t_n) \to q$. If $\varepsilon > 0$ and $T > 0$ are given, then there exists N such that

$$\rho(f(p, t_N), q) < \varepsilon, \quad t_N > T.$$

Conversely, let for every $\varepsilon > 0$ and $T > 0$ (1.22) holds. Let then $\varepsilon_n \to 0$, $\varepsilon_n > 0$ and $T_n \to \infty$. For every n we have

$$\rho\left(f\left(p, t_n\right), q\right) < \varepsilon_n, \qquad t_n > T_n.$$

Passing to the limit $n \to \infty$, we get $\lim\limits_{n \to \infty} f\left(p, t_n\right) = q$ that is $q \in \Omega_p$. The proof is complete.

This theorem points to the fact that if the motion $f(p, t)$ possesses ω-limit point, then whatever the given neighborhood of this point may be, there exists a moment such that the motion get into this neighborhood. Moreover, if the neighborhood is chosen, we can define the monotonically increasing time moments in which the motion will pass through this neighborhood. Thus, the motion possesses the property of some recurrence in a neighborhood of its ω-limit point. The case is possible when the point p is ω -limit point of the motion $f(p, t)$.

In this case the motion $f(p, t)$ returns in any neighborhood of the initial point p. Such motions are called stable according to Poisson. A strict definition and properties of motions stable according to Poisson will be given below.

We now establish the important properties of the sets Ω_p and A_p such as the completeness and the invariance.

Theorem 1.15. *The sets Ω_p and A_p are closed.*

Proof. Let $q \in \bar{\Omega}_p$. Then there exists a sequence $q_n \to q$, $q_n \in \Omega_p$. We show that $q \in \Omega_p$. Let us take any ε_0–neighborhood of the point q. Then there exists $N\left(\varepsilon_0\right)$ such that

$$q_n \in O_{\varepsilon_0}\left(q\right), \qquad n \geq N\left(\varepsilon_0\right).$$

Further, there exists ε_1 such that $O_{\varepsilon_1}\left(q_N\right) \subset O_{\varepsilon_0}\left(q\right)$. Since $q_N \in \Omega_p$, for every $T > 0$, there exists $t^* > T$ such that $f\left(p, t^*\right) \in O_{\varepsilon_1}\left(q_N\right)$ $t^* > T$. But then $f\left(p, t^*\right) \in O_{\varepsilon_0}\left(q\right)$ and consequently by Theorem 1.14 $q \in \Omega_p$. (Notice that the completeness of Ω_p follows immediately from (1.21).)

Theorem 1.16. *The sets Ω_p and A_p are invariant.*

Proof. Let $q \in \Omega_p$. Then there exists a sequence $\{t_n\}$ such that $f\left(p, t_n\right) \to q$, $t_n \to +\infty$. But, for every $t \in R$ we have $t_n + t \to +\infty$. Therefore

$$\lim_{n \to \infty} f\left(p, t_n + t\right) = \lim_{n \to \infty} f\left(f\left(p, t_n\right), t\right) = f\left(q, t\right).$$

Thus, there exists a sequence $\{t_n + t\}$ such that

$$\{f\left(p, t_n + t\right)\} \to f\left(q, t\right), \qquad t_n + t \to +\infty.$$

Consequently, the point $f\left(q, t\right)$ belongs to Ω_p for every $t \in R$, that is $f\left(q, R\right) \subset \Omega_p$ and the set Ω_p consists of trajectories (i.e. with its every point q the set contains whole its trajectory $f\left(q, R\right)$).

Nonemptyness of ω-limit sets is connected with the stability of motions according to Lagrange. Consider the stability according to Lagrange.

Definition 1.8. The motion $f(p,t)$ (and the point p) is said to be positively (negatively) stable according to Lagrange if the closure of positive (negative) semitrajectory is a compact set, i.e. a positive (negative) semitrajectory $f(p, R_+)$ $(f(p, R_-))$ is precompact.

A motion which is at the same time positively and negatively stable according to Lagrange is called stable according to Lagrange. In these cases we shall speak about the L^+-stability, the L^--stability and the L-stability of motion. From the above definition the proposition follows.

Proposition 1.9. If the phase space X of dynamical system $f(p,t)$ is *compact*, then all motions in the phase space are stable according to Lagrange.

In the finite-dimensional case, when $X = E_n$ is Euclidian n-dimensional space the stability according to Lagrange means the boundedness of the solution (trajectory), i.e. those motions are stable according to Lagrange, whose trajectories are located in a bounded portion of the phase space.

Theorem 1.17. *If the motion $f(p,t)$ is positively (negatively) stable according to Lagrange, then:*

1) the set Ω_p is not empty (the set A_p is not empty);

2) Ω_p is compact (A_p is compact);

3) $f(p,t) \to \Omega_p$ as $t \to +\infty$, $(f(p,t) \to A_p$ as $t \to -\infty)$.

Proof. 1. Let the motion $f(p,t)$ be L^+-stable and $t_n \to +\infty$. Then, evidently,

$$f(p, t_n) \in f(p, R_+) \in \overline{f(p, R_+)} \equiv \overline{\gamma_t(p)}.$$

By compactness of $\overline{\gamma_+(p)}$ from the sequence $f(p,t_n)$ there can be chosen a convergent subsequence $f(p,t_{n_j}) \to q$. But then q will be ω-limit point of the motion, that is $q \in \Omega_p$. Thus, Ω_p is not empty.

2. Since $\Omega_p \subset \overline{f(p, R_+)}$, by virtue of (1.20) and since Ω_p is closed, it is compact.

3. We need to prove that $\rho(f(p,t), \Omega_p) \to 0$ as $t \to +\infty$, i.e. for every $\varepsilon > 0$ there exists $T(\varepsilon)$ such that $\rho(f(p,t)\,\Omega_p) < \varepsilon$, for $t > T$. If this is not the case, then there exists ε_0 such that whatever T may be, there exists $t > T$ such that

$$\rho(f(p,t), \Omega_p) \geq \varepsilon_0. \qquad (1.23)$$

Let $\{T_n\}$ be a sequence of numbers such that $T_n \to +\infty$ as $n \to \infty$. In this case, by (1.23), for every n, there exists $t_n > T_n$ such that

$$\rho(f(p,t_n), \Omega_p) \geq \varepsilon_0. \qquad (1.24)$$

Since $\{f(p, t_n)\} \subset \overline{f(p, R_+)}$, this sequence may be thought of as convergent. Let q be the limit of the sequence.

Then passing to the limit in (1.24), we get $\rho(q, \Omega_p) \geq \varepsilon_0$. But by the definition q is ω-limit point of motion $f(p, t)$, $q \in \Omega_p$. We have arrived at a contradiction and the theorem is proved.

We can demonstrate the converse. Here the following theorem takes place.

Theorem 1.18. *If the motion $f(p, t)$ is to be positively stable according to Lagrange, it is necessary and sufficient that the following conditions are fulfilled:*

1) Ω_p *is not empty;*

2) Ω_p *is compact;*

3) $\lim\limits_{t \to +\infty} \rho(f(p, t), \Omega_p) = 0$.

Proof. The necessity was proved in the preceding theorem. We prove now the sufficiency.

Let the conditions of the theorem be fulfilled. Then the semitrajectory $f(p, R_+)$ is precompact in X. Take any sequence of the points $\{q_n\} \subset f(p, R_+)$, $q_n = f(p, t_n)$, $t_n \geq 0$.

If the sequence $\{t_n\}$ is bounded, there can be chosen from it a convergent subsequence. Then the corresponding subsequence of the points of $\{q_n\}$ will be convergent.

Let $t_n \to +\infty$ as $n \to \infty$. Then, by virtue of the condition 3 $\rho(q_n, \Omega_p) \to 0$ as $n \to \infty$. This means that there exists a sequence of points $\{r_n\} \subset \Omega_p$ such that $\rho(q_n, r_n) \to 0$ as $n \to \infty$. Since Ω_p is compact by the condition of the theorem, the sequence $\{r_n\}$ may be thought of as convergent to the point $r \in \Omega_p$. But then $q_n \to r$. Hence, $f(p, R_+)$ is precompact , i.e. its closure is compact, that is the motion $f(p, t)$ is L^+-stable.

Remark. If the phase space X is locally compact then the condition 3 in the proven theorem may be omitted, i.e. in locally compact spaces (for example, in Euclidian n-dimensional spaces), if the motion is to be stable according to Lagrange, it is necessary and sufficient that Ω_p is not empty and compact. It can be shown that if the motion $f(p, t)$ is L^+-stable, then its Ω_p-set is connected.

Theorem 1.19. *Let Ω_p be ω-limit set of the L^+-stable motion. Then all motions in Ω_p are stable according to Lagrange, i.e. they are L-stable.*

Proof. Let $f(p, t)$ be L^+-stable and the point $q \in \Omega_p$. The set Ω_p is invariant; that is it contains the trajectory $f(q, R) \subseteq \Omega_p$. By completeness of Ω_p it follows that $\overline{f(q, R)} \subseteq \Omega_p$.

By virtue of the L^+-stability the set Ω_p is compact. But then $\overline{f(q, R)}$ is also compact, that is the motion $f(q, t)$ is stable according to Lagrange. The theorem is proved.

We now coming to the consideration of such motions $f(p, t)$, for which the point p is ω-limit. As noted, such motions arbitrary often return in any given neighborhood of the initial point p.

These motions are said to be the motions stable according to Poisson. We shall give the corresponding definitions.

Definition 1.9. The point p is called positively stable according to Poisson (P^+-stable), if for any neighborhood U of the point p and for any $T > 0$, there can be found a value $t \geq T$ such that $f(p, t) \in U$.

If there can be found a value $t \leq -T$ such that $f(p, t) \subset U$ then the point p is negatively stable according to Poisson (P^--stable).

A point stable according to Poisson both as $t \to +\infty$ and as $t \to -\infty$ is called simply stable according to Poisson.

Let $O_\varepsilon(p)$ be an arbitrary ε-neighborhood of the point p and $\{T_n\}$ is the sequence such that $T_n \to +\infty$ as $n \to \infty$. Then by the above definition there exists a sequence $t_n \geq T_n$ such that $f(p, t_n) \in O_\varepsilon(p)$. This means that the motion $f(p, t)$ arbitrary often returns in the given neighborhood of the initial point p.

The given definition is equivalent to the following one.

Definition 1.10. The point p is positively stable according to Poisson, if there exists a sequence $\{t_n\}, t_n \to +\infty$ as $n \to \infty$ such that $f(p, t_n) \to p$ as $n \to \infty$, i.e. p is ω-limit point of the motion $f(p, t)$ (and $p \in \Omega_p$).

Prove this. Let $f(p, t_n) \to p$, $t_n \to +\infty$ as $n \to \infty$. Then for any neighborhood U of the point p and for any T there exists $t_N \geq T$ such that $f(p, t_N) \in U(p)$, that is P^+-stability of the point p is established.

Conversely, let the point p be positively stable according to Poisson. Then, for any sequence $\varepsilon_1 > \varepsilon_2 > \ldots > \varepsilon_n > \ldots, > \varepsilon_n \to 0$ as $n \to \infty$, there exist the numbers $t_n > n$ such that

$$\rho(f(p, t_n), p) < \varepsilon_n.$$

As $n \to \infty$, hence $f(p, t_n) \to p$ for $t_n \to +\infty$. By the definition of the ω-limit point $p \in \Omega_p$, i.e. p is ω-limit point.

Theorem 1.20. *If the point p is positively stable according to Poisson, then every point of its trajectory $f(p, R)$ is also P^+-stable.*

Proof. Let $q \in f(p, R)$ be any point of the trajectory. Then $q = f(p, t_q)$. We have to prove that there exists a sequence $t_n \to +\infty$ as $n \to \infty$ such that $f(q, t_n) \to q$ as $n \to \infty$.

By stability of the point p there exists a sequence $t_n \to +\infty$ such that $f(p, t_n) \to p$. Let $f(p, t_n) = p_n$. Then it follows from the condition $p_n \to p$ that

$$f(p_n, t_q) \to f(p, t_q) = q, \quad n \to \infty,$$

or

$$f(f(p, t_n), t_q) = f((p, t_q), t_n) = f(q, t_n) \to q, \quad n \to \infty.$$

Taking into account this theorem, we shall speak about the motions of positively (negatively) stable according to Poisson and the motions simple stable according to Poisson.

Corollary 1.1. If the motion $f(p,t)$ is positively stable according to Poisson, then all points of semitrajectory $f(p, R_+)$ are ω-limit, that is $f(p, R_+) \subset \Omega_p$.

By completeness of Ω_p it follows that $\overline{f(p, R_+)} \subset \Omega_p$. Since $\Omega_p \subset \overline{f(p, R_+)}$ always, $\overline{f(p, R_+)} = \Omega_p$.

Similar reasoning may be applied to the P^--stable and the P-stable motions. The results may be represented in form of the following theorem.

Theorem 1.21. *For the P^+-stable motion $A_p \subset \Omega_p = \overline{f(p, R_+)}$, where A_p may be empty.*

For the P^--stable motion $\Omega_p \subset A_p = \overline{f(p, R_-)}$, where Ω_p may be empty. For the P-stable motion

$$A_p = \Omega_p = \overline{f(p, R)}. \tag{1.25}$$

We can show that if the above equation holds then the motion is P-stable.

Theorem 1.22. *If the motion $f(p,t)$ is to be stable according to Poisson, it is necessary and sufficient that the equalities (1.25) hold.*

In the applications we are forced to consider the stability according to Poisson for the discrete time, that is $t = n$, $n = 1, 2, \ldots$.

Let us take the following definition.

Definition 1.11. If the sequence of points $\{f(p, n)\}$ for $n = 1, 2, \ldots$, possesses the point p as its limit point then p is P^+-stable.

It can be shown that if there is a P^+-stability of the point p at the continuous time, then there is also a P^+-stability of the point at the discrete time. We shall use this fact to find all P^+-stable points.

Demand for this purpose that the phase space of dynamical system be a metric space with a countable base. For example, X may be a compact space. Let $\{V_n\}$, $n = 1, 2, \ldots$, be a basis of the space X.

We now define $V_n^* = V_n \setminus V_n \cap (\bigcup_{m=1}^{\infty} f(V_n, -m))$. The sets V_n^* satisfy the relationships:

$$f(V_n^*, m) \cap V_n = \emptyset, \quad f(V_n^*, m) \cap V_n^* = \emptyset, \quad m = 1, 2, \ldots \tag{1.26}$$

Define two sets V^+ and E^+, suggesting

$$V^+ = \bigcup_{n=1}^{\infty} V_n^*, \quad E^+ = X \setminus V^+. \tag{1.27}$$

Theorem 1.23. *Let X be a metric space with a countable base. Then V^+ is the set of all P^+-unstable points, while E^+ is the set of all P^+-stable points of dynamical system which acts in X.*

Proof. Let $p \in E^+$ and let V_N be any neighborhood containing this point. Then the point p does not belong to V_N^* by virtue of the definition of the set E^+. Consequently, there exists $k \geq 1$ such that $p \in V_N \cap f(V_N, -k)$. Hence $f(p, k) \in f(V_N, k) \cap V_N \subset V$. But this is the definition of the P^+-stability for the discrete case.

If $p \in V^+$, then there exists a neighborhood V_M of this point such that $p \in V_M^*$. But then $f(V_M^*, m) \cap V_M = \emptyset$, $m = 1, 2, \ldots$, according to (1.26).

Thus $f(p, m) \cap V_M = \emptyset$, $m = 1, 2, \ldots$, that is the point p leaves its V_M neighborhood forever. The theorem is proved.

Similarly we can introduce the sets

$$V^- = \bigcup_{n=1}^{\infty} V_n^{**}, \quad V_n^{**} = V_n \setminus V_n \cap (\bigcup_{k=1}^{\infty} f(V_n, k)), \quad E^- = X \setminus V^-.$$

Then $E^+ \cap E^-$ is the set of all stable according to Poisson points of dynamical system.

Theorem 1.24. *The set of all P^+-stable and (P^-)-stable points of dynamical system is invariant.*

Proof. If the point $p \in F$, where F is the set of all P^+-stable points, then every point of the trajectory $f(p, R)$ is P^+-stable, i.e. $f(p, R) \subset F$. Hence, F is invariant.

Stability according to Lyapunov. In conclusion we give the definition of the stability according to Lyapunov.

The point p and the motion $f(p, t)$ are said to be positively stable according to Lyapunov if for any $\varepsilon > 0$ there can be found $\delta > 0$ such that if $\rho(p, q) < \delta$ and $t \in R_+$, then

$$\rho(f(p, t), f(q, t)) < \varepsilon.$$

The point $p \in X$ and the motion $f(p, t)$ are said to be positively stable according to Lyapunov with respect to the set $B \subseteq X$ if $p \in B$ and for any $\varepsilon > 0$ there exists $\delta > 0$ such that if $\rho(p, q) < \delta$, $q \in b$, $t \in R_+$ then the above inequality fulfills.

It is clear that the positive stability according to Lyapunov is the condition of the integral continuity for $T = +\infty$. If $B = X$, then from this definition the positive stability according to Lyapunov follows. If the point p is positively stable according to Lyapunov with respect to the set B then on the trajectory $f(p, R)$ there can exist some points positively unstable according to Lyapunov with respect to B.

One can prove that the set of points which are positively stable according to Lyapunov with respect to the set B is invariant.

The set $A \subseteq X$ is called positively stable according to Lyapunov with respect to the set $B \subseteq X$, if every point of the set A is positively stable according to Lyapunov with respect to the set B ($A \subseteq B$).

1.5   Recurrence of Domains

The motions of dynamical systems may be classified in two different manners:

1) with respect to the properties of dynamically limit sets Ω_p and A_p;

2) with respect to the properties of recurrence of motions in the neighborhood of their initial state.

Consider both of these classifications. We shall take the following classification of motions with respect to the sets Ω_p and A_p.

Definition 1.12. The motion $f(p,t)$ and the point p are called departing as $t \to +\infty$ ($t \to -\infty$), if the set $\Omega_p\,(A_p)$ is empty.

Definition 1.13. The motion $f(p,t)$ and the point p are called positively asymptotic, if $\Omega_p\,(A_p)$ is not empty, while an intersection $f(p, R_+)(f(p, R_-))$ with $\Omega_p\,(A_p)$ is empty.

Definition 1.14. The motion $f(p,t)$ and the point p are called positively (negatively) stable according to Poisson, if $\Omega_p\,(A_p)$ is not empty and the intersection $f(p, R_+)(f(p, R_-))$ with $\Omega_p\,(A_p)$ is also not empty.

As is easy to see, this definition of the stability according to Poisson is equivalent to the above definition.

Theorem 1.25. *For any dynamical system the following sets are invariant:*

1) *the set of all departing as $t \to +\infty$ (as $t \to -\infty$) points;*

2) *the set of all positively (negatively) asymptotic points;*

3) *the set of all P^+-stable and (P^-)-stable points.*

Proof. Let $p \in F$, where F is the set of all departing as $t \to +\infty$ points. Then each point of the trajectory $f(p, R)$ will be departing. Indeed, if $q \in f(p, R)$ then $\Omega_p = \Omega_q = \emptyset$, i.e. $f(p, R) \subset F$. Hence, F is invariant as the set consisting of entire trajectories.

Let $p \in G$, where G is the set of positively asymptotic points. Then every point of the trajectory $f(p, R)$ will be asymptotic, i.e. $f(p\,R) \subset G$, therefore G is invariant. The invariance of the set of all P^+-stable points was proved in the Theorem 1.20.

We are coming now to the classification of motions on the recurrence in the neighborhood of their initial state. From physical point it is natural to select and investigate such motions which possess some regularity, for example, they posses one or other of the properties of recurrence or periodicity. Among these motions are: 1) stationary states, 2) periodic motions, 3) almost periodic motions, 4) recurrent motions, 5) stable according to Poisson motions, 6) nonwandering motions.

We shall give now the corresponding definitions.

If $f(p,t) = p$ for all $t \in R$ then point p is said to be a *rest point* or a *stationary point*. We shall give some statements related to the stationary points.

Theorem 1.26. *The set of stationary points is a closed set.*

Proof. Let $\{p_n\}$ for $n = 1, 2 \ldots$, be a sequence of stationary points and let $p_n \to p_0$ as $n \to \infty$. Then passing to the limit in the equality $f(p_n, t) = p_n$, we get $f(p_0, t) = p_0$ for all $t \in R$.

Theorem 1.27. *No motion enters the stationary point for a finite time t.*

Proof. If a motion $f(p, t)$ enters the stationary point $q \neq p$ at $t = t_0$, i.e. if $f(p, t_0) = q$, then $f(p, R) = f(q, R) = q$ and from this it follows that $p = q$. The contradiction so obtained proves the theorem.

If for certain $\tau > 0$

$$f(p, t + \tau) = f(p, t) \tag{1.28}$$

for all $t \in R$, then the motion $f(p, t)$ is said to be *periodic* with the period τ. Notice that then the motion $f(p, t)$ admits also the periods $n\tau, n = \pm 1, 2, \ldots$. The smallest positive number τ satisfying the condition (1.28) is called the period of the motion $f(p, t)$.

In order to give the definition of almost periodic function, we introduce the concept of the relatively dense set of numbers on the real axis.

Definition 1.15. The set $M \subset R$ of real numbers is called relatively dense on the real axis R, if there exists a number $l > 0$ such that any interval $a \leq x \leq a + l$, $a \in R$ of length l contains at least one point of the set M.

The number τ is called ε-almost period of the function $f(t)$ if for every $t \in R$

$$\mid f(t + \tau) - f(t) \mid < \varepsilon.$$

Definition 1.16. The motion $f(p, t)$ is called almost periodic, if for any $\varepsilon > 0$ there exists the number $l(\varepsilon) > 0$ such that on any time interval of length $l(\varepsilon)$ there is the number τ for which

$$\rho(f(p, t + \tau), f(p, t)) < \varepsilon, \quad \forall t \in R. \tag{1.29}$$

The set of numbers τ (of ε-almost periods) satisfying the condition (1.29) forms the relatively dense set on the real axis R. We denote the points of the set by τ_n. Then (1.29) can be written in the form

$$\rho(f(p, t + \tau_n), f(p, t)) < \varepsilon,$$

where $\{\tau_n\}$ is the relatively dense set of numbers on R. The set will be depend on ε, but it exists for every $\varepsilon > 0$.

The main property of almost periodic motions, which will be important in the sequel, holds that: if $f(p, t)$ is almost periodic motion, then the closure of its trajectory $f(p, R)$ is *compact*, i.e. the set $f(p, R)$ is *precompact* in X. (X is a complete metric space.)

Another (equivalent) definition of almost periodic function has been given by Bochner: the continuous function $f(t)$, $t \in R$ is almost periodic if from any sequence of its translations $f(p, t + h_n)$, $h_n \in R$ it is possible to choose an uniformly convergent on R subsequence, that is the family of functions $f(t + h)$, $h \in R$, is precompact by virtue of the uniform convergence [91]. For almost periodic function $f(t)$ we can define the spectral function

$$a(\lambda) = \lim_{T \to \infty} \frac{1}{T} \int_0^T f(t) e^{i\lambda t} dt.$$

The values of λ, such that $a(\lambda) \neq 0$, are a finite or countable sequence of real numbers $\lambda_1, \lambda_2, \ldots$. The numbers are called the Fourier exponents (or spectrum) of the function $f(t)$, while the numbers $a(\lambda_n) = A_n$ are the Fourier coefficients.

The quasiperiodic (or conditionally periodic) functions are the particular case of almost periodic functions. These functions have the form

$$f(t) = f(\omega_1 t, \omega_2 t, \ldots, \omega_n t),$$

where f is 2π-periodic in every argument $\omega_k t$, $k = \overline{1, n}$, while the numbers ω_k are rationally independent.

Almost periodic motion possesses the recurrence property, i.e. it arbitrary often returns in a given neighborhood not only of the initial point, but in that of every point of its trajectory. The recurrence is regular in the sense that it is determined by the relatively dense set $\{\tau_n\}$. The following class of motions possessing the recurrence property is the class of the recurrent motions.

Definition 1.17. The motion $f(p, t)$ is said to be *recurrent* if for any $\varepsilon > 0$ there can be found an interval $T(\varepsilon) > 0$ such that any arc $f(p, [t, t + T])$ of the trajectory approximates all trajectory with a precision to within ε, i.e.

$$f(p, R) \in O_\varepsilon(f(p, [t, t + T])), \quad \forall t \in R. \tag{1.30}$$

The definition requires some explanations. Introduce the concept of an approximation of a set by a trajectory of a motion.

Definition 1.18. The trajectory $f(p, R)$ uniformly approximates the set $M \subset X$ if for every $\varepsilon > 0$ there exists the number $T > 0$ such that the arc of the trajectory $f(p, R)$ of length T approximates the set M with a precision to within ε, i.e. $M \subseteq O_\varepsilon(f(p, [t_0, t_0 + T]))$ for every $t_0 \in R$.

Taking into account the above definition, one can say that the motion is recurrent if its trajectory uniformly approximates itself.

It follows from the definition 1.17 that whatever the numbers s and t may be, there exists a number τ such that $t < \tau < t + T$ and

$$\rho(f(p, s), f(p, \tau)) < \varepsilon. \tag{1.31}$$

In other words, the recurrent motions possess the following property. They arbitrary often return in every neighborhood of every point of their trajectory. Unlike almost periodic motions, such a returning is not regular. Therefore the class of recurrent motions is more large than the class almost periodic ones, but it is narrower than the class of the stable according to Poisson motions. We now prove this statement.

Theorem 1.28. *Every recurrent motion is stable according to Poisson.*

Proof. Let $\varepsilon > 0$ and $t_0 > 0$. By the recurrence there exist the numbers t_1 and t_2 such that $t_0 \le t_1 \le t_0 + T(-t_0 - T \le t_2 \le -t_0)$ and $\rho(p, f(p, t_i)) < \varepsilon$, $i = 1, 2$. But this is a P^+ (P^-)-stability, since t_0 may be taken as large as one likes.

Theorem 1.29. *Every almost periodic motion is recurrent.*

Proof. For the recurrent motion for given $\varepsilon > 0$ there can be found an interval of length $T(\varepsilon)$ such that the arc of the trajectory $f(p, [a, a+T])$ for every a approximates the every point q of trajectory with a precision to within ε: there exists $t_0 \in (a, a + T)$ such that

$$\rho(q, f(p, t_0)) < \varepsilon.$$

If $q \in f(p, R)$, then $q = f(p, t_q)$. Let $f(p, t)$ be almost periodic motion.

Let in the definition of almost periodicity the number $l(\varepsilon)$ be equal to $T(\varepsilon)$ and ε-almost period be $\tau \in (a - t_q, a - t_q + T)$. By the definition of almost periodic motion

$$\rho(f(p, t_q), f(p, t_q + T)) < \varepsilon$$

or assuming that $t_q + \tau = t_0$, $a < t_0 < a + T$, we find

$$\rho(f(p, t_q), f(p, t_0)) < \varepsilon,$$

this inequality is the condition of recurrence (1.31).

From the above results it follows that we have the next "chain" of more and more large classes of the motions possessing some recurrence property:

Stationary state $\subset$ periodic motions $\subset$ almost periodic motions $\subset$ recurrent motions $\subset$ stable according to Poisson motions.

Finally, we shall define the most large class of motions possessing the recurrence property. Those are nonwandering motions.

Definition 1.19. The point p is called nonwandering in X if for every T and every neighborhood $U(p)$ of the point p there can be found a moment $t > T$ such that

$$U(p) \cap f(U(p), t) \ne \emptyset. \tag{1.32}$$

The point p of the phase space of dynamical system $f(p,t)$ is called *wandering* if there exist a neighborhood $U(p)$ of it and $T > 0$ such that

$$f(U(p), t) \cap U(p) = \emptyset, \quad \forall\, t \geq T, \tag{1.33}$$

i.e. there exists $T > 0$ such that the motion $f(p,t)$, $t > T$ never will return in U-neighborhood of the initial point p.

We denote by W the set of all wandering points and by W_N the set of all nonwandering points of the phase space. Then

$$X = W \cup W_N.$$

For the set of nonwandering points we introduce the following notation

$$C_0 = W_N(f, X),$$

pointing out that this set is the set of nonwandering points of dynamical system f given on the phase space X .

Theorem 1.30. *Sets of wandering and nonwandering points are invariant.*

Proof. Let $p \in W$. Then there exists a number T and a neighborhood $U(p)$ such that for all $t > T$ we have the equality (1.33). Applying the transformation $f(\cdot, t_0)$, for all $t > T$, to (1.33) we obtain

$$f(U(p), t + t_0) \cap f(U(p), t_0) = \emptyset$$

or

$$f(f(U(p), t_0), t) \cap f(U(p), t_0) = \emptyset.$$

Introduce a notation

$$f(U(p), t_0) = U_*(f(p, t_0)),$$

where U_* is the neighborhood of point $f(p, t_0)$, i.e. the open set as the image of the open set $U(p)$. Then for all $t > T$ the previous equality becomes

$$f(U_*(f(p, t_0)), t) \cap U_*(f(p, t_0)) = \emptyset.$$

This equality shows that the point $f(p, t_0)$ is wandering, therefore $f(p, t_0) \in W$ for every $t_0 \in R$. Hence all points of the trajectory $f(p, R) \subseteq W$ are wandering. Consequently, W is invariant.

The invariance of the set of nonwandering points W_N follows from the equality $W_N = X \setminus W$.

Theorem 1.31. *The set of wandering points is open, while the set nonwandering points is closed.*

Proof. It follows from (1.33) that together with the point $p \in W$ all points of its neighborhood $U(p)$ are wandering, i.e. W is open. Since $W_N = X \setminus W$, it follows that W_N is closed.

Theorem 1.32. *If the phase space of dynamical system is compact,
then the set of its nonwandering points is not empty.*

Proof. Let set W_N be empty. Then every point p of the phase
space X is wandering and for the point there can be found a neighborhood $U(p)$ satisfying, for $t > T$, the equality (1.33). By compactness
of X from all such neighborhoods of wandering points there can be
chosen a finite number of neighborhoods $U_j, \, j = \overline{1, N}$ so that

$$X = \bigcup_{j=1}^{N} U_j.$$

Let $T_j, \, j = \overline{1, N}$ be the numbers associated with these neighborhood. Let the point p belongs to the neighborhood U_{n_1}. Then in view
of (1.33) after the expiration of time (more or equal to) T_{n_1} the point
will leave this neighborhood forever and get into a new neighborhood
U_{n_2}, that one the point will leave after a time interval more or equal to
T_{n_2} and so on. Finally, after a time interval $T \geq T_{n_1} + T_{n_2} + \ldots + T_{n_N}$
there will be nowhere for the point to go and it will be forced to enter
one of those neighborhoods, in which it was already, i.e. the point p
is nonwandering.

Theorem 1.33. *If dynamical system possesses at least one L^+ (or
L^-)-stable motion, then the set of nonwandering points is not empty.*

Proof. Let $f(p, t)$ be positively stable according to Lagrange.
Then set Ω_p is not empty and compact. By invariance the set Ω_p can
be considered as a new phase space. It is compact and by the previous
theorem the set of nonwandering points in it is not empty, hence the
set W_N of the space X is also not empty. This proves the theorem.

Here the question arises, what a behavior have wandering motions?
The following theorem answers the question.

It is found that if a phase space is compact, then wandering motions
spend practically all their time in a neighborhood of nonwandering
points, that is all motions in such a space, in some sense, are tending
to be recurrent (returning) ones.

Theorem 1.34. *Let phase space of dynamical system be compact
and $C_0 = W_N(f, X)$ is the set of its nonwandering points. Then any
wandering motion $f(p, t)$ spends only a finite time not exceeding a
certain $T(\varepsilon)$ outside ε-neighborhood of the set C_0.*

Proof. Since X is compact, while $O_\varepsilon(C_0)$ is open then set
$Y = X \setminus O_\varepsilon(C_0)$ is closed, therefore it is compact. It consists of only
from wandering points. Therefore for every point p of this set there
exist a neighborhood $U(p)$ and $t > T$ such that the equality (1.33)
holds. Let us cover the set Y by the finite number of open sets (of
neighborhoods) $U_1, U_2, \ldots, U_N$. Let $T_1, T_2, \ldots, T_N$ be the numbers
associated with these sets.

Then, as by the proving Theorem 1.32, we see that the time which the point p spends in Y does not exceed $T = T_1 + \ldots + T_N$. The theorem is proved.

Thus, for dynamical systems with a compact phase space the set C_0 of nonwandering points is *not empty, closed and invariant*, consequently, it is *compact*.

Therefore C_0 may be taken as a new phase space for dynamical system f. Again we get the dynamical system with the compact metric phase space. All points of this space C_0 will be divided on nonwandering and wandering ones. We obtain the set of nonwandering points

$$C_1 = W_N(f, C_0).$$

Continuing this procedure by induction we define the sets

$$C_{n+1} = W_N(f, C_n), \quad n = 0, 1, \ldots.$$

It has been proved that the intersection of all sets C_j is not empty. Denote the intersection by C_ω.

Definition 1.20. The set C_ω is called a center of dynamical system f. The center is *the greatest closed invariant set*, all points of which are *nonwandering* with respect to this set. The motions on the central set are said to be the *central* motions.

In the set of nonwandering points C_0 (as in the set of the central motions) the recurrence of sets may take place. We shall give the corresponding definition.

Definition 1.21. Dynamical system $f(p, t)$ possesses the property of recurrence of sets (domains) if for any domain $G \subset X$ and any $T > 0$ there can be found $t > T$ such that $G \cap f(G, t) \neq \emptyset$.

Minimal sets. Every closed invariant set which does not contain its closed invariant subset is called a minimal set. Every closed invariant compact set F contains some minimal set. If the space X is compact then it contains some minimal set.

In general case, the dynamical system given on the compact metric space does not possesses the property of recurrence of sets. Only the dynamical systems with an invariant measure possess the recurrence property. We proceed now to the consideration of such systems. For that purpose it will be helpful if we recall some information from the theory of measure.

1.6   Measure. Krylov–Bogolyubov Theorem

1. *Measure.* The measure is a natural generalization of the concepts of length, area and volume for the sets having sufficiently complicated

structure. Just as in elementary geometry we assign to certain geometrical object certain number called its length, area or volume, so we can assign to some complicate set a number called measure of this set. The set is said to be measurable if we can assign to this set the measure. If the set is some geometrical figure, then its area or volume and measure are coincide. A collection of measurable sets forms a system of sets called σ-algebra. The measure itself is treated as a function given on this system of the sets and which to any set of the system puts in correspondence some number, i.e. the measure of this set. We now give the strict definitions.

Thus, the measure is defined always on a system of the sets, i.e. its domain of definition are sets which have been selected according to some common property. Usually these sets form σ-algebra. Recall the definition of σ-algebra. Let X be arbitrary set. The system Σ of subsets of the set X is said to be σ-algebra if the following conditions are fulfilled:

1) $X \in \Sigma$;

2) if $A \in \Sigma$ then $X \setminus A \in \Sigma$;

3) if $A_j \in \Sigma$, $j = 1, 2, \ldots$, then $\bigcup_{j=1}^{\infty} A_j \in \Sigma$.

From these conditions it follows that if $A \in \Sigma$ and $B \in \Sigma$, then the sets $A \cup B$, $A \cap B$, $A \setminus B$, $A \triangle B$ also belong to Σ.

Further, if $A_j \in \Sigma$, $j = 1, 2, \ldots$, then $\bigcap_{j=1}^{\infty} A_j \in \Sigma$. The simplest example of σ-algebra is the totality of all subsets of a given set X. Often σ-algebra is constructed from certain given system S of the sets. If in the some manner certain system S of the sets is selected, then there always exists σ-algebra $\Sigma(S)$ containing S and contained in any σ-algebra which contains S. This σ-algebra is called a minimal σ-algebra containing S. If X is a metric space the minimal σ-algebra generated by the class of the open sets of X is called *Borel* σ-algebra, while its elements are called the Borel set. On the set X with selected σ-algebra Σ of its subsets one can define a measure. We shall define the finite σ-additive measure.

A function μ defined on the sets of the system Σ is called a finite countable-additive measure if:

1) μ takes the real non-negative values on the sets of the system Σ, i.e. $\mu(\omega) \geq 0$, $\forall \omega \in \Sigma$;

2) the function μ is countable-additive, that is for any sequence of the mutually disjoint sets ω_j, $j = 1, 2, \ldots$ of Σ the following equality takes place

$$\mu\left(\bigcup_{j=1}^{\infty} \omega_j \right) = \sum_{j=1}^{\infty} \mu(\omega_j), \ \omega_j \in \Sigma, \ j = 1, 2, \ldots, \ \omega_i \cap \omega_j = \emptyset, \ i \neq j;$$

3) $\mu(X) < \infty$.

The number $\mu(\omega)$ is called a measure of the set ω. If $\mu(X) = 1$, then the measure μ is called *normalized* or *probability* measure. The triple (X, Σ, μ) is said to be a measure space, and if $\mu(X) = 1$, then this space is called a probability space.

There are different ways of the choose of the system Σ for a given set X and there are different ways of representation the measure μ on this set.

Practically the construction of a measure begins with the construction of so called outer measure. If the outer measure is constructed, then by the use of it there can be chosen σ-algebra of sets, on which the outer measure will be a countable-additive function of sets, i.e. it will be a measure. Such a procedure of the construction of a measure was proposed by Karatheodory.

We shall draw on this fact to construct the measure by the use of linear continuous functionals. Now we shall describe the properties of the outer measure and the Karatheodory's procedure of the construction of the measure using the outer measure [85, 108].

Non-negative function μ^* of sets defined on all subsets Σ of the space X is said to be a outer measure on Σ if:

1) $\mu^*(A) \geq 0, \quad \forall A \subset X, \quad \mu^*(\emptyset) = 0$;

2) if $A \subset B$, then $\mu^*(A) \leq \mu^*(B)$;

3) for any countable sequence of sets $\{A_j\}$, $j = 1, 2, \ldots$ the inequality holds

$$\mu^* \left(\bigcup_{j=1}^{\infty} A_j \right) \leq \sum_{j=1}^{\infty} \mu^*(A_j);$$

4) if $\rho(A, B) > 0$, then $\mu^*(A \cup B) = \mu^*(A) + \mu^*(B)$, where

$$\rho(A, B) = \inf_{x \in A, y \in B} \rho(x, y);$$

5) for every sets A it is supposed that

$$\mu^*(A) = \lim_{Q \supset A} \inf \mu^*(Q),$$

where Q are the open sets containing A.

We shall consider that the outer measure μ^* is finite. Having the outer measure one can to introduce the concept of a measurable sets. By the definition the set $A \subset X$ is said to be *a measurable* set, if for any set $W \subset X$ the following equality holds:

$$\mu^*(W) = \mu^*(W \cap A) + \mu^*(W \setminus W \cap A).$$

We shall denote by μ the measure of the measurable set A and set $\mu(A) = \mu^*(A)$. From this definition and the properties of the outer measure the following propositions emerge:

1) all open and all closed sets are measurable sets;
2) if A is measurable set, then $X \setminus A$ is also measurable;
3) if A and B are measurable sets, then their intersection $A \cap B$ and union $A \cup B$ are also measurable; if the intersection is empty, then

$$\mu(A \cup B) = \mu(A) + \mu(B);$$

4) the intersection and the union of countable number of the measurable sets are measurable, thereby if the measurable sets A_j, $j = \overline{1, \infty}$ are mutually disjoint, then

$$\mu(\bigcup_{j=1}^{\infty} A_j) = \sum_{j=1}^{\infty} \mu(A_j).$$

Thus, all the μ-measurable sets form σ-algebra. It can be shown also that, in fact, it is sufficient to give the outer measure μ^* only on the open sets of the space X.

2. *Measurable functions. Integral.* Let $\varphi : X \to R$ be a real function given on the metric space X with values in R. Function $\varphi(p)$ is said to be μ-measurable function, if for any real number a the set

$$E = \{p : \varphi(p) > a\}.$$

is measurable. If $\varphi(p)$ is μ-measurable function, then the following sets are μ-measurable:

$$E_1 = \{p : \alpha < \varphi(p) < \beta\}, \quad E_2 = \{p : \varphi(p) = b\},$$

where α, β and b are some real numbers. Let $\varphi(p)$ be a bounded μ-measurable function, i.e. $m \leq \varphi(p) \leq M$. Let us divide the segment $[m, M]$ into n parts by the points $m = l_0 < l_1 \ldots < l_n = M$. Denote by E_i and E_j^* the sets of the form

$$E_i = \{p : l_i < \varphi(p) < l_{i+1}\}, \quad i = 0, 1, \ldots, n - 1,$$
$$E_j^* = \{p : \varphi(p) = l_j\}, \quad j = 0, 1, \ldots, n.$$

Compose the integral sum

$$\sum_{i=0}^{n-1} l_i \, \mu \, E_i + \sum_{j=0}^{n} l_j \, \mu \, E_j^*.$$

When $n \to \infty$ and the greater of differences $l_{i+1} - l_i$ tends to zero, this sum has unique limit called Lebesgue–Radon integral and denoted by

$$\int_X \varphi(p) \mu(dp) \text{ or } \int_X \varphi(p) \, d\mu.$$

Thus, by the definition

$$\int_X \varphi(p)\,\mu(dp) = \lim_{n\to\infty}\left\{\sum_{i=0}^{n-1} l_i\,\mu\,E_i + \sum_{j=0}^{n} l_j\,\mu\,E_j^*\right\},\ \max\mid l_{i+1} - l_i\mid\ \to\ 0.$$

For the unbounded functions the integral is defined as follows. If $\varphi(p)$ is not bounded and if $\varphi(p) \geq 0$ and it is also μ-measurable function, then one can introduce the functions $\varphi_n(p) = \varphi(p)$, if $0 \leq \varphi(p) \leq n$ and $\varphi(p) = n$, if $\varphi(p) > n$. By the definition we suppose that

$$\int_X \varphi(p)\,\mu(dp) = \lim_{n\to\infty}\int_X \varphi_n(p)\,\mu(dp).$$

The limit in the left side exists always, but it may be equal to the infinity. If the limit is not equal to $+\infty$, then $\varphi(p)$ is called μ-summable function. The integral of the negative measurable function is defined analogously.

Any μ-measurable function can be presented in the following view: $\varphi(p) = \varphi_+(p) + \varphi_-(p)$, where $\varphi_-(p) = 0$, if $\varphi(p) \geq 0$ and it is equal to $\varphi(p)$, if $\varphi(p) < 0$; analogously $\varphi_+(p) = \varphi(p)$, if $\varphi(p) > 0$ and it is equal to 0, if $\varphi(p) \leq 0$. According to the definition, we set

$$\int_X \varphi(p)\,\mu(dp) = \int_X \varphi_+(p)\,\mu(dp) + \int_X \varphi_-(p)\mu(dp).$$

If both integrals in the right side are finite, then the function $\varphi(p)$ is said to be μ-summable function. Let X be a metric space with a countable base and let R be real axis. Introduce the space $X \times R$. The points (p, t) this space are the points of X and the numbers of R. The space $X \times R$ will become a metric one, if we introduce in it the distance

$$\rho\left((p_1, t_1), (p_2, t_2)\right) = \sqrt{\rho^2(p_1, p_2) + (t_1 - t_2)^2}.$$

If $U_j \subset X$ and $\Delta_j \subset R$, then on $U_i \times \Delta_j$ we define the measure ν assuming $\nu(U_i \times \Delta_j) = \mu\,U_i\,\mathrm{mes}\,\Delta_j$, where *mes* is Lebesgue measure on real axis. Here $\{U_i\}$ is the base of the space X, while Δ_j is the set of open intervals with rational edges on the real axis. The measure ν can be extended on the other sets of the space $X \times R$.

Here the following Fubini's theorem takes place: if non-negative function $F(p,t)$ is ν-measurable in $X \times R$, then

$$\int_R dt \int_X F(p, t)\,\mu(dp) = \int_X \mu(dp)\int_R F(p, t)\,dt.$$

For the function $F(p, t)$ of arbitrary sign we set $F = F_+ + F_-$. The following statement on approximation is often very helpful.

Let X be a compact metric space and $\varphi(p)$ be μ-summable function. Then for any $\varepsilon > 0$ there can be found a continuous and uniformly bounded on X function $f_\varepsilon(p)$ such that

$$\int_X |\varphi(p) - f_\varepsilon(p)| \, \mu(dp) < \varepsilon.$$

Finally, if the equality

$$\int_X \varphi(p) \, \mu_1(dp) = \int_X \varphi(p) \, \mu_2(dp)$$

holds for any continuous uniformly bounded on R function $\varphi(p)$, then $\mu_1 = \mu_2$, where μ_1 and μ_2 are measures.

Indeed, this equality by the statement on approximation will hold also for the function $\varphi_G(p)$, which is a characteristic function of the open set G, i.e.

$$\int_X \varphi_G(p) \, \mu_1(dp) = \int_X \varphi_G(p) \, \mu_2(dp),$$

since $\varphi_G(p)$ is μ_1 and μ_2-summable function. Further,

$$\int_X \varphi_G(p) \, \mu_1(dp) = \mu_1 G, \quad \int_X \varphi_G(p) \, \mu_2(dp) = \mu_2 G,$$

hence $\mu_1 G = \mu_2 G$ for any open set G. In the ordinary way we can now to prove that the measures μ_1 and μ_2 coincide on the other sets also.

3. *Construction of a measure by the use of linear functional. Riesz–Radon Theorem.*

Let X be a compact metric space. On the sets of all continuous functions $\{\varphi(p)\}$ given on X we define a linear continuous functional $L(\varphi)$, i.e. $|L(\varphi)| \leq \max |\varphi| \, \mu X$. Let $L(\varphi)$ be a positive functional, i.e. $L(\varphi) \geq 0$ if $\varphi \geq 0$ and $L(\varphi) = 1$, if $\varphi \equiv 1$.

It is seen that every such functional generates the normalized measure, while the functional itself is expressed by the Lebesgue–Radon integral on this measure. This result constitute the content of the Riesz–Radon theorem [108].

Theorem 1.35. *Every linear continuous positive functional $L(\varphi)$ defined on the sets of all continuous functions $\{\varphi(p)\}$ of a compact metric space X such that $L(1) = 1$ generates a normalized measure μ on σ-algebra of Borel sets of the space X, thereby*

$$L(\varphi) = \int_X \varphi(p) \, \mu(dp). \tag{1.34}$$

Proof. Let us extend the domain of definition of the functional $L(\varphi)$. Let G is any open set of X and φ_G is its characteristic function. Now give a sequence of the continuous functions $\{\varphi_n(p)\}$ convergent to φ_G, i.e. let, for example,

$$\varphi_n(p) = \begin{cases} 1, & \rho(p, R \setminus G) \geq 1/n, \\ 0, & p \in R \setminus G, \\ n\rho(p, R \setminus G), & 0 \leq \rho(p, R \setminus G) \leq 1/n. \end{cases}$$

It is evident that

$$\lim_{n \to \infty} \varphi_n(p) = \varphi_G(p),$$

$$0 < \varphi_1(p) \leq \varphi_2(p) \leq \ldots \leq \varphi_n(p) \leq \ldots \leq 1.$$

Therefore,

$$L(\varphi_1) \leq L(\varphi_2) \leq \ldots \leq L(\varphi_n) \leq \ldots \leq L(1) = 1,$$

i.e. the sequence of numbers $L(\varphi_j)$, $j = 1, 2, \ldots$ converges

$$\lim_{n \to \infty} L(\varphi_n) = g = L(\varphi_G) \leq 1.$$

(We notice that if $\varphi \geq \psi$, then $L(\varphi) \geq L(\psi)$, since $\varphi - \psi \geq 0$, $L(\varphi - \psi) = L(\varphi) - L(\psi) \geq 0$). Now we shall call the number

$$\mu G = L(\varphi_G).$$

the measure of the open set G. Clear, $0 \leq \mu G \leq 1$, $\mu X = 1$; if $G_1 \subset G_2$, then $\mu G_1 \leq \mu G_2$, because $\varphi_{G_1} \leq \varphi_{G_2}$, $L(\varphi_{G_1}) \leq L(\varphi_{G_2})$. It follows from the fact that if $G_1 \subset G_2$, then

$$\varphi_n^{(1)}(p) \to \varphi_{G_1}, \quad \varphi_n^{(2)}(p) \to \varphi_{G_2}, \quad \varphi_n^{(1)}(p) \leq \varphi_n^{(2)}(p), \quad n \to \infty.$$

Let $\{G_j\}$, $j = 1, 2, \ldots$ be a sequence of open sets and $G_i \cap G_j = \emptyset$, $i \neq j$. Then $\cup_{j=1}^{\infty} G_j$ is an open set and

$$\mu \left(\bigcup_{j=1}^{\infty} G_j \right) = \sum_{j=1}^{\infty} \mu(G_j).$$

Indeed, the set

$$G = \bigcup_{j=1}^{\infty} G_j$$

is an open set, therefore μG has a sense according to the above-said. Notice that if $p \in G_1 \cup G_2$, then either $p \in G_1$ or $p \in G_2$. Hence $\varphi_{G_1 \cup G_2} = \varphi_{G_1} + \varphi_{G_2}$. It is seen that

$$\varphi_{\cup_{j=1}^{\infty} G_j} = \sum_{j=1}^{\infty} \varphi_{G_j}, \quad G_i \cap G_j = \emptyset \quad \forall p \in \bigcup_{j=1}^{\infty} G_j.$$

Consequently,

$$\mu\left(\bigcup_{j=1}^{\infty} G_j\right) = L\left(\varphi_{\bigcup_{j=1}^{\infty} G_j}\right) = L\left(\sum_{j=1}^{\infty} \varphi_{G_j}\right) = \sum_{j=1}^{\infty} L\left(\varphi_{G_i}\right) = \sum_{j=1}^{\infty} \mu\, G_j.$$

Thus, the measure μ on the open sets is σ-additive and satisfies the Karatheodory's conditions for the outer measures. Hence this measure can be extended on the Borel sets of the space X. Now we shall show that the functional $L(\varphi)$ can be presented as integral (1.34).

Notice that $m \le \varphi(p) \le M$. Let us devide the segment $[m, M]$ into n parts by the points $m = l_0, l_1, \dots, \quad l_N = M$. Let F_i and G_i are the sets of the form

$$F_i = \{p : \varphi(p) = l_i\}, \quad i = \overline{0, N},$$
$$G_j = \{p : l_j < \varphi(p) < l_{j+1}\}, \quad j = \overline{0, N-1}.$$

Introduce the function $\varphi_N(p)$ as follows

$$\varphi_N(p) = \sum_{i=0}^{N} l_i\, \varphi_{F_i} + \sum_{j=0}^{N-1} l_i\, \varphi_{G_i}, \ |\, L(\varphi) - L(\varphi_N)\,| \le \frac{M-m}{N}.$$

Then

$$L(\varphi_N) = \sum_{i=0}^{N} l_i\, L(\varphi_{F_i}) + \sum_{j=0}^{N-1} l_j\, L(\varphi_{G_j}) = \sum_{i=0}^{N} l_i\, \mu\, F_i + \sum_{j=0}^{N-1} l_j\, \mu\, G_j.$$

Passing to the limit as $N \to \infty$, we get

$$L(\varphi) = \int_R \varphi(p)\, \mu(dp).$$

4. *Convergence of measures.* In what follows we shall deal with the sequences of measures $\{\mu_n\}$ and their convergence. By the convergence of measures will be meant always the weak convergence.

Definition 1.22. The sequence of measure $\{\mu_n\}$ weakly converges to the measure μ, if for any continuous function $\varphi(p)$ we have

$$\lim_{n \to \infty} \int_X \varphi(p)\, \mu_n(dp) = \int_X \varphi(p)\, \mu(dp). \qquad (1.35)$$

We define the metric on the sets of continuous functions $\{\varphi(p)\}$, which are given usually on the compact metric space X

$$\rho(\varphi_1, \varphi_2) = \{\max |\, \varphi_1(p) - \varphi_2(p)\,|, \quad p \in X\}.$$

This space will be separable one, i.e. it contains a countable everywhere dense set of functions. Using this fact we can demonstrate the following fundamental theorem.

Theorem 1.36. *Let $\{\mu_n\}$, $n = 1, 2, \dots$ be any sequence of normalized measures given on a compact metric space. Then from this sequence one can always choose the subsequence $\{\mu_{n_j}\}$ weakly convergent to certain normalized measure μ, i.e. there exists a limit*

$$\lim_{n_j \to \infty} \int_X \varphi(p)\, \mu_{n_j}(dp) = \int_X \varphi(p)\, \mu(dp).$$

This property can be briefly formulated as follows. The set of normalized measures in a compact metric space is weakly precompact.

Proof.   From the countable everywhere dense set of continuous functions we choose a countable sequence $\{\varphi_j(p)\}$ such that $|\varphi_j(p)| \le 1$. Then the set of numbers

$$\int_X \varphi_1(p)\mu_n(dp), \quad n = 1, 2, \dots$$

will lie on the segment $[-1, +1]$. We choose from this set a convergent sequence

$$\int_X \varphi_1(p)\mu_n^{(1)}(dp), \quad n = 1, 2, \dots$$

Consider now the sequence of numbers

$$\int_X \varphi_2(p)\mu_n^{(1)}(dp),$$

also lying on the segment $[-1, +1]$. From the above sequence we choose again a convergent subsequence

$$\int_X \varphi_2(p)\mu_n^{(2)}(dp).$$

Continuing this procedure we get the sequences of measures $\{\mu_n^{(k)}\}$ such that there exists a limit

$$\lim_{n \to \infty} \int_X \varphi_k(p)\mu_n^{(k)}(dp), \quad k = 1, 2, \dots$$

We chose from the matrix $\{\mu_n^{(k)}\}$ the diagonal elements $\mu_k^{(k)} \equiv \mu^{(k)}$. They will possess the following property: for any function φ_n, $n = 1, 2, \dots$ there exists a limit

$$\lim_{k \to \infty} \int_X \varphi_n(p)\mu^{(k)}(dp) = \lim_{k \to \infty} L^{(k)}(\varphi_n), \tag{1.36}$$

where $L^{(k)}(\varphi)$ linear functional of the form

$$L^{(k)}(\varphi) = \int_X \varphi(p)\mu^{(k)}(dp).$$

By the continuity it can be shown that the limit (1.36) exists and $|\varphi| \leq 1$ (does not necessarily enter the countable everywhere dense set). Denote the limit by $L\varphi$. Thus, for any continuous function $\varphi(p)$, $|\varphi| \leq 1$ we have

$$\lim_{k\to\infty} L^{(k)}(\varphi) = \lim_{k\to\infty} \int_X \varphi(p)\,\mu^{(k)}(dp) = L(\varphi).$$

So obtained functional may be defined on any continuous function φ, if we set

$$L(\varphi) = \max |\varphi| \cdot L\left(\frac{\varphi}{\max |\varphi|}\right).$$

It is evident that $L(\varphi)$ is continuous, $L(1) = 1$ and $L(\varphi) \geq 0$ if $\varphi \geq 0$. Hence, by virtue of the theorem of Riesz–Radon there exists a normalized measure μ such that

$$L(\varphi) = \int_X \varphi(p)\,\mu(dp).$$

Thus,

$$\lim_{k\to\infty} \int_X \varphi(p)\,\mu^{(k)}(dp) = \int \varphi(p)\,\mu(dp)$$

for any continuous function $\varphi(p)$, i.e. $\mu^{(k)} \to \mu$ in the weak sense.

In conclusion we notice that the phase space X always has a normalized measure. For example, let $p \in X$ be any point. We define the measure m_p associated with this point, setting for any set $A: m_p A = 1$ if $p \in A$, and $m_p A = 0$ if $p \in X \setminus A$.

It can be shown that the measures $\{m_p\}$ possess the property: any normalized measure μ is a weak limit of linear combinations

$$\sum_{j=1}^n \alpha_j m_{p_j}, \quad \sum_{j=1}^n \alpha_j = 1, \quad \alpha_j > 0, \quad j = \overline{1, n}.$$

Let us proceed now to the presentation of the Krylov–Bogolyubov theorem on the existence of invariant probability measure of dynamical system acting in compact phase space [85].

Theorem 1.37. *A dynamical system acting in a compact metric space possesses an invariant probability measure.*

Proof. Let $f(p,t)$ be a dynamical system acting in a compact metric space X and let m be any probability measure in X. For any T and any continuous function $\varphi(p)$ we define a linear functional

$$L_T(\varphi) = \frac{1}{T} \int_0^T dt \int_X \varphi(f(p,t)) \, m(dp).$$

The functional is continuous and positive one. By the theorem of Riesz–Radon it defines the normalized measure m_T, thereby

$$\frac{1}{T} \int_0^T dt \int_X \varphi(f(p,t)) \, m(dp) = \int_X \varphi(p) \, m_T(dp).$$

By Theorem 1.36 the set of measures $\{m_T\}$ is compact. We can choose from it some convergent (weakly) sequence $m_{T_n} \to \mu$, $T_n \to +\infty$, $n \to \infty$. Then the equalities will take place

$$\lim_{n\to\infty} \frac{1}{T_n} \int_0^{T_n} dt \int_X \varphi(f(q,t)) m(dq)$$
$$= \lim_{n\to\infty} \int_X \varphi(p) \, m_{T_n}(dp) = \int_X \varphi(p) \, \mu(dp). \qquad (1.37)$$

So constructed measure μ will be invariant one. If the measure is to be invariant, it is necessary and sufficient that for any t_0 and any continuous function $\varphi(p)$ the equality holds

$$\int_X \varphi(p) \, \mu(dp) = \int_X \varphi(f(p,t_0)) \mu(dp). \qquad (1.38)$$

In fact, as it was shown above, the functional (1.37) has also a sense when φ is characteristic function φ_G of the open set G. If (1.38) holds then for any open set G we have

$$\int_X \varphi_G(p) \, \mu(dp) = \mu G = \int_X \varphi_G(f(p,t_0)) \mu(dp) = \mu f(G, -t_0).$$

That is, $\mu G = \mu f(G, -t_0)$. Further, in the ordinary way one can prove that for any measurable set $E \subset X$ we have $\mu E = \mu f(E,t)$, i.e. μ is an invariant measure. Let us prove (1.38). By virtue of (1.37)

$$\int_X \varphi(f(p,t_0)) \, \mu(dp)$$
$$= \lim_{n\to\infty} \frac{1}{T_n} \int_0^{T_n} dt \int_X \varphi(f(q, t_0 + t)) \, m(dq). \qquad (1.39)$$

Hence by virtue of (1.38) it is sufficient to demonstrate that the integral in left side of (1.37) is equal to the integral in the right side of (1.39). For that purpose we estimate the difference

$$
\left| \frac{1}{T_n} \int_0^{T_n} \varphi\left(f(p,t)\right) dt - \frac{1}{T_n} \int_0^{T_n} \varphi\left(f(q, t_0 + t)\right) dt \right|
$$

$$
\leq \frac{1}{T_n} \left(\left| \int_0^{t_0} \varphi(fq, t)) \, dt \right| + \left| \int_{T_n}^{T_n + t_0} \varphi(f(q, t)) \, dt \right| \right) \leq \frac{2\,M \,\mid t_0 \mid}{T_n},
$$

where $\{\mid \varphi(q) \mid \leq M, q \in X\}$. For the sufficiently large T_n this value is small. Consequently, it is proved that integrals are equal one to another.

As the initial probability measure in the Krylov – Bogolyubov theorem one can use the measure m_p connected with the point $p \in X$. Using this measure we construct the measures m_{pT} according to the equality

$$
\frac{1}{T} \int_0^T dt \int_X \varphi\left(f(q, t)\right) m_p(dq) = \int_X \varphi(q)\, m_{pT}(dq).
$$

Since for the measure m_p the following equality holds

$$
\int_X \varphi(f(q, t))\, m_p(dq) = \varphi(f(p, t)),
$$

then the measures m_{pT} will be defined by the relation

$$
\frac{1}{T} \int_0^T \varphi(f(p, t))\, dt = \int_X \varphi(q)\, m_{pT}(dq). \tag{1.40}
$$

Now we shall not choose the sequence $T_n \to \infty$, for which integral in the left side of (1.40) converges, we consider those points p of the space X, for which this limit exists as $T \to \infty$. In this case the sequence of measures $\{m_{pT}\}$ will be convergent as $T \to +\infty$ weakly to the invariant measure μ_p, that is

$$
\lim_{T \to \infty} \frac{1}{T} \int_0^T \varphi(f(p, t))\, dt = \lim_{T \to \infty} \int_X \varphi(q)\, m_{pT}(dq) \tag{1.41}
$$

$$
= \int_X \varphi(p)\, \mu_p(dp).
$$

An invariant measure μ_p is called *individual*, while the points p, for which there holds (1.40), are called *quasiregular*. From the above discussion it follows that a dynamical system in a compact metric space possesses the invariant measures.

All the totality of invariant measures admitted by a given dynamical system will not be considered here. Notice only that any invariant measure μ is connected with an individual measure μ_p by the next equality [108]:

$$\mu E = \int_U \mu_p(E)\mu(dp).$$

In the sense of applicability of the Krylov–Bogolyubov theorem the question arises: what invariant measure of a given dynamical system is *essential*, i.e. has a physical sense adequate to the problem under consideration.

1.7 Dynamical Systems with Invariant Measure

Let X be a metric space and $f(p,t)$ be a dynamical system on it. Let, further, μ be a normalized (probability) measure invariant with respect to $f(p,t)$, i.e.

$$\mu A = \mu f(A, t), \quad \mu X = 1, \quad \forall A \in \sigma,$$

where σ is σ-algebra of subsets of the space X.

Systems with invariant measure possess the properties which set off them from the other dynamical systems. The main property of these systems is that they return in a given neighborhood of their initial point.

Theorem 1.38. (Poincaré, recurrence of sets). *Let $(X\ \sigma,\ \mu,\ f)$ be a dynamical system with invariant normalized measure μ.*

Let $A \in \sigma$ and $\mu A = a > 0$ ($\mu X = 1$). Then there always can be found $t = t^*$ such that

$$\mu(A \cap f(A, t^*)) > 0.$$

Proof.   Notice that since $A \subset X$ one has $\mu A < \mu X = 1$, that is, $0 < a < 1$. We consider the images of the set A at moments $t = 0, \pm 1, \pm 2, \ldots$, i.e.

$$A_n = f(A, n), \quad A_0 \equiv A, \quad n = 0, \pm 1, \pm 2, \ldots$$

By the invariance of measure μ we have

$$\mu A = \mu f(A, n) = \mu A_n = a > 0, \quad n = 0, \pm 1, \pm 2, \ldots$$

If we assume that the sets $A_0, A_1, \ldots, A_k$ are mutually disjoint (or the measure of set of their intersection is equal to zero), then

$$\mu\left(\bigcup_{j=0}^{k} A_j\right) = \sum_{j=0}^{k} \mu A_j = k\,a.$$

Since

$$\bigcup_{j=o}^{k} A_j \subset X,$$

one has

$$k\,a = \mu\big(\bigcup_{j=0}^{k} A_j\big) < \mu\,X = 1,$$

i.e. there must be $k\,a < 1$. If we take $k > [1/a]$, then we get a contradiction

$$\mu\big(\bigcup_{j=1}^{k} A_j\big) > 1.$$

Consequently, if $k \le [1/a]$, then among sets $A_0, A_1, \ldots, A_k$ there must be found a pair of sets A_j and A_i, whose intersection measure is greater than zero, i.e.

$$\mu\,(A_i \cap A_j) > 0, \quad i \le k, \quad j \le k.$$

Let $0 \le i < j \le k$. Let us apply to $A_i \cap A_j$ the transformation $f\,(\cdot, -i)$. Noticing that $f\,(A_m, -m) = A_0$, we find

$$f\,(A_i \cap A_j, -i) = f\,(A_i, -i) \cap f\,(A_j, -i) = A_0 \cap f\,(A_0, j - i).$$

Hence,

$$\mu\,(A_i \cap A_j) = \mu\,f\,(A_i \cap A_j, -i) = \mu\,(A_0 \cap f\,(A_0, j - i)) > 0, \quad j - i \ge 1.$$

Thus, for $t^* = j - i \ge 1$ the sets A_0 and $f(A_0, t^*)$ are intersect and the measure of their intersection is greater than zero, moreover

$$1 \le j - i \le 1 + [1/a], \quad 1 \le t^* \le 1 + [1/a].$$

Applying the transformation $f(\cdot, -j)$ to $A_i \cap A_j$, we find that

$$\mu\,(A_0 \cap f\,(A_0, i - j)) > 0, \quad i - j \le -1.$$

It must be noted that the values of t, for which $\mu(A \cap f(A, t)) > 0$, may be as large as one likes.

Indeed, let $T > 0$ be any number and $N > T$ be an integer. We choose the sequence $A_0, A_N, A_{2N}, \ldots, A_{kN}$. Then, for $t = N(j - i)$ we obtain

$$\mu\left(A_0 \cap f\,(A_0, N\,(j - i))\right) > 0, \quad N\,(j - i) \ge N > T.$$

Theorem of Poincaré does not indicate how much the number $\mu(A \cap f(A, t))$ is greater than zero and how is the set of values of t, for which the recurrence occurs. On this matter there is theorem of Khintchine. Before presenting this theorem we shall prove the following proposition.

Let μ be a normalized measure ($\mu X = 1$) and let E_k, $k = 1, 2, \ldots$ be measurable sets $E_k \in \sigma$ having the same measure: $\mu E_k = a$, $k = 1, 2, \ldots$, $0 < a < 1$. Then among the sets $\{E_k\}$ there can be found at least two sets E_i and E_j such that for any $0 < \lambda < 1$ the inequality holds

$$\mu(E_i \cap E_j) > \lambda a^2.$$

In fact, let $\varphi_{E_k}(p)$ be a characteristic function of the set E_k. Denote

$$f_n(p) = \sum_{k=1}^{a} \varphi_{E_k}(p).$$

Assume that the proposition is false, i.e. $\mu(E_i \cap E_j) \leq \lambda a$ for any E_i and E_j. Hence

$$(n\,a)^2 = \sum_{k=1}^{n} (\mu E_k)^2 = \left(\sum_{k=1}^{n} \int_X \varphi_{E_k}(p)\mu(dp) \right)^2 = \left(\int_X f_n(p)\mu(dp) \right)^2$$

$$\leq \int_X f_n^2(p)\,\mu(dp) \int_X 1^2\,\mu(dp) = \sum_{k=1}^{n} \mu E_k$$

$$+ \sum_{i,j=1}^{n} \mu(E_i \cap E_j) \leq n\,a + n\,(n-1)\,\lambda\,a^2, \quad i \neq j.$$

Thus, $(n\,a)^2 \leq n\,a + n\,(n-1)\,\lambda\,a^2$, that will not take place for large n. The proposition is proved.

Now we give the Khintchine theorem [108].

Theorem 1.39. *Under conditions of Poincaré's theorem on recurrence of sets for any μ-measurable set A such that $\mu A = a > 0$, and for any $0 < \lambda < 1$ the inequality*

$$\mu(A \cap f(A, t)) > \lambda a^2$$

holds for a relatively dense set of values of t. The theorem is proved by reductio ad absurdum [108].

Let us proceed now to the theorem of recurrence of points. Let (X, σ, μ, f) be a dynamical system with invariant normalized measure μ and $A \in \sigma$, $\mu A = a > 0$. Introduce the notations

$$T A = f(A, 1), \quad T^n A = f(A, n), \quad n = \pm 1, \pm 2, \ldots$$

A point $p \in A$ is said to be *recurrent* into the set A for $t > 0$, if $T^n p \in A$ for some $n \geq 1$. Similarly, a point $p \in A$ is said to be recurrent into A for $t < 0$, if $T^{-n} p \in A$ for some $n \geq 1$. (Sometimes this is called *a recurrence in the sense of Poincaré.*) Let W_+ denote the set of non-recurrent for $t > 0$ (*wandering*) points of the set A.

Similarly, we shall denote by W_- the set of wandering for $t < 0$ points of the set A. The set $W = W_+ \cup W_-$ is the set of wandering points of the set A. It is evident that if $p \in W \subset A$, then one has $T^n p \notin A$ for $n = \pm 1, \pm 2, \ldots$. If the point p returns into A, then it returns into A arbitrary often.

Theorem 1.40. $\mu W = 0$. (*The measure of wandering points of the set A is equal to zero, this means that almost all points of A are recurrent.*)

Proof. We show that $\mu W_+ = 0$ (the proof of the fact that $\mu W_- = 0$ is similar). Since $W_+ \subset A$, one has $\mu W_+ \leq \mu A$.

If $p \in W_+$, then $T^n p \notin A$ for $n = 1, 2, \ldots$. Moreover $T^n p \notin W_+$, that is $p \notin T^{-n} W_+$. Consequently,

$$W_+ \cap T^{-n} W_+ = \emptyset, \quad n = 1, 2, \ldots$$

But then the sets $W_+, T^{-1} W_+, T^{-2} W_+, \ldots$ are mutually disjoint, because for $0 < k < m$ we have

$$T^{-k} W_+ \cap T^{-m} W_+ = T^{-k} (W_+ \cap T^{-(m-k)} W_+) = \emptyset.$$

Therefore,

$$1 \geq \mu \left(\bigcup_{n=0}^{\infty} T^{-n} W_+ \right) = \sum_{n=0}^{\infty} \mu (T^{-n} W_+) = \sum_{n=0}^{\infty} \mu (W_+) = \infty,$$

i.e. $\mu (W_+) = 0$. It is noteworthy that

$$W_+ = A \setminus \bigcup_{n=1}^{\infty} (A \cap T^{-n} A), \quad W_- = A \setminus \bigcup_{n=1}^{\infty} (A \cap T^n A).$$

In fact, if $p \in W_+$, then $T^n p \notin A$ for $n = 1, 2, \ldots$. Let

$$p \in A \setminus \bigcup_{n=1}^{\infty} (A \cap T^{-n} A),$$

Then $p \in A$, $p \notin A \cap T^{-n} A$, $n = 1, 2, \ldots$, i.e. $p \notin T^{-n} A$, $n = 1, 2, \ldots$. Hence, $T^n p \notin A$, $n = 1, 2, \ldots$, i.e. p is a wandering point.

Theorem 1.41. (Poincaré, recurrence of points). *Let (X, σ, μ, f) be a dynamical system with invariant normalized measure μ and let X be a compact metric space.*

Then almost all points $p \in X$ in the sense of measure μ are stable according to Poisson, i.e. for any neighborhood $U(p)$ of the point p one can find $t = t^*$ such that

$$f(p, t^*) \in U(p).$$

Proof. By the compactness of the space X there exists a countable base of the neighborhoods $\{U^{(n)}\}$, $n = 1, 2, \ldots$.

Let $W^{(n)} = W_+^{(n)} \cup W_-^{(n)}$ be a set of wandering points of U^n. Since $\mu W^{(n)} = 0$, $n = 1, 2, \ldots$, one has

$$\mu W = \mu \left(\bigcup_{n=1}^{\infty} W^{(n)} \right) = 0.$$

Let $p = X \setminus W$, then there exists a neighborhood $U^{(k)}$ such that $p \in U^{(k)}$ and p is nonwandering point of the set $U^{(k)}$. Therefore, there exists m such that $p \in f(U^{(k)}, m)$. From this it follows that $f(p, -m) \in U^{(k)}$, where m may be as positive so negative integer. Since any neighborhood $U(p)$ of the point p is an open set, we have

$$U(p) = \bigcup_{n=1}^{N} U^{(n)}, \quad N \leq \infty.$$

The theorem is proved.

Remark. The closure of the set of points stable according to Poisson forms the set C of *central motions*. From the above theorem it follows that the complement $X \setminus C$ has a zero measure, while the set C is not the everywhere dense one in X, i.e. it does not fill the whole space . If the system has an integral invariant, then C fills the whole space X [108].

It may be noted that two solutions $f(p, t)$ and $f(q, t)$ emanating from the nearby points p and q lying in the small neighborhood U, after a certain time will return again in the neighborhood U, i.e. if $p \in U$, $q \in U$, $\rho(p, q) < \delta$, then

$$\rho(f(p, t^*), f(q, t^{**})) < \varepsilon.$$

However, the difference $|t^* - t^{**}|$ may be large. Poincaré's theorem holds also in the case when X need not to be compact. The theorem is true for the case when X is locally compact space with a countable base, similar to Euclidian finite-dimensional space R_n.

However, the invariant measure in R_n may be not the normalized one, i.e. $\mu R_n = \infty$. Hopf had generalized the theorem of Poincaré for the case when $\mu X = \infty$.

Theorem 1.42. (Hopf). *Let there be given a locally compact metric space X with a countable base and $\mu X = \infty$.*

Let there be defined in it an invariant measure having the following properties: but for any compact set $F \subset X$ the measure μ is finite. Then almost all points $p \in X$ as $t \to +\infty$ are either stable according to Poisson or departing.

In many problems there is the necessity to know how much time the solution $f(p, t)$ spends in one or another portion of the phase space X, for example, how much time the solution spends within the set A. Let $\varphi_A(p)$ be a characteristic function of the set A and $t \in [0, T]$.

Then the number of entrances of the solution $f(p, t)$ into A in a interval $0 \leq t \leq T$ will be equal to

$$\tau = \int_0^T \varphi_A \left(f \left(p, t \right) \right) dt.$$

The relation τ/T represents the relative time which the solution spends within the set A in a time interval $[0, T]$. Clearly, $0 \leq \tau/T \leq 1$. If there exists a limit

$$\lim_{T \to \infty} \frac{\tau}{T} = \lim_{T \to \infty} \frac{1}{T} \int_0^T \varphi_A \left(f \left(p, t \right) \right) dt,$$

then it called *the probability of occurrence the point within the set A* as $t \to \infty$. The question arises: does there exist this limit, and if it exists, to what value it is equal. The answer is provided by Birkhoff theorem. Usually, instead of the characteristics function φ_A one takes any μ-measurable function φ. In this case the limit

$$\lim_{T \to \infty} \frac{1}{T} \int_0^T \varphi \left(f(p, t) \right) dt$$

is called the mean of a function along the solution $f(p, t)$.

Theorem 1.43. (Birkhoff). *Let (X, σ, μ, f) is a dynamical system with invariant normalized measure μ. Then for almost all (in the sense of μ-measure) points $p \in X$ there exists a mean time (along the solution $f(p, t)$)*

$$\lim_{T \to \infty} \frac{1}{T} \int_0^T \varphi \left(f(p, t) \right) dt = \psi(p)$$

for any μ-summable function $\varphi(p)$.

A limit function $\psi(p)$ possesses the following properties:
1) *it is summable function;*
2) $\psi(p) = \psi(f(p, t))$, *i.e. ψ is an invariant function;*
3) $\int_X \varphi(p) \mu(dp) = \int_X \psi(p) \mu(dp)$.

The proof of this theorem is very cumbersome [108]. We omit it here. We shall give only some comments.

Let, again, $\varphi = \varphi_A$ be a characteristic function of the set A. Then

$$\lim_{T \to \infty} \frac{1}{T} \int_0^T \varphi_A \left(f \left(p,\, t \right) \right) dt = \psi \left(p,\, A \right),$$

where $\psi \left(p,\, A \right)$ will be equal to the probability of occurrence of the point p within set A. For the different points p, i.e. for different solutions we get different probabilities of occurrence within a given set A. If the function $\psi \left(p,\, A \right)$ does not depend on the point p, i.e. $\psi \left(A \right) = c = \text{const}$, then according to the property 3 we obtain

$$\int_X \varphi_A \left(p \right) \mu \left(dp \right) = \mu \left(A \right) = \int_X c \, \mu \left(dp \right) = c \, \mu \, X = c,$$

i.e. $c = \mu \left(A \right)$. In this case, the probability of occurrence of any point of the space X in the set A is the same and is equal to $\mu \left(A \right)$.

Let us consider the discrete case. Let

$$T p = f \left(p,\, 1 \right), \quad T^n p = f \left(p,\, n \right), \quad n = 1,\, 2,\, \dots$$

Then the theorem of Birkhoff can be reformulated as follows. For almost all points $p \in X$ there exists a mean

$$\lim_{n \to \infty} \frac{1}{n} \sum_{n=0}^{n-1} \varphi \left(T^n p \right).$$

If $\varphi = \varphi_A$, then $\dfrac{1}{n} \sum_{n=0}^{n-1} \varphi_A \left(T^n p \right)$ is the frequency of the entrances of the point p into set A. The theorem of Birkhoff asserts that there exists with the probability equal to 1 a limit of this frequency of entrances as $n \to \infty$, i.e. this is the statement similar to the low of large numbers

Remark. The law of large numbers (Chebyshev's theorem). Let $x_1,\, x_2,\, \dots,\, x_n$ are independent random values, whose dispersions do not exceed the number c. Then for any $\varepsilon > 0$ we have

$$\lim_{n \to \infty} P \left(| \frac{1}{n} \sum_{j=1}^n x_j - \frac{1}{n} \sum_{j=1}^n M \left(x_j \right) | < \varepsilon \right) = 1,$$

where $M \left(x_j \right)$ is mathematical expectation x_j.

Ergodic systems and measures. Let us now examine the ergodic dynamical systems. Such systems possess a number of important properties. Some of these properties will be described below. For instance, if the system is ergodic, then its invariant measure is unique.

Definition 1.23. Let there be given a system $\left(X,\, \sigma,\, \mu,\, f \right)$. The dynamical system $f(p, t)$ is said to be ergodic with respect to the

measure μ, if X cannot be presented as two mutually disjoint invariant
sets of positive measure. This means that there cannot take place
an equality $X = A \cup B$, $A \cap B = \emptyset$, where $\mu A > 0$, $\mu B > 0$,
$A = f(A, t)$, $B = f(B, t)$.

Hence it is seen that if the system is ergodic with respect to μ and
$\mu A > 0$, $A = f(A, t)$, then $\mu(X \setminus A) = 0$, since $X \setminus A$ is invariant.

If the system is ergodic with respect to μ, then the measure of any
invariant set is either equal to 0 or 1.

Theorem 1.44. *If the measure μ is invariant, but is not ergodic
with respect to $f(p, t)$, then using it one can construct so much invariant measures for $f(p, t)$ as one likes.*

Proof.   Let μ be not ergodic, but let it be invariant; this means
that there exists an invariant set A, $\mu(A) > 0$ and $\mu(X \setminus A) > 0$.
Define the system of measures $\mu_\alpha(B)$, $\forall B \in \sigma$, setting

$$\mu_\alpha(B) = \alpha \frac{\mu(A \cap B)}{\mu(A)} + (1 - \alpha) \frac{\mu((X \setminus A) \cap B)}{\mu(X \setminus A)}, \quad 0 < \alpha < 1.$$

Measures μ_α are invariant ones. In fact, direct calculations show that
$\mu_\alpha(B) = \mu_\alpha(f(B, t))$.

Let us make a remark relatively to the ergodic measures.

The measure μ_1 is said to be *absolutely continuous* with respect to
the measure μ_2 if by virtue of $\mu_2(A) = 0$, $A \in \sigma$, we have $\mu_1(A) = 0$.

Measures μ_1 and μ_2 are called *singular* with respect to one another
if X may be broken into two sets X_1 and X_2 having no points in
common and such that

$$\mu_1(X_1) = 0, \quad \mu_2(X_2) = 0, \quad X_1 \in \sigma, \quad X_2 \in \sigma.$$

Singularity of measures means that the measure μ_1 is not equal to zero
only on subsets of some set μ_2 with zero measure and conversely.

We shall give some statements from the theory of ergodic systems.

Let there be given on the space (X, σ) a dynamical system $f(p, t)$,
$p \in X$ and two normalized invariant with respect to $f(p, t)$ measures
μ_1 and μ_2. Then:

1) if the measures μ_1 and μ_2 are ergodic, then either $\mu_1 = \mu_2$, or
measures μ_1 and μ_2 are mutually singular;

2) if μ_1 is ergodic, while μ_2 is absolutely continuous with respect
to μ_1, then μ_2 is ergodic and $\mu_1 = \mu_2$.

This statement practically shows that in the ergodic case the invariant measure is unique.

If system $f(p, t)$ is ergodic (with respect to the measure μ), then
any invariant function $g(p)$, $p \in X$ is equal to almost everywhere
constant one.

Remark. Let there be given a system $(X \sigma, \mu, f)$ and let $E \in \sigma$ be any measurable set, $\mu(E) > 0$. This set can always be regarded as a space with measure μ_E defined as follows

$$\mu_E(A) = \frac{\mu(A \cap E)}{\mu(A)}, \quad \forall A \subseteq E, \quad A \in \sigma.$$

We get the probability space (E, σ_E, μ_E). Let now the system be ergodic (i.e. measure μ is not only invariant, but it is ergodic one). Then the limit function $\psi(p)$ must be (almost everywhere) constant by virtue of its invariance, that is $\psi = \text{const} = c$. Then, from the condition 3) of Theorem 1.43 it follows that

$$c = \frac{1}{\mu(X)} \int_X \varphi(p) \mu(dp),$$

i.e. c is equal to the mean of function φ over the whole space X. Now we have the equality

$$\lim_{T \to \infty} \frac{1}{T} \int_0^T \varphi(f(p, t)) \, dt = \frac{1}{\mu(X)} \int \varphi(p) \mu(dp).$$

This equality means that for ergodic systems the average along the solution of the system is equal to the phase space average. All the above-said one can formulate as follows.

Theorem 1.45. *If a system is ergodic and a measure μ is invariant and normalized then the time mean*

$$\lim_{T \to \infty} \frac{1}{T} \int_0^T \varphi(f(p, t)) \, dt$$

has the same value for almost all points $p \in X$.

We regard the particular case, when in the theorem there is $\varphi = \varphi_E$ which represents a characteristic function of the set $E \in \sigma$. Then

$$\lim_{T \to \infty} \frac{1}{T} \int_0^T \varphi_E(f(p,t)) \, dt = \mu(E).$$

But in this case, the time mean is the mean time which the solution $f(p, t)$ spends within the set E. As it is seen, in the ergodic case this time does not depend on the solution $f(p, t)$ and it is the same for different solutions. Let us present it as follows. The mean time which almost all trajectories of the ergodic system spend within the set E is proportional to the measure $\mu(E)$ of this set.

In other words, the probability of finding of almost any trajectory in the set E is equal to $\mu(E)$.

To put it differently, if the system is ergodic, then the trajectory of almost every point of the phase space enters any set of positive measure and spends within it a time proportional to the measure of this set. If in equality

$$\lim_{T \to \infty} \frac{1}{T} \int_0^T \varphi(f(p, t)) \, dt = \int_X \varphi(p) \mu(dp)$$

we set $\varphi = \varphi_A$ and multiply this equality by $g = g_B$ and integrate over X, then we obtain the following relation

$$\lim_{T \to \infty} \frac{1}{T} \int_0^T \mu(f(A, t) \cap B) \, dt = \mu(A) \mu(B),$$

where it is assumed that the measures of the sets A and B are greater than zero. If we require the fulfillment of more strong condition, namely,

$$\lim_{t \to \infty} \mu(f(A, t) \cap B) = \mu(A) \mu(B),$$

then such a system is called a system with *intermixing*.

Consider the dynamical system (X, σ, μ, f), where μ is a normalized measure invariant with respect to f.

Before it was shown that for any set $E \in \sigma$ there is a set W of measure zero (the set of wandering points) such that for any $p \in E \setminus W$ there exists a positive integer $n(p)$ satisfying the condition $T^{n(p)} p \in E$, where $T p = f(p, 1)$, $T^k p = f(p, k)$. In other words, almost all points of any set $E \in \sigma$ return into this set after a certain time.

It is natural in this case to pose a question about the estimate of the return time of the point $p \in E - W$ into the set E.

We introduce the concept of the return time $\tau_E(p)$ i.e. the time of recurrence of the point p into the set E.

Let E be a measurable set (that is $E \in \sigma$) and $\mu(E) > 0, \mu$ be a normalized invariant measure. By $\tau_E(p)$ we shall denote, for each $p \in E$, the least positive integer such that

$$T^{\tau_E(p)} p \in E, \quad T p = f(p, 1), \quad T p^k = f(p, k).$$

In other words,

$$\tau_E(p) = \left\{ \min n : T^n p \in E \, n \geq 1 \right\}.$$

The function $\tau_E(p)$ is called a return time of the point p into the set E. It is evident that the function $\tau_E(p)$ is defined almost everywhere on the set E. We shall prove that this function is summable.

Theorem 1.46. *Function $\tau_E(p)$ is summable and*

$$\int_E \tau_E(p)\,\mu(dp) = \mu\Big(\bigcup_{n=0}^{\infty} T^n E\Big) \leq 1.$$

Proof. Let us define the sets

$$E_n = \{p \in E : \tau_E(p) = n\}$$

as the sets of points of E, which return into the set E after n iterations.

Then the sets $T^j E_n$, $0 \leq j < n$ are mutually disjoint. Now we notice that

$$\bigcup_{n=0}^{\infty} T^n E = \bigcup_{n=1}^{\infty}\Big(\bigcap_{j=0}^{n-1} T^j E_n\Big) = E_1 \cup [E_2, \cap T\, E_2] \cup \ldots$$

Since

$$\bigcup_{n=0}^{\infty} T^n E \subset X,$$

one has

$$\mu\Big(\bigcup_{n=0}^{\infty} T^n E\Big) \leq \mu X = 1.$$

Therefore,

$$1 \geq \mu\Big(\bigcup_{n=0}^{\infty} T^n E\Big) = \mu\Big(\bigcup_{n=1}^{\infty}\bigcup_{j=0}^{n-1} T^j E_n\Big)$$

$$= \sum_{n=1}^{\infty}\sum_{j=0}^{n-1} \mu(T^j E_n) = \sum_{n=1}^{\infty} n\,\mu(E_n) = \int_E \tau_E(p)\,\mu(dp).$$

(Here we used the definition of integral of function $\tau_E(p)$). Hence it follows that $\tau_E(p) \in L_1(X, \sigma, \mu)$, if we assume $\tau_E = 0$ on $X \setminus E$.

Theorem 1.47. *Let there be given a system (X, σ, μ, f), where μ is an invariant normalized, while f is ergodic with respect to this measure. Let $E \in \sigma$ and $\mu(E) > 0$. Then*

$$\int_E \tau_E(p)\,\mu(dp) = 1.$$

Proof. By virtue of the above-said, it remains only to show that if system is ergodic, then one has

$$\mu\Big(\bigcup_{n=0}^{\infty} T^n E\Big) = 1.$$

We have

$$B = \bigcup_{n=0}^{\infty} T^n E = \bigcup_{n=0}^{\infty} f(E, n) = f(E, \bigcup_{n=0}^{\infty} n),$$

since

1) $f(B, 1) = f(f(E, \bigcup_{n=0}^{\infty} n), 1) = f(E, \bigcup_{n=0}^{\infty} (n+1)) \subset B$;

2) $\mu(T B) = \mu(B) \geq \mu(E)$

by the invariance of measure.

Characteristic function φ_B of the set B will be invariant. Now one can find the invariant set B_* such that $\mu(B \triangle B_*) = 0$. By the ergodicity it follows that $\mu(B_*) = 1$, that is, $\mu(B) = 1$.

1.8   Nonlinear Dissipative Systems

We now turn our attention to the semidynamical systems, i.e. such dynamical system whose solutions are determined for $t \geq 0$. The general definition of semidynamical systems will be given in Chapter 3. We now consider the class of continuous semidynamical systems.

Definition 1.23. A semidynamical system is a family of mappings $S(t)$, $t \geq 0$, which maps the metric space X into itself and satisfies the conditions:

1) $S(t + s) = S(t)S(s)$, $\forall s, t \geq 0$;

2) $S(0) = I$;

3) $S(t)$ are continuous nonlinear operators from X into X, for any $t \geq 0$;

4) for any $u \in X$ the mapping $t \to S(t)u$ is continuous.

As noted at the beginning of this chapter, if there is given an autonomous system of equations

$$\partial_t u = F(u), \quad u \mid_{t=0} = u \tag{1.42}$$

and its solution $u = u(t, u)$ is determined for all $t \geq 0$ and continuous in the pair of variables (t, u_0), then the solution of such a system may be written as:

$$u(t) = u(t, u_0) = S(t) u_0,$$

where $S(t)$ is the family of nonlinear operators acting in the corresponding functional space, in which the problem (1.42) is solvable. Hence, investigating the semidynamical systems, we study the properties of the solutions of the equations (1.42).

Many concepts such as the concepts of the invariant sets, ω-limit sets and other ones of dynamical systems can be extended to the semidynamical systems also.

For example, the set $A \subset X$ is called invariant (positively invariant, negatively invariant), if

$$S(t) A = A, \ t \geq 0, \quad (S(t) A \subset A, \ t > 0, \quad S(t) A \supset A, \ t > 0).$$

For the point $u_0 \in X$ and for the set $A \subset X$ ω-limit sets are defined as follows:

$$\omega(u_0) = \bigcap_{s>0} \overline{\bigcup_{t \geq s} S(t) u_0}, \quad \omega(A) = \bigcap_{s>0} \overline{\bigcup_{t \geq s} S(t) A},$$

where the overline denotes the closure in X. Notice that in this definition it can be considered that $s \geq s_0$, $s_0 > 0$.

The main characteristic property of the ω-limit sets is as follows.

The property **F**: $u \in \omega(A)$ if and only if there exist some sequences $u_n \in A$ and $t_n \to +\infty$ such that $S(t_n) u_n \to u$ as $n \to \infty$. The point $u_0 \in X$ is *the stationary* point of $S(t)$ if $S(t) u_0 = u_0$ for all $t \geq 0$. The set

$$\gamma^+(u_0) = \bigcup_{t \geq 0} S(t) u_0.$$

is called the positive semitrajectory (or the trajectory starting at the point u_0.)

Let M be an invariant set. Then each point $u \in M$ defines a positive semitrajectory $\gamma^+(u)$ lying within M. We show that this semitrajectory can be extended to the complete trajectory.

Let $u = v_0$. For the point v_0 there can be found $v_1 \in M$ such that $S(1) v_1 = v_0$. For the point v_1 there can be found a point v_2 such that $S(1) v_2 = v_1$, and so forth. In other words, for the point v_n we can find the point v_{n+1} such that $S(1) v_{n+1} = v_n$, $n = 0, 1, \ldots$. It is evident that $v_0 = S(k) v_k$, $k = 1, 2, \ldots$. Let $v_k = S(-k) v_0$.

We determine the operators $S(-t)$ for all $t > 0$, setting

$$S(-k + t) v_0 = S(t) v_k, \quad t \in [0, 1), \quad k = 1, 2 \ldots$$

Thus, the negative semitrajectory $\gamma^-(u)$ is

$$\gamma^-(u) = \{S(t) v_k, \quad t \in [0, 1), \, k = 1, 2, \ldots, \}, \qquad (1.43)$$
$$v_k = S(-k) v_0, \quad k = 1, 2, \ldots, \quad v_0 = u.$$

Therefore the semidynamical system on an invariant set can be extended to the dynamical system, if the operator $S(t)$ is one-to-one on the invariant set.

We shall call the function such that

$$f(s + t) = S(t)f(s), \quad s \in R, \quad t \in R_+.$$

the complete trajectory of the semigroup $S(t)$, $t \geq 0$.

We shall establish now the theorem on the properties of the ω-limit set $\omega(A)$, which will be very important in the sequel.

Theorem 1.48. *Let for the nonempty set $A \subset X$ for certain $t_0 > 0$ the set $G(t_0) = \bigcup_{t \geq t_0} S(t)A$ is a precompact set in X. Then $\omega(A)$ is a nonempty, compact and invariant set. In particular, if the positive semitrajectory $\gamma^+(u_0)$ is precompact, then $\omega(u_0)$ is a nonempty, compact and invariant set.*

Proof. It is evident that the sets $\overline{G(s)}$ for every $s \geq t_0$ are nonempty, compact sets which decrease as s increases. Their intersection, for every $s \geq t_0$ is then a nonempty, compact set and is equal to $\omega(A)$.

Let us prove the invariance of $\omega(A)$, i.e. that $S(t)\omega(A) = \omega(A)$. Let $u \in S(t)\omega(A)$. Then $u = S(t)v$, $v \in \omega(A)$ and by virtue of the Property **F** there exist some sequences $v_n \in A$ and $t_n \to +\infty$ as $n \to \infty$ such that $S(t_n)v_n \to v$ as $n \to \infty$. Therefore,

$$S(t)S(t_n)v_n = S(t + t_n)v_n \to S(t)v = u,$$

that is u is the limit of the sequence $S(t + t_n)v_n$, where $v_n \in A$, $t + t_n \to +\infty$ as $n \to \infty$. Consequently, $u \in \omega(A)$.

Let $u \in \omega(A)$. Then there exist the above sequences v_n and t_n such that $S(t_n)v_n \to u$ as $n \to \infty$. We notice that for $t_n \geq t$ the sequence $S(t_n - t)v_n$ is precompact, i.e. from it there can be chosen a convergent subsequence $S(t_{n_j} - t)v_{n_j} \to v$, hence $v \in \omega(A)$. In the other hand, $S(t_{n_j})v_{n_j} \to u$ as $n_j \to \infty$ by the choose of the sequence (t_n, v_n). Hence $S(t_{n_j})v_{n_j} = S(t)S(t_{n_j} - t)v_{n_j} \to S(t)v$, that is $u = S(t)v$ and $u \in S(t)\omega(A)$.

The principal condition of this theorem is the requirement that the set $\bigcup_{t \geq t_0} S(t)(A)$ be precompact. In the finite-dimensional case this means that the set is bounded. In the infinite-dimensional case it is a common practice to demonstrate the boundedness of this set in some space which is compactly embedded in X.

Let us come now to the consideration of nonlinear dissipative systems, which can be defined as the semidynamical systems possessing an absorbing set.

Attracting set. Global attractor. Absorbing set. The word "attractor" in a general sense is used as the notation for the mathematical object, which attracts. With respect to the character of the object itself and the fact what and how the object attracts there is a variety of the definitions of the attractor.

We dwell on the concept of the global attractor because it is the only attractor we shall deal with in the sequel.

Two main characteristics of attractor are attraction and invariance. If by the definition of the attractor we emphasize the property of attraction, then it can be seen that the attractor is *the least* set among the attracting sets. If we emphasize the invariance, then the attractor will be the *largest* (maximal) set among the invariant sets.

We give now the strict definitions.

Let X be the metric space with the metric $\rho = \rho(x, y)$. By $O_\varepsilon(A)$ we denote ε-neighborhood of the set A.

Let $S(t)$ for $t \geq 0$ be the semigroup of nonlinear operators acting in X.

Definition 1.24. The set $F \subset X$ is said to be *the attracting* set if for any $\varepsilon > 0$ and for any *bounded* set $B \subset X$ there can be found the number $T(\varepsilon, B)$ such that

$$S(t) B \subset O_\varepsilon(F), \quad \forall \, t \geq T(\varepsilon, B).$$

Notice that: 1) the set F attracts every bounded set of the space X; 2) this attraction is uniform with respect to all points of the set B, i.e. for $t \geq T(\varepsilon, B)$ all points of the set B will enter a given ε-neighborhood of the attracting set F. This may be presented as follows

$$\lim_{t \to +\infty} \sup_{b \in B} \rho(S(t) b, F) = 0.$$

Since each point $x \in X$ is a bounded set, the set F attracts also each point of the space X, i.e. there exists $T(\varepsilon, x_0)$ such that

$$S(t) x_0 \subset O_\varepsilon(F), \quad \forall \, t \geq T(\varepsilon, x_0),$$

or

$$\lim_{t \to +\infty} \rho(S(t) x_0, F) = 0.$$

From the above definition it follows that the attracting set F possesses the property of the global attraction, i.e. it attracts any bounded sets (and each points) of the space X, wherever they are in X and however "far apart" they are from the set F.

To emphasize also the fact that the set F attracts the bounded sets $B \subset X$, one says that F is the global B-attracting set. It is natural to choose from the global B-attracting sets the least (nonempty) set. According to the Ladyzhenskaya O.A. [87, 88], we introduce the following definition.

Definition 1.25. The minimal global B-attractor of the semigroup $S(t)$, $t \geq 0$ is the least closed nonempty set attracting any bounded set B of the space X (i.e. it is the least closed global B-attracting set). We shall denote this attractor by A_b.

Definition 1.26. The minimal global attractor of the semigroup $S(t)$, $t \geq 0$ is the least closed nonempty set attracting each point of the space X. We denote this attractor by A_p. It is evident that $A_p \subset A_b$.

On the practice the attractor often is not only the closed set, it is the compact one also. Apart from the property of attraction the attractor possesses another important property, that is the invariance. Taking this into account, we introduce the following definition of the *global* attractor, which we shall use repeatedly in the sequel.

Definition 1.27. The set $A \subset X$ is said to be the global attractor (of the semigroup $S(t)$, $t \geq 0$), if:

1) A is a compact set,

2) A is an invariant set, i.e. $S(t)(A) = A$, $\quad \forall\, t \geq 0$,

3) A attracts every bounded set $B \subset X$.

It will be shown that, the global attractor is also the minimal global B-attractor.

Thus, *the global attractor A having the property of attraction* possesses yet another important property, it is *invariant*. Among the invariant sets of the semigroup $S(t)$, $t \geq 0$ the global attractor is *the largest* set, i.e. if M is any bounded invariant set, that is $S(t)M = M$, $t \geq 0$, then $M \subset A$. Indeed, since M is bounded, one has $S(t)M \subset O_\varepsilon,(A)$ for $\forall\, t \geq T(\varepsilon, M)$ and for any $\varepsilon > 0$.

Then by invariance $M \subset O_\varepsilon(A)$. As $\varepsilon \to 0$ we get $M \subseteq A$. By virtue of the above-said *the global attractor* is called sometime *the maximal* (or universal) attractor. We shall not use this term further.

Remark. The more general concept of the global (F, D)-attractor may be used [4]. We now introduce this definition. Let F and D are spaces, therefore $D \subset F$. A set $A \subset D$ is said to be global (F, D)-attractor of the semigroup $S(t)$, $t \geq 0$ if:

1) A is an invariant set, i.e. $S(t)A = A$,

2) the set A is compact in D and bounded in F,

3) A attracts any bounded set of the space F in the topology of the space D, i.e. it is (F, D)-attracting set.

In particular, according this terminology the global attractor is the global (X, X)-attractor.

We shall introduce the concept of the *absorbing* set.

Definition 1.28. The set $B_a \subset X$ is said to be an *absorbing* set, if for any bounded set $B \subset X$ there can be found $T(B)$ such that

$$S(t)B \subset B_a, \quad \forall t \geq T(B),$$

i.e. any bounded set of X enters after a certain time the absorbing set B_a and will remain in it forever. It is evident that each point of X will also enter the absorbing set B_a and remains in it forever.

If the semigroup $S(t)$, $t \geq 0$ possesses the global attractor A, then any neighborhood of the attractor A is the absorbing set.

Indeed, let $O_\varepsilon(A)$ be a given ε-neighborhood of global attractor A and let B be any bounded set. Then there can be found $T(\varepsilon, B)$ such that $S(t)B \subset O_{\varepsilon/2}(A)$, $\forall\, t \geq t(\varepsilon, B)$. The converse is not true. In order that the system possessing the absorbing set has the global attractor, there are needed some additional conditions.

Theorem 1.49. *Let the semigroup $S(t)$, $t \geq 0$ has the compact attracting set $K \subset X$. Then the set $G = \bigcup_{t \geq t_0} S(t)B$ is a precompact set, where B is any bounded set and $\omega(B) \subset K$.*

Proof. Let $\{g_n\}$, $n = 1, 2, \ldots$ be a sequence of points of G. We need to prove that from it there can be always chosen a convergent subsequence. It is evident that for any g_n there exist $t_n \geq 0$ and $b_n \in B$ such that $g_n = S(t_n)b_n$. Since K is the attracting set one has $\rho(S(t)b_n, K) \to 0$ as $t \to +\infty$ hence $\rho(S(t + t_n)b_n, K) \to 0$ as $t \to +\infty$, that is $\rho(S(t)g_n, x_n) \to 0$ as $t \to +\infty$, for $n = 1, 2, \ldots$. By the compactness of K from $\{x_n\}$ there can be chosen a convergent subsequence $x_{n_j} \to x_0$.

Consequently, there exists a limit

$$\lim_{n \to \infty} \lim_{t \to +\infty} \rho(S(t)g_{n_j}, x_{n_j}) = 0.$$

By the continuity of $S(t)$ and $\rho(x, y)$ it follows that there exists a limit g_{n_j} as $n_j \to \infty$.

We prove that $\omega(B) \subset K$. Let $x \in \omega(B)$. Then there exist some sequences $b_n \in B$ and $t_n \to +\infty$ such that $S(t_n)b_n \to x$. On the other hand, $\rho(S(t_n)b_n, K) \to 0$ as $n \to \infty$, i.e. there is a point $y \in K$ such that $S(t_n)b_n \to y$. Consequently, $x = y \in K$.

Remark. It may be shown that if K is a closed, bounded and attracting set, then $\omega(B) \subset K$, i.e. K need not to be compact. The prove is the same as the one of the above theorem.

From the proved theorem and Theorem 1.48 it follows that if the semigroup $S(t)$, $t \geq 0$ has the compact attracting set, then for any bounded set B its ω-limit set $\omega(B)$ is a nonempty, compact and invariant set. We prove that in this case $\omega(B)$ attracts the set B.

Theorem 1.50. *Let the semigroup $S(t)$, $t \geq 0$ possesses the compact attracting set $K \subset X$ and let B be any bounded set. Then $\omega(B)$ attracts set B and $\omega(B) \subset K$.*

Proof. Let $\omega(B)$ does not attracts B. Then there exist two sequences $b_n \in B$ and $t_n \to +\infty$ such that for some $\delta > 0$ we shall have

$$\rho(S(t_n)b_n, \omega(B)) \geq \delta, \quad n = 1, 2, \ldots \tag{1.44}$$

But since K is an attracting set, one has $\rho(S(t_n)b_n, K) \to 0$ as $n \to \infty$. This means that there exists a point $x \in K$ such that $S(t_n)b_n \to x$ as $n \to \infty$. In this case $x \in \omega(B)$, that contradicts the inequality (1.44). The following theorem provides the conditions of the existence of the global attractor [71].

Theorem 1.51. *Let semigroup* $S(t) : X \to X$ *possesses the compact attracting set* K, *that is*

$$\lim_{t \to +\infty} \sup_{b \in B} \rho\left(S(t)b, K\right) = 0$$

for any bounded set $B \subset X$. *Then there exists a unique set* A *possessing the following properties:* 1) A *is a compact set;* 2) A *is an attracting set, i.e.*

$$\lim_{t \to +\infty} \sup_{b \in B} \rho\left(S(t)b, A\right) = 0$$

for any bounded set $B \subset X$;
3) A *is invariant, i.e.* $S(t)A = A$, $t \geq 0$;
4) *if* F *is any bounded closed attracting set, then* $A \subset F$, *i.e.* A *is the least among the attracting sets;*
5) $A = \{f(s), s \in R\}$, *where* $f(s)$ *is complete bounded trajectory of the semigroup* $S(t)$.

Proof. Let u_0 is any point in X. We define now the closed balls $B_n(x_0)$ centered at this point with radius n:

$$B_n(u_0) = \{u \in X : \rho(u, u_0) \leq n\}.$$

Let

$$A = \overline{\bigcup_{n \in N} \omega(B_n)}.$$

We show that A possesses the properties mentioned in the theorem. Evidently, $\omega(B_n) \subset K$, $n = 1, 2, \ldots$ by Theorem 1.50. Hence $A \subset K$ and A are compact, because by the definition A is closed.

Further, for any bounded set B there can be found a ball $B_N(u_0)$ such that $B \subset B_N(u_0)$. Then $\omega(B) \subset \omega(B_N)$ and A is the attracting set. Since $\omega(B_n)$ are invariant, and their union and closure are also invariant, the set A is invariant.

Let the set F is closed, bounded and attracting. Then $\omega(B_n) \subset F$ for all $n = 1, 2, \ldots$. Hence $A \subset F$. If A_* is another compact attracting and minimal set, then we must have $A_* \subset A$ and $A \subset A_*$, i.e. A having the above properties is unique.

Thus, A is the global attractor of the semigroup $S(t)$. It may be noted that the proof is true for any point u_0.

We give now yet another theorem on the existence of the global attractor [125].

Theorem 1.52. *Let the semigroup* $S(t)$, $t \geq 0$ *has the bounded absorbing set* B_a *and for any bounded set* B *there exists* $t_0 \geq 0$ *such that* $\bigcup_{t \geq t_0} S(t)B$ *is a precompact set. Then the semigroup* $S(t)$, $t \geq 0$ *possesses a global attractor* A *which is* ω-*limit set of the absorbing set* B_a, *i.e.* $A = \omega(B_a)$.

Proof. By Theorem 1.48 the set $\omega\left(B_a\right)$ is a nonempty, compact and invariant set. We need to prove the property of the attraction. We suppose that $\omega\left(B_a\right)$ does not possess the property of attraction of the bounded sets. Let a bounded set B_0, such that $\operatorname{dist}_X\left(S\left(t\right)B_0, A\right)$ does not tend to zero as $t \to +\infty$. Then there exists $\delta > 0$ and sequences $b_n \in B_0$ and $t_n \to +\infty$ such that

$$\rho\left(S\left(t_n\right)b_n, \quad \omega\left(B_a\right)\right) \geq \delta, \quad n = 1, 2, \ldots$$

By the precompactness it may be supposed that the sequence $S\left(t_n\right)b_n$ converges to a certain point b. Passing to the limit in this inequality we obtain

$$\rho\left(b, \omega\left(B_a\right)\right) \geq \delta > 0. \tag{1.45}$$

On the other hand,

$$b = \lim_{n \to \infty} S\left(t_n\right)b_n = \lim_{n \to \infty} S\left(t_n - T\right)S\left(T\right)b_n.$$

Since $a_n \equiv S\left(T\right)b_n \in B_a$ for sufficiently large T and $t_n - T \to +\infty$, we shall have $b \in \omega\left(B_a\right)$, that is in contradiction with the inequality (1.45). The attractor $A = \omega\left(B_a\right)$ is maximal. In fact, let $A_* \supseteq A$ is another, larger attractor.

Then $A_* \subseteq B_a$ because for the sufficiently large values of t it must enter absorbing set, by virtue of the fact that $S\left(t\right)A^* = A^*$. Have $A_* = \omega\left(A_*\right) \subset \omega\left(B_a\right) = A$, we have arrived at a contradiction.

Dimension of attractors. The above attractors are compact sets in some functional spaces. These sets in the general case are not manifolds and in order to define their dimension we need to use the concepts of Hausdorff dimension and fractal dimension of the sets.

The estimate of the dimension of attractor is important by virtue of the following fact. Since all the solutions tend to the attractor, the final motions of the system for $t > T$ will occur either on the attractor or in its neighborhood. If the dimension of the attractor is finite, then infinite-dimensional dynamical system for large t falls into this finite-dimensional set and its dynamics is, in fact, finite-dimensional. The number of degrees of freedom of this finite-dimensional system will be in order of its attractor dimension.

We shall give the definitions of Hausdorff and fractal dimensions of sets. Let X is compact set in the metric space H. Cover the set by the balls $B_{r_j}\left(x_j\right)$ of radius $r_j \leq \varepsilon$ centered at the point x $x_i \in X$. Given covering will be denoted by $U\left(\varepsilon\right)$. It is seen that

$$X \subseteq U\left(\varepsilon\right) = \bigcup_{j=1}^{M\left(\varepsilon, U\right)} B_{r_j}\left(x_j\right).$$

Let

$$\mu_H\left(X, \varepsilon, U\right) = \sum_{j=1}^{M\left(\varepsilon, U\right)} r_j^d,$$

where $d \geq 0$ is a parameter.

This expression defines "the volume" of the set U consisting of the balls of radius $r_j \leq \varepsilon$ with the dimension d, "covering" X.

Now, from all coverings of the set X by the d-dimensional balls of radius $r_j \leq \varepsilon$ we chose the covering containing the least number of the balls. Let

$$\mu_H\left(X, d, \varepsilon\right) = \inf_U \mu_H\left(X, \varepsilon, U\right) = \inf_U \sum_{j=1}^{M} r_j^d = \sum_{j=1}^{N} r_j^d.$$

Definition 1.29. A number

$$\mu_H\left(X, d\right) = \lim_{\varepsilon \to 0} \mu_H\left(X, d, \varepsilon\right) = \lim_{\varepsilon \to 0} \sum_{j=1}^{N\left(\varepsilon\right)} r_j^d. \tag{1.46}$$

is said to be Hausdorff measure of the dimension d of the set X. As $\varepsilon \to 0$ the number of terms in the last expression will increases and every term r_j^d decreases. Since (1.46) may represent the infinite series, whose convergence depends on d. It is natural to chose the least d for which this series converges.

Definition 1.30. A number

$$d_H\left(X\right) = \dim_H X = \{d : \inf_d \mu_H\left(X, d\right) < \infty\}$$

is said to be the Hausdorff dimension of the set X. Notice that the Hausdorff dimension of finite and countable sets is equal to zero.

If we shall cover the set X by the balls of one and the same radius $r = \varepsilon$, then we may introduce the concept of the fractal dimension. If we denote by $N_\varepsilon\left(X\right)$ the least number of the balls of radius $r = \varepsilon$ which cover X, then for the definition of the fractal dimension $d_F\left(X\right)$ we obtain the following formula:

$$d_F\left(X\right) = \lim_{\varepsilon \to 0} \sup \frac{\ln N_\varepsilon\left(X\right)}{\ln 1/\varepsilon}.$$

Always $d_H\left(X\right) \leq d_F\left(X\right)$.

Theorem 1.53 [125]. *Let H be a Hilbert space and $S\left(t\right)$, $t \geq 0$ be the nonlinear continuous semigroup generated by the equation*

$$\partial_t u = F\left(u\right), \quad u\left(0\right) = u_0.$$

Further, let $X \subset H$ be the compact invariant set $(S(t)X = X,$ $t \geq 0)$. Assume that the mapping $S(t) : u_0 \rightarrow u(t)$ is continuously differentiable in H and its differential is the linear operator $L(t, u_0)$: $\xi \in H \rightarrow U(t) \in H$, where $U(t)$ is the solution of the variation equation

$$\partial_t U = F'(S(t)u_0)U, \quad U(0) = \xi.$$

For any integer $N \geq 1$ we find the numbers q_N, setting

$$q_N = \lim_{t \to \infty} \sup \sup_{u_0 \in X} \sup_{\|\xi_i\| \leq 1} \frac{1}{t} \int_0^t Tr \, F' \, S(\tau) \, u_0 \, Q_N(\tau) dt,$$

$$\xi_i \in H, \, i = \overline{1, N},$$

where $Q_N(\tau)$ is orthoprojector in H on Span $\{U_1(\tau), \ldots, U_N(\tau)\}$, and $U_j(t)$ is the solution of the variation equation with the initial condition $U_j(0) = \xi_j$. Let $N_0 \geq 1$ be such that $q_N < 0$ for all $N \geq N_0$. Then the Hausdorff dimension of the set X is less or equal to N_0, that is $\dim_H X \leq N_0$.

This theorem will be used in the sequel to obtain the attractor dimension estimate of climate models.

Lyapunov Exponents and Dimension of Attractors. The Lyapunov numbers and Lyapunov exponents are introduced for the description of the evolution of the elementary m-dimensional volume containing the point $u_0 \in H$ for the large t under acting of the semigroup $S(t)$, $t \geq 0$. Notice that these numbers exist not for each point u_0.

Lyapunov numbers (if they exist) of the semigroup $S(t)$, $t \geq 0$ at the point u_0 are numbers

$$\lambda_k(u_0) = \lim_{t \to \infty} [\alpha_k(L(t, u_0))]^{1/t}, \quad k = 1, 2, \ldots,$$

where

$$\alpha_k(L(t, u_0)) = \sup_{F \subset H} \inf_{\varphi \in F} \|L(t, u)\varphi\|_H, \, \dim F = k, \, \|\varphi\|_H = 1,$$

and the linear operator $L(t, u_0)$ is the Freshet derivative of $S(t)u_0$. Lyapunov exponents μ_k are numbers

$$\mu_k(u_0) = \ln \lambda_k(u_0).$$

Let us give more detailed description of the Lyapunov exponents for the finite-dimensional case. Let $x = \varphi_t(x_0)$ be the solution of the problem

$$\frac{dx}{dt} = F(x), \quad x \mid_{t=0} = x_0, \quad x \in R^n,$$

and $U(t, x_0)$ be fundamental matrix of the solutions of the variation equation

$$\frac{du}{dt} = \frac{\partial F}{\partial x}(\varphi_t(x_0))\, u, \quad \frac{\partial F}{\partial x} = \left(\frac{\partial F_i}{\partial x_j}\right), \quad i, j = \overline{1, n}.$$

The trajectory of the solution $\varphi_t(x_0)$ in the phase space will be denoted by $\gamma(x_0)$. Let $\{a_1, a_2, \ldots, a_n\}$ be the system of linearly independent vectors at the point x_0. We chose from them k arbitrary vectors $\{e_1, e_2, \ldots, e_k\}, 1 \le k \le n$, therefore it need not be $e_j = a_j$. Further we construct on them parallelepiped which will be denoted by $P_k = \|e_1 \wedge e_2 \wedge \ldots \wedge e_k\| = |\det A|^{1/2}$, $A = (a_{ij})$, $a_{ij} = (e_i, e_j)$. In time t the point x_0 will go into the point $\varphi_t(x_0)$, and the transformation of translation $\varphi_t(x_0)$ will map the vectors $\{e_1, \ldots, e_k\}$ into vectors $\{U(t, x_0)e_1, \ldots, U(t, x_0)e_k\}$. We denote the volume of parallelepiped constructed on these vectors by $\varphi_t(P_k) = \|U e_1 \wedge U e_2 \ldots \wedge U e_k\|$. Let there exists a limit

$$\lambda_k(x_0) = \lim_{t \to \infty} \sup \frac{1}{t} \ln \frac{\|U(t, x_0) e_1 \wedge \ldots \wedge U(t, x_0)\|}{\|e_1 \wedge e_2 \wedge \ldots \wedge e_k\|}.$$

The number $\lambda_k(x_0)$ is called k-dimensional Lyapunov exponent of the trajectory $\gamma(x_0)$.

There are the following properties of Lyapunov exponents:

1) there can be no more than n of different one-dimensional Lyapunov exponents $\lambda_1 \ge \lambda_2 \ge \ldots \ge \lambda_n$;

2) there can be no more than $\binom{n}{k}$ of the different k-dimensional Lyapunov exponents therefore every of them is equal to the sum of k-one-dimensional Lyapunov exponents;

3) for the arbitrarily chosen independent vectors $\{e_1, \ldots, e_k\}$ the limit defining $\lambda_k(x_0)$, converges usually to the maximal k- dimensional Lyapunov exponent λ_k^{max}.

The dimension of the attractor of the finite dimensional dissipative system can be found by use of Lyapunov exponents (Lyapunov dimension of attractor). We give now the formulas for the estimate of the attractor dimension by Lyapunov exponents.

Let $\lambda_1 \ge \lambda_2 \ge \ldots \ge \lambda_n$ are one-dimensional Lyapunov exponents of some finite dimensional nonlinear dissipative system which are arranged in the order of decreasing.

Formula of Kaplan–Yorke:

$$K_L = j + |\lambda_{j+1}|^{-1} \sum_{k=1}^{j} \lambda_k,$$

Here the number j is defined from the conditions: $\lambda_1 + \lambda_2 + \ldots + \lambda_j \ge 0$, and $\lambda_1 + \lambda_2 + \ldots + \lambda_j + \lambda_{j+1} < 0$.

Formula of Mori:

$$M_L = k + \frac{\sum_{i=1}^{m} \lambda_i^+}{\sum_{j=1}^{l} \mid \lambda_j^- \mid},$$

where λ_i^+ are positive Lyapunov exponents and m is their number, λ_j^- is negative Lyapunov exponents, and l is their number, k is the number of non-negative Lyapunov exponents.

Formula of Yang:

$$Y_L = k + \mid \lambda_n \mid^{-1} \sum_{i=1}^{k} \lambda_i,$$

where k is the number of non-negative Lyapunov exponents. Evidently, $M_L \leq Y_L \leq K_L$.

By way of example let us consider the dissipative system with three Lyapunov exponents $\lambda_1 \geq \lambda_2 \geq \lambda_3$. If all them are negative $(-,-,-)$, then the attractor is the stationary point; if one of them is equal to zero, while two others are negative $(0,-,-)$, then attractor is limit cycle; if two exponents are equal to zero, while one is negative $(0,0,-)$, then the attractor is two-dimensional torus.

Finally, if the first exponent is positive, the second is equal to zero, while the third is negative $(+,0,-)$, then the attractor is the strange attractor.

In the first case the given formulas are not applicable. In the second case we have $M_L = Y_L = 1$, and the formula for K_L is not applicable. In the third case $M_L = Y_L = 2$, the formula for K_L is not applicable again. In the fourth case $K_L = M_L = Y_L = 2 + \lambda_1 / \mid \lambda_3 \mid$. (Here the condition of the dissipativity $\lambda_1 + \lambda_2 + \lambda_3 < 0$ is taken into account.)

Statistical solutions and invariant measures. As is known, the real fields used by the numerical computations as the initial conditions may be given with certain error.

Because of this it is reasonable to consider the problems, when instead of the precisely given initial conditions $u_0 \in H$ we know only the probability distributions of the initial data (H is the complete metric space).

By $\mathcal{B}(H)$ we denote the σ-algebra of the Borel sets of H. Probability measure $\mu_0(\omega)$, $\omega \in \mathcal{B}(H)$ is σ- additive nonnegative function given on $\mathcal{B}(H)$ and satisfying the condition $\mu_0(H) = 1$. Probability measure $\mu_0(\omega)$, $\omega \in \mathcal{B}(H)$ is called concentrated on the Borel set ω^*, if $\mu_0(\omega^*) = 1$.

In this case for every $\omega \in \mathcal{B}(H)$, $\omega \cap \omega^* = \emptyset$ we have $\mu_0(\omega) = 0$. (Here the null index in the notation of the measure $\mu_0(\omega)$ indicates only the fact that this measure is given on the space of the initial conditions H, $u_0 \in H$).

Statistical solution is the family of measures $\mu(t, \omega), \omega \in \mathcal{B}(H)$ defined by

$$\mu(t, \omega) = \mu_0(S(t)^{-1}\omega), \quad \forall \, \omega \in \mathcal{B}(H),$$

where $S(t)^{-1}\omega$ is the complete inverse image of the set ω by the mapping $S(t)$, that is

$$S(t)^{-1}\omega = \{u \in H : S(t)u \in \omega\}.$$

Statistical solution $\mu(t, \omega)$ is said to be stationary solution if the family of measures $\mu(t, \omega)$ does not depend on t.

Probability measure $\bar{\mu}_0(\omega)$, $\omega \in \mathcal{B}(H)$ is said to be *invariant* with respect to the semigroup $S(t)$, if for any $t > 0$ we get

$$\bar{\mu}((S(t))^{-1}\omega) = \bar{\mu}(\omega), \quad \forall \omega \in \mathcal{B}(H).$$

The measure $\bar{\mu}$ is invariant if and only if $\bar{\mu}$ is stationary statistical solution.

The principal property of the nonlinear dissipative system possessing the attractor means that any initial probability measure $\mu_0(\omega)$ generates the invariant measure $\hat{\mu}(\omega)$, concentrated on the attractor of this system. (In the general case the measure $\hat{\mu}(\omega)$ can be not unique. In other words, the system "forgets" its initial distribution and tends to some probability distribution on the attractor. The uniqueness of the limit measure $\hat{\mu}(\omega)$ is connected with the ergodicity of the system under consideration. The sense of the statistical solution $\mu(t, \omega)$ is that the solution gives possibility to find the system in the set ω at the moment t.

The dependence of the attractor on the systems parameters. In the real situations the parameters of the considered system may vary (or they may be known not precisely). Hence it is natural to pose the question: what happens with the attractor if the parameters of the system vary smoothly (continuously). In the general case there is occurred some continuity (upper semicontinuity) of the attractor in the parameters.

However, we must keep in mind that there can be possible such a situation when the parameters can approach to certain critical values, for which such a continuous dependence will be broken. Then in the system the bifurcation will be happen and the initial attractor disappears and instead of it a new one arises, it will have another dimension and possesses other properties and structure.

Let $S_\lambda(t) u_0$ be the solution of the considered dissipative system, where $S_\lambda(t)$, $t \geq 0$ is the semigroup of nonlinear operators, which acts in the some complete metric space $X(u_0 \in X)$ and depends on the parameters λ.

We shall assume that λ belongs to some complete metric space Λ. In other words, let $(S_\lambda(t), t \geq 0, X, \Lambda)$ be semidynamical system and A_λ be its global attractor depending on the parameter λ. Further, let λ_0 be nonisolated point Λ, while ρ is the metric of X.

Here the following theorem takes place [76].

Theorem 1.54. *Let the conditions be fulfilled:*

1) *if $\lambda_k \to \lambda_0$, then set $K = \cup_{k=1}^\infty A_{\lambda_k}$ is a precompact set;*

2) *if $\lambda_k \to \lambda_0$ and $u_k \to u_0$, then $S_{\lambda_k}(t)u_k \to S_{\lambda_0}(t)u_0$ for some $t > 0$ and $u_k \in A_{\lambda_k}$.*

Then for any neighborhood $O(A_{\lambda_0})$ of the attractor A_{λ_0} there can be found $\delta > 0$ such that $A_{\lambda_k} \subset O(A_{\lambda_0})$ for all $\lambda_k \in \Lambda$ such that $d(\lambda_k \lambda_0) < \delta$, where d is the metric of Λ.

Proof. Suppose that the statement of the theorem is false. Then for some $\varepsilon > 0$ there exists a sequence $\lambda_k \in \Lambda$ convergent to λ_0 and a sequence $u_k \in A_{\lambda_k}$, $k = 1, 2, \ldots$ such that $\rho(u_k, A_{\lambda_0}) \geq \varepsilon$ for all $k = 1, 2, \ldots$.. By the condition of the theorem the set $\bar{K}$ is compact. Therefore it can be considered that u_k converges to some point $u_0 \in \bar{K}$, hence $\rho(u_0, A_{\lambda_0}) \geq \varepsilon$. We shall prove that $u_0 \in A_{\lambda_0}$. Then we arrive at a contradiction with this inequality.

By virtue of the fact that $S_{\lambda_k}(t, u_k) \to S_{\lambda_0}(t, u_0)$, $t > 0$ and $S_{\lambda_k}(t, u_k) \in A_{\lambda_k}$ by the invariance of A_{λ_k}, we get $S_{\lambda_k}(t, u_0) \in \bar{K}$ for all $n = 1, 2, \ldots$.. Again, by invariance of A_{λ_k} for every $u_k \in A_{\lambda_k}$ there can be found $u_k(-t) \in A_{\lambda_k}$ such that $S_{\lambda_k}(t, u_k(-t)) = u_k$. By the compactness $\bar{K}$ there can be assumed that $u_k(-t)$ converges to some $u_0(-t) \in \bar{K}$. But then $u_k = S_{\lambda_k}(t, u_k(-t)) \to S_{\lambda_0}(t, u_0(-t))$. Taking into account $u_k \to u_0$, we get $S_{\lambda_0}(t, u_0(-t)) = u_0$. Repeating these reasonings we find $u_0(-nt)$, $n = 2, 3, \ldots$, such that $u_0(-nt) \in \bar{K}$ and $S_{\lambda_0}(t, u_0(-nt)) = u_0(-nt + t)$.

In other words, through the point u_0 there goes a complete trajectory of the semigroup $S_{\lambda_0}(\tau)$, $\tau = mt$, $t > 0$, $m = 1, 2, \ldots$ completely lying within $\bar{K}$. Thus, if the attractor A_{λ_0} of the semigroup $S_{\lambda_0}(t)$ is the union of complete trajectories, then the constructed trajectory must belong to it and, consequently, $u_0 \in A_{\lambda_0}$.

Here we used the following property of semidynamical system: the global attractor of it does not vary, when we consider the discrete case, i.e. if t will take the values $0, t_0, 2t_0, \ldots$, where $t_0 > 0$.

Local attractors. Attractors of mappings. Scenario of passing to the chaos. Above we have considered only the global attractors, i.e. such attractor, whose basin of attraction is the whole phase space of the dynamical system.

The global attractor, if it exists, is unique. However, one can consider also the system possessing the local attractors, whose basins of attraction are some neighborhoods of these attractors.

This often occurs in the discrete systems of the view

$$x_{n+1} = f_\lambda(x_n), \quad n = 0, 1, 2 \ldots,$$

where λ is a parameter, with the change of which the character of the mapping may be changed.

Let us consider, for example, Newtonian algorithm of the calculation of the equation roots $z^3 = 1$ at the complex plain:

$$z_{n+1} = z_n - (z_n^3 - 1)/3\, z_n^2. \tag{1.47}$$

The roots of the equation are

$$\bar{z}_1 = 1, \quad \bar{z}_2 = -1/2 + i\sqrt{3}/2, \quad \bar{z}_3 = -1/2 - i\sqrt{3}/2$$

and they are the local attractors of the mapping (1.47). The boundary separating the basins of attraction of these local attractors is the fractal set (called the set of Julia).

We shall give the general definition of the attractor. Let in the phase space X there is semidynamical system representing the semigroup of nonlinear mappings $S(t)$ and let $A \subset X$. The set A is called the attractor of the semigroup $S(t)$ (the attractor of the semidynamical system), if there exists such a λ-neighborhood, $O_\lambda(A) = \Lambda$, of the set A then $A = \omega(\Lambda)$, i.e. the attractor A is a ω-limit set of some its λ-neighborhood.

The basin of attraction A is called the set

$$W = \{x \in X : \omega(x) \subset A\},$$

where $\omega(x)$ is the ω-limit set of the point x.

If the basin of attraction of the attractor is the whole phase space (precisely, there are attracted every bounded sets), then the attractor A is global.

Another branch of the theory of mappings is the scenario of passing of the system to the chaos through the sequence of the bifurcations by the changing of some parameter of mappings. The mapping of the form

$$x_{n+1} = \lambda x_n(1 - x_n), \quad x \in [0, 1], 0 < \lambda < 4$$

is the typical one. If $0 < \lambda \leq 1$, then the mapping will have unique fixed point $x = 0$. For $1 < \lambda \leq 3$ the point $x = 0$ looses its stability, but at the same time there appears another fixed point $x_1 = 1 - 1/\lambda$. If $3 < \lambda \leq 1 + \sqrt{6}$, then the point x_1 becomes unstable and there appears stable double cycle. As the further increasing of λ double cycle looses its stability and the stable four cycle arises and so forth. (There happens the bifurcation of the doubling of the periods of the attracting cycle).

The detailed description of the scenario of passing to the chaos can be found in the literature. We shall not consider questions concerned the local attractors of mappings.

In conclusion we give the Ladyzhenskaya theorem [88].

Theorem 1.55. *Let $V_t : X \to X$, $t \in R_+$ be a semigroup, where V_t is a compact operator for all $t > 0$, while X is the whole metric space. Then:* 1) *if the set $\gamma^+(B)$ is bounded, then $\omega(B)$ is nonempty invariant compact set, which attracts B;* 2) *if for every bounded set B the set $\gamma^+(B)$ is bounded and there exists a bounded attracting set B_a, then the minimal global B-attractor A_b of the semigroup V_t is the nonempty compact invariant set and every invariant bounded set A, in particular, the complete bounded trajectories, lies within A_b.* 3) *If V_t for $t > 0$ is the one-to-one mapping on the set $\omega(B)$ (or on A_b), then the inverse operators V_t^{-1} are continuous on $\omega(B)$ (or on A_b) and the semigroup V_t, $t \in R$ on $\omega(B)$ (or on A_b) is extended to the group V_t, $t \in R$.*

It should be noted that if X is bounded, then as the set B_a there can be taken X.

1.9   Inertial Manifolds of Dissipative Systems

For certain class of nonlinear dissipative systems one can establish the existence of more simple objects called *the inertial manifolds*. The concept of inertial manifold was introduced in paper [57].

The inertial manifolds represent the finite-dimensional Lipschitz manifolds. They possess the property of the strong exponential attraction of all solutions of dissipative system. The solution which falls into the inertial manifold will remain on it forever.

The global attractor lies in the inertial manifold. As is known, attractors of dissipative systems, in the general case, are not manifolds and may have a very complicated structure. In this regard the inertial manifold is more simple object because it itself is a smooth manifold. Since the solutions of the system enter at the fast speed the neighborhood of the inertial manifold, the dynamics of such a system is actually finite-dimensional. We can project the initial system on the inertial manifold and in such a way to obtain a finite system of ordinary differential equations (inertial equations), which will adequately describe the behavior of the infinite-dimensional dissipative system under study. Physically, the inertial manifolds offer an approach to the description of a turbulence. The existence of the inertial manifold means the existence of the low of the interaction between the large-scale structures and the small-scale ones occurred in the turbulent flow.

For the solutions lying on the inertial manifold the small and the large whirls are connected between them by the manifold equation.

Since any solution exponentially approaches to the inertial manifold, this latter, in fact, determines the regime of the system.

The main condition providing for the possibility to establish the existence of the inertial manifold represents so-called the spectral gap condition, what means that the distance between two next different eigenvalues of the linear operator of the problem must grow as the number of eigenvalue grows.

We are coming now to the consideration of the systems, for which we can establish the existence of the inertial manifolds following the works [18, 19] and using the notations accepted in them.

Consider the evolution equation

$$\frac{d\,u}{d\,t} + A\,u = f\,(u), \quad u\,(0) = u_0. \tag{1.47}$$

Let A be linear unbounded positive selfadjoint operator in the Hilbert space H with the norm $|\cdot|$. Let A has a compact resolvente, so that the eigenvectors of A form the orthonormal basis $\{w_j\}$, $j \in N$ in H, while its eigenvalues are positive.

$$0 < \lambda_1 \le \lambda_2 \le \ldots \le \lambda_j \le \ldots, \quad \lambda_j \to \infty, \; j \to \infty.$$

The powers A^s, $s \ge 0$ of the operator A are determined routinely. The norm in $D\,(A^s)$ is denoted by $|\cdot|_{D\,(A^s)} = |\cdot|_s = |\,A^s\cdot\,|$.

The nonlinear function f belongs to C^1 and acts from $D\,(A^{\alpha+\gamma})$ into $D\,(A^\gamma)$ for certain $\gamma \ge 0$ and $\alpha \in [0, 1)$. To shorten the writing we introduce the notations $D\,(A^{\alpha+\gamma}) = E$, $D\,(A^\gamma) = F$. We shall assume that

$$|\,f\,(u)\,|_\gamma \le M_0, \quad |\,f\,(u_1) - f\,(u_2)\,|_\gamma \le M_1\,|\,u_1 - u_2\,|_{\alpha+\gamma}. \tag{1.48}$$

The fulfillment of these conditions in many cases is unreal. They are satisfied actually not on the whole space E, but only in certain bounded domain of the space, for example,

$$|\,f\,(u_1) - f\,(u_2)\,|_\gamma \le M_1\,(r)\,|\,u_1 - u_2\,|_{\alpha+\gamma}, \tag{1.49}$$
$$|\,f\,(u)\,|_\gamma \le M_0\,(r), \quad |\,u\,|_{\alpha+\gamma} \le r, \quad |\,u_j\,|_{\alpha+\gamma} \le r, \quad j = 1, 2.$$

We suppose that the semigroup $S\,(t)$, $t \ge 0$, associated with the problem (1.47), possesses an absorbing set in the space E. Let R be the radius of the ball in E which contains the absorbing set. We determine the function

$$g\,(u) = \theta\left(\frac{|\,u\,|_{\alpha+\gamma}^2}{R^2}\right)\,f\,(u).$$

Here $\theta\,(x)$ is the smooth function such that $\theta : R_+ \to [0, 1]$, therefore $\theta\,(x) = 1$, if $0 \le x \le 1$, and $\theta\,(x) = 0$, if $x \ge 2$.

The function $g(u)$ will satisfy the condition (1.48). If we replace in the equation (1.47) the function f by g, the behavior of the solutions as $t \to \infty$ of the obtained equation will be similar to that of the initial equation. Therefore we shall suppose that the conditions (1.48) are fulfilled in the whole space E.

We shall denote by $P \equiv P_n$ the projector onto the subspace spanned by eigenfunctions of the operator A which correspond to eigenvalues $\lambda_1, \lambda_2, \ldots, \lambda_n$. Let $Q \equiv Q_n = I - P_n$. Spaces PH and QH are invariant with respect to the semigroup e^{-At} for all $t \geq 0$, and the semigroup e^{-At}, $t \geq 0$ on PH can be extended to the group.

Notice that $e^{-At} F \subset E$ for all $t > 0$. We introduce the notations: $\lambda = \lambda_n$, $\Lambda = \lambda_{n+1}$. The following equalities will take place here [65,78,79]. If $t \leq 0$, then

$$\|e^{-At} P\|_{\mathcal{L}(E)} \leq e^{-\lambda t}, \quad \|e^{-At} P\|_{\mathcal{L}(E,F)} \leq \lambda^\alpha e^{-\lambda t}. \qquad (1.50)$$

If $t > 0$, then

$$\|e^{-At} Q\|_{\mathcal{L}(E)} \leq e^{-\Lambda t}, \quad \|e^{-At} Q\|_{\mathcal{L}(E,F)} \leq (t^{-\alpha} + \Lambda^\alpha) e^{-\Lambda t}. \qquad (1.51)$$

We shall assume that if $u_0 \in E$, then the problem (1.47) has an unique solution $u(t) \in C([0, +\infty), E)$.

From here on we shall use the spectral gap condition, whose formulation is given below.

Let us come to the description of the algorithm of the construction of the inertial manifold [18, 19]. Setting $u = y + z$, $y \in PE$, $z \in QE$, we write the initial system (1.47) as follows

$$\frac{dy}{dt} + Ay = P f(y + z), \quad \frac{dz}{dt} + Az = Q f(y + z). \qquad (1.52)$$

We shall look for the inertial manifold of the form $z = \Phi(y)$, where $\Phi(y)$ and y must be determined. In this case (1.52) takes the form

$$\frac{dy}{dt} + Ay = P f(y + \Phi(y)), \quad \frac{d\Phi(y)}{dt} + A\Phi(y) = Q f(y + \Phi(y)).$$

We write the last equation in the integral form

$$\Phi(y(t)) = e^{-A(t-t_0)} \Phi(y(t_0)) + \int_{t_0}^{t} e^{-A(t-s)} Q f(y(s) + \Phi(y(s))) ds.$$

Tending t_0 to $-\infty$, we find

$$\Phi(y(t)) = \int_{-\infty}^{t} e^{-A(t-s)} Q f(y(s) + \Phi(y(s))) ds.$$

Setting in this equation $t = 0$, we obtain

$$\Phi\left(y_0\right) = \int_{-\infty}^{0} e^{As} Q f\left(y\left(s\right) + \Phi\left(y\left(s\right)\right)\right) ds. \qquad (1.53)$$

where $y_0 = y\left(0\right)$. The equation (1.53) must be considered together with the equation

$$\frac{dy}{dt} + Ay = Pf\left(y + \Phi\left(y\right)\right), \quad y\left(0\right) = y_0. \qquad (1.54)$$

We shall seek the function Φ as fixed point of the next mapping T

$$T\Psi\left(y_0\right) = \int_{-\infty}^{0} e^{As} Q f\left(y\left(s\right) + \Psi\left(y\left(s\right)\right)\right) ds,$$

acting on the set of functions $\Psi : PH \rightarrow QH, \quad \Psi \in F_{l,b}$, where

$$F_{l,b} = \{\Psi : \mid \Psi \mid_{\infty} = \sup_{y \in PH} \mid \Psi\left(y\right)\mid_{E} \leq b, \quad \text{Lip } \Psi \leq l\}.$$

The set $F_{l,b}$ is a complete metric space with the metric

$$\rho\left(\Psi_1, \Psi_2\right) = \sup_{y \in PH} \mid \Psi_1\left(y\right) - \Psi_2\left(y\right) \mid_{E} \equiv \mid \Psi_1 - \Psi_2 \mid_{\infty}.$$

The set $F_{l,b}$ is the set of bounded Lipschitz-functions.

In the sequel we shall need the following inequality: if for $t < 0$ the inequality holds

$$u\left(t\right) \leq a e^{-\delta t} + b \int_{t}^{0} e^{-\gamma\left(t-s\right)} u\left(s\right) ds,$$

where $u\left(t\right) \geq 0$ and the constants a, b, γ are positive, while $\gamma + b > \delta > 0$, then

$$u\left(t\right) \leq a \frac{\gamma - \delta + 2b}{\gamma - \delta + b} e^{-\left(b+\gamma\right)t}. \qquad (1.55)$$

In particular, if $\delta = \gamma$, then $u\left(t\right) \leq 2 a e^{-\left(b+\gamma\right)t}$. To establish this inequality, we set

$$v\left(t\right) = \int_{t}^{0} e^{\gamma s} u\left(s\right) ds.$$

Hence

$$-v' \leq a e^{\left(\gamma-\delta\right)t} + bv, \quad v\left(0\right) = 0.$$

Applying to this inequality the Gronwall's lemma, we get (1.55). We now proceed to the presentation of the strong results [18, 19].

Lemma 1.1. *If $\Psi \in F_{l,b}$, then $| T \Psi (y_0) |_E \leq b$ for any $y_0 \in PH$.*
Proof.   We have

$$| T \Psi (y_0) |_E \leq \int_{-\infty}^{0} | e^{As} Q \; f (y (s) + \Psi (y (s))) |_E \; ds$$

$$\leq M_0 \int_{-\infty}^{0} (| s |^{-\alpha} + \Lambda^{\alpha}) e^{\Lambda s} \; ds = M_0 (1 + \gamma_{\alpha}) \Lambda^{\alpha - 1} \leq b,$$

where

$$\gamma_{\alpha} = \int_{0}^{\infty} t^{-\alpha} e^{-t} \, dt, \quad \Lambda^{1 - \alpha} \geq M_0 (1 + \gamma_{\alpha}) b^{-1}. \tag{1.56}$$

The last inequality places the condition on Λ.

Lemma 1.2. *Let $\Psi \in F_{l,b}$. Then for any y_0^1 and y_0^2 of PH takes place the inequality*

$$| T \Psi (y_0^1) - T \Psi (y_0^2) |_E \leq l \; | y_0^1 - y_0^2 |_E \; .$$

Proof.   Notice that for $t < 0$, we have:

$$y^i (t) = e^{-At} y_0^i + \int_{0}^{t} e^{-A (t-s)} P f (y^i (s) + \Psi (y^i (s))) \, ds, \; i = 1, 2.$$

Hence for the difference $y (t) = y^1 (t) - y^2 (t)$ with respect to (1.55) we obtain the estimate

$$| y (t) |_E \leq 2 \; | y_0^1 - y_0^2 |_E \; e^{-(\lambda + \sigma) t}, \quad \sigma = M_1 (1 + l) \lambda^{\alpha}.$$

Under condition $2 M_1 (1 + l)(1 + \gamma_{\alpha}) \Lambda^{\alpha} (\Lambda - \lambda - \sigma)^{-1} \leq l$, it can be easily shown that

$$| T \Psi (y_0^1) - T \Psi (y_0^2) |_E$$

$$\leq M_1 (1 + l) \int_{-\infty}^{0} (| s |^{-\alpha} + \Lambda^{\alpha}) e^{\Lambda s} \; | y^1 (s) - y^2 (s) |_E \; ds$$

$$\leq 2 M_1 (1 + l)(1 + \gamma_{\alpha}) \Lambda^{\alpha} (\Lambda - \lambda - \sigma)^{-1} \; | y_0^1 - y_0^2 |_E \leq l \; | y_0^1 - y_0^2 |_E,$$

This condition will be satisfied if we require the fulfillment of the inequality

$$\Lambda - \lambda \geq \left[M_1 (1 + l) + 2 M_1 (1 + \gamma_{\alpha}) \frac{1 + l}{l} \right] (\Lambda^{\alpha} + \lambda^{\alpha}). \tag{1.57}$$

From two lemmas given above it follows that T maps the set $F_{l,b}$ into $F_{l,b}$ if the inequalities (1.56) are (1.57) hold.

Lemma 1.3. *For any* Ψ_1 *and* Ψ_2 *of* $F_{l,b}$ *the following inequality is true*

$$| T \Psi - T \Psi_2 |_\infty \leq \delta \ | \Psi_1 - \Psi_2 |_\infty .$$

Proof. As before, for the difference $y(t) = y_1(t) - y_2(t)$ we get the estimate

$$| y(t) |_E \leq M_1 (1+l) \int_t^0 e^{-\lambda(t-s)} \ | y(s) |_E \ ds + M_1 \lambda^{\alpha-1} \ | \Psi_1 - \Psi_2 |_\infty .$$

Hence we find

$$| y(t) |_E \leq 2 \, M_1 \, \lambda^{\alpha-1} \, e^{-(\lambda+\sigma)t} \ | \Psi_1 - \Psi_2 |_\infty .$$

Further, on some transformations, we get

$$| T \, \Psi_1 (y_0) - T \, \Psi_2 (y_0) |_E$$
$$\leq M_1 \int_t^0 (| \, s \, |^{-\alpha} + \Lambda^\alpha) e^{\Lambda s} \Big[(1+l) \ | y(s) |_E$$
$$+ \ | \Psi_1 - \Psi_2 |_\infty \Big] ds \leq \delta \ | \Psi_1 - \Psi_2 |_\infty,$$

where

$$M_1 (1 + \gamma_\alpha) \lambda^{\alpha-1} (2 \, M_1 (1+l)(\Lambda - \lambda - \sigma)^{-1} \Lambda^\alpha + 1) \leq \delta.$$

From this lemma it follows that if $\delta < 1$, then the mapping T is strictly contracted and then the function Φ sought will represent the fixed point of this mapping.

Lemma 1.4. *If* $f \in C^1$, *i.e. it is continuously differentiable, then* Φ *is also continuously differentiable.*

The proof is given in [19].

Lemma 1.5. *The graph* Φ *is positively invariant.*

Proof. The function Φ satisfies the following conditions:

$$\Phi(y_0) = \int_{-\infty}^0 e^{As} Q f (y(s) + \Phi(y(s))) \, ds,$$

$$\frac{dy}{dt} + A y = P f (y + \Phi(y)), \ y(0) = y_0. \tag{1.58}$$

It is evident that the solution $\bar{y}$ of the equation

$$\frac{d\bar{y}}{dt} + A \bar{y} = P(f(\bar{y} + \Phi(\bar{y}))), \quad \bar{y}(0) = y(t_0), \quad t_0 > 0$$

has the form $\bar{y}(t) = y(t + t_0)$.

Further,

$$\Phi\left(y\left(t_0\right)\right) = \int_{-\infty}^{0} e^{-At} Q f\left(\bar{y}\left(s\right) + \Phi\left(\bar{y}\left(s\right)\right)\right) ds$$

$$= \int_{-\infty}^{t_0} e^{-A\left(t-t_0\right)} Q f\left(y\left(s\right) + \Phi\left(y\left(s\right)\right)\right) ds.$$

Differentiating this equation on t_0 and replacing then t_0 by t, we find

$$\frac{d\,\Phi\left(y\right)}{d\,t} + A\,\Phi\left(y\right) = Q\,f\left(y + \Phi\left(y\right)\right), \tag{1.59}$$

where $y = y\left(t\right)$, $\Phi\left(y\left(t\right)\right)\big|_{t=0} = \Phi\left(y_0\right)$.

Thus, the inertial manifold is determined by the system of equations (1.58) – (1.59), from which it follows that

$$\frac{d\left(y + \Phi\left(y\right)\right)}{d\,t} + A\left(y + \Phi\left(s\right)\right) = f\left(y + \Phi\left(y\right)\right), \tag{1.60}$$

$$\left(y + \Phi\left(y\right)\right)\big|_{t=0} = y_0 + \Phi\left(y_0\right).$$

On the other hand, the solution of the equation (1.47) with the initial condition $u\big|_{t=0} = y_0 + \Phi\left(y_0\right)$ will be $u\left(t\right) = S\left(t\right)\left(y_0 + \Phi\left(y_0\right)\right)$. Comparing (1.47) and (1.60), we find by virtue of the uniqueness theorem

$$y\left(t\right) + \Phi\left(y\left(t\right)\right) = S\left(t\right)\left(y_0 + \Phi\left(y_0\right)\right).$$

Lemma 1.6. *The next estimate of the closeness of the solution* $S\left(t\right)u_0$ *of the problem* (1.47) *and the inertial manifold* M *takes place:*

$$\mathrm{dist}_E\left(S\left(t\right)u_0,\, M\right) \leq \left|\, z\left(t\right) - \Phi\left(y\left(t\right)\right)\,\right|_E$$

$$\leq C\,e^{-2^{-1}At}\,\left|\, z_0 - \Phi\left(y_0\right)\,\right|_E. \tag{1.61}$$

Proof. Let $u = y + z$, $y = P\,u$, $z = Q\,u$ is the solution of the problem (1.47). Then

$$\frac{d\left(z - \Phi\left(y\right)\right)}{d\,t} = \frac{d\,z}{d\,t} - \Phi'\left(y\right)\frac{d\,y}{d\,t}$$

$$= -A\,z + Q\,f\left(y + z\right) - \Phi'\left(y\right)\left(-A\,y + P\,f\left(y + z\right)\right)$$

$$= -A\left(z - \Phi\left(y\right)\right) + Q\,f\left(y + z\right) - Q\,f\left(y + \Phi\left(y\right)\right)$$

$$-\Phi'\left(y\right)\left(P\,f\left(y + z\right) - P\,f\left(y + \Phi\left(y\right)\right)\right)$$

$$\equiv -A\left(z - \Phi\left(y\right) + R\left(y,\, z\right)\right). \tag{1.62}$$

Here we use the equality

$$\Phi'\left(y\right)\left(-A\,y + P\,f\left(y + \Phi\left(y\right)\right)\right) + A\,\Phi\left(y\right) = Q\,f\left(y + \Phi\left(y\right)\right).$$

Writing (1.62) in the integral form, we find

$$z(t) - \Phi y(t)) = e^{-At}(z_0 - \Phi(y_0)) + \int_0^t e^{-A(t-s)} R(y(s), z(s)) ds.$$

Making the corresponding estimates, we get

$$\mid z(t) - \Phi(y(t) \mid_E \leq \mid z_0 - \Phi(y_0) \mid_E e^{-\Lambda t} + M_1(1+l) \int_0^t [(t-s)^{-\alpha}$$
$$+ \Lambda^\alpha] e^{-\Lambda(t-s)} \mid z(s) - \Phi(y(s)) \mid_E ds. \tag{1.63}$$

We introduce the notation

$$K(t) = \max_{s \in [0, t]} e^{\Lambda s/2} \mid z(s) - \Phi(y(s)) \mid_E .$$

Then from (1.63) it follows that

$$K(t) \leq \mid z_0 - \Phi(y_0) \mid_E + c_0 K(t), c_0 = M_1(1+l)\left[\left(\frac{\Lambda}{2}\right)^{\alpha-1}\gamma_\alpha + 2\Lambda^{\alpha-1}\right].$$

We assume that the following inequality holds

$$\lambda^{1-\alpha} \geq 2 M_1(1+l)(2 + 2^{1-\alpha}\gamma_\alpha).$$

Hence,

$$\text{dist}_E(S(t)u_0, M) \leq \mid z(t) - \Phi(y(t)) \mid_E \leq 2 e^{-2^{-1}\Lambda t} \mid z_0 - \Phi(y_0) \mid_E .$$

It can be shown that all restrictions placed on λ and Λ in lemmas 1.1–1.6, will be met, if $\delta = 1/2$ and

$$\Lambda - \lambda \geq A_1 M_1(\Lambda^\alpha + \lambda^\alpha), \quad \Lambda^{1-\alpha} \geq A_2(M_0 + M_1),$$

where

$$A_1 = 2(1+l) + 2(\gamma_\alpha + 1)\max(8, \frac{1+l}{l}),$$
$$A_2 = \max\{b^{-1}, 2^{2-\alpha}(1+l), 6\}(1+\gamma_\alpha).$$

We present now yet another theorem on the existence of the inertial manifold.

Let nonlinear term $f(u)$ in (1.47) for certain real α is locally Lipschitz mapping from $D(A^\alpha)$ into $D(A^{\alpha-1/2})$, i.e.

$$\mid f(u) \mid_{\alpha-1/2} \leq C_1(r), \quad \mid f(u) - f(v) \mid_{\alpha-1/2} \leq C_2(r) \mid u - v \mid_\alpha$$

for any $u, v \in D(A^\alpha)$ such that $\mid u \mid_\alpha \leq r, \mid v \mid_\alpha \leq r$.

We consider the equation

$$\frac{d\,u}{d\,t} + A\,u + f_\theta\,(u) = 0, \tag{1.64}$$

alongside with (1.47), where $f_\theta(u) = \theta_\rho(|A^\alpha(u)|)f(u)$, $\forall\,u \in D(A^\alpha)$. Here $\theta_\rho(x) = \theta(x/\rho)$ $\rho > 0$, and $\theta(x) = 1$, as $x \in [0, 1]$ and is equal to 0, as $x \geq 2$, and is equal to $\sup |\;\theta(x)\;| \leq 1$, as $x \geq 0$. It is assumed that $\theta(x) \in C^\infty(R_+)$.

Let $S\,(t)$ be the semigroup generated by the problem (1.47). From the above-said the next theorem follows [125].

Theorem 1.56. *Let the following conditions are satisfied:*

1) *for any $u_0 \in D\,(A^\alpha)$ the problem (1.47) has a unique solution $u\,(t)$ determined on $R_+ = [0, \infty)$, satisfying the condition $u\,(0) = u_0$ and such that*

$$u \in C\,(R_+\,, D\,(A^\alpha)) \cap L^2\,(0,\,T;\,D\,(A^{\alpha+1/2}))$$

for any $T > 0$;

2) *the mapping $S\,(t) : u_0 \rightarrow u\,(t)$ is continuous in $D\,(A^\alpha)$ for any $t > 0$;*

3) *the semigroup $S\,(t)$ has the absorbing set $B_\alpha \in D\,(A^\alpha)$, which is positively invariant;*

4) *ω-limit set of the set B_α is the global attractor A of the semigroup $S\,(t)$ in $D\,(A^\alpha)$;*

5) *let $\rho > 0$ be such that B_α and, consequently, the attractor A contained in the ball centered at 0 of radius $\rho/2$ in $D\,(A^\alpha)$, while the ball of radius ρ is the absorbing set for (1.64);*

6) *there exists an integer $N > 0$ such that $\lambda_{N+1} > C_3^2\,(20 + 8e^{-1/2})^{1/2}$, $\lambda_{N+1}^{1/2} - \lambda_N^{1/2} > 18C_3$, where $C_3 = 2C_1(2\rho)\rho^{-1} + C_2(2\rho)$.*

Then the equation (1.47) possesses in $D\,(A^\alpha)$ the N-dimensional inertial manifold, which represents the graph of certain mapping Φ acting from $P_N\,D\,(A^\alpha)$ into $Q_N\,D\,(A^\alpha)$, where P_N is the orthoprojector in H on span $\{w_1,\ldots,w_N\}$, with support belonging to the ball of radius 2ρ of $P_N\,D\,(A^\alpha)$.

The application of this theorem to the barotropic vorticity equation on the sphere will be given in Chapter 3 in which we shall prove the existence of the inertial manifolds if the dissipative operator has the degree greater than 2.

Using the results of the works [86, 126] it is possible to prove the existence of the inertial manifold for the case when the degree of the dissipative operator is equal to 2. Many other theorems on the existence of inertial manifolds for nonlinear dissipative systems can be found in [9,11,13-15, 115,134].

Chapter 2

Non-autonomous Dissipative Systems, Their Attractors and Averaging

2.1 Introduction

In this chapter we consider non-autonomous nonlinear dissipative dynamical systems. Such systems appear, for example, when external forcing explicitly depends on time and when coefficients or parameters of the model vary with time.

Autonomous systems are associated to semigroup of nonlinear operators, while non-autonomous ones are associated to the process and the *family of processes.* The process is a natural generalization of the concept of semigroup.

Considering the attractors of non-autonomous dissipative systems, we have to center upon *attraction* and *minimality* instead of the invariance of attractor. A similar situation holds when we deal with inertial manifolds. For non-autonomous systems, there are used the concepts of a kernel and its section. The kernel consists of the collection of all bounded complete trajectories of the system (i.e. trajectories defined for all t), while the section of kernel at certain time is a set of points lying on the above trajectories at this time. The kernel and the section of the kernel are natural generalizations of the concept of a maximal invariant set to non-autonomous nonlinear dissipative systems. Some relation between the autonomous dissipative systems and non-autonomous dissipative ones can be established by the use of averaging methods. Certain classes of non-autonomous nonlinear dissipative systems one can average over a time explicitly occurred in them and hence the study of non-autonomous systems may be connected with the study of the correspondent autonomous averaged system.

There can be averaged some systems of a standard form

$$\partial_t u = \varepsilon F(u, t),$$

where $\varepsilon > 0$ is a small parameter. In particular, the systems with fast oscillating external forcing, the systems with fast oscillating coefficients and those with small nonlinearity and other can be reduced to the standard form.

Together with the above equation one can consider an averaged equation:

$$\partial_t v = \varepsilon \bar{F}(v),$$

where

$$\bar{F}(v) = \lim_{T \to \infty} \frac{1}{T} \int_0^T F(v, t) \, dt.$$

On the averaging, we arrive at an autonomous nonlineary dissipative system. Our purpose is to establish the closeness between solutions, attractors and inertial manifolds of an original system and those of an averaged system. The chapter is dedicated to this problem.

The main impediment to the extension of the averaging methods to the partial differential equations is that the nonlinearity must be a globally Lipschitzian one. This is not always the case. However, for dissipative systems possessing an absorbing set one can overcome this impediment by taking the initial data for the original and averaged systems inside of the absorbing set or by modification the nonlinearity in the special way as it is done in the theory of inertial manifolds (where the nonlinearity is multiplied by a smooth function, which is equal to 1, when the solution is within an absorbing set and which is equal to zero, if the solution is outside the absorbing set).

We now proceed to the more detailed description of non-autonomous dissipative systems, which will be considered in this chapter. Here we shall follow [8], the interested reader is referred to this work for the details. It is convenient to write non-autonomous evolutionary equations, which will be considered in what follows, in the form

$$\partial_t u = F(u, \sigma(t)), \tag{2.1}$$

where the dependence of the right side of the equation on t is expressed in terms of the function

$$\sigma(t) = (\sigma_1(t), \sigma_2(t), \ldots, \sigma_n(t)).$$

A vector–function $\sigma(t)$ may represent, say, a collection of time-depending coefficients of a system, an external nonstationary forcing and others parameters of system under consideration. In applications there were the cases when the coefficients and external forcings are periodic, quasiperiodic or almost periodic time-functions.

We assume that values of $\sigma(t)$ belong to a complete metric space X and let $\sigma(t) \in C_b(R, X)$, where $C_b(R, X)$ denotes the space of continuous bounded functions on R with values in X. We denote by

$$\rho_{C_b}(f_1, f_2) = \sup_{t \in R} \rho_X(f_1(t), f_2(t)),$$

the metric of $C_b(R, X)$, where ρ_X is a metric in X, while f_1 and f_2 are functions of $C_b(R, X)$. Usually, the function $\sigma(t)$ belongs to some closed set $\Sigma \subset C_b(R, X)$.

A set Σ is called a symbol space, while a function $\sigma(t)$ is called the time symbol (or simply symbol) of the equation (2.1).

We make a useful for what follows remark. The equation (2.1) can be considered either for a given fixed function $\sigma(t)$ or if $\sigma(t)$ is an arbitrary element of some given family of functions. In the first case we shall deal with a fixed process, in the second case we shall concern with a family of processes of the equation (2.1)

In many cases, as such a family of functions there is taken a set Σ of all functions which are obtained from $\sigma(t)$ by a shift by h of the argument of the symbol $\sigma(t)$.

By way of example, let us consider the case when $\sigma(t)$ is almost periodic function. Such a function possesses the following property: a set of all its translations $\sigma(t + h) = T(h)\sigma(t)$, $h \in R$ is a precompact set in the space $C_b(R, X)$.

A hull $H(\sigma)$ of the function $\sigma(t)$ is the closure in $C_b(R, X)$ of the set of its translations. In this case, the hull H of almost periodic function σ is taken as a symbol space, i.e. $\Sigma = H(\sigma)$.

Let us consider three Banach spaces E_1, E and E_0 that are continuously embedded one another: $E_1 \subseteq E \subseteq E_0$.

We shall assume that for fixed t the nonlinear operator $F(u, \sigma(t))$ acts from E_1 into E_0. We shall consider the equation (2.1) with the initial condition

$$u\mid_{t=\tau} = u_\tau, \ u_\tau \in E, \ t \geq \tau, \ \tau \in R. \tag{2.2}$$

and make the following assumption.

Assumption A. *Let, for any symbol $\sigma(t) \in \Sigma$, the problem (2.1)-(2.2) be uniquely solvable for any $\tau \in R$ and any $u_\tau \in E$, while its solution $u(t)$ belongs to E for any $t \geq \tau$. The solution of the problem (2.1)-(2.2) will be written as follows:*

$$u(t) = U_\sigma(t, \tau) u_\tau \quad t \geq \tau, \ \tau \in R, \tag{2.3}$$

where $U_\sigma(t, \tau)$ is two-parametric (parameters t and τ) family of operators, which act from E into E.

From Assumption **A** one can obtain the following properties of the family of operators $U_\sigma(t,\tau)$. The solution of (2.3) for $t = s$ has the form

$$u(s) = U_\sigma(s,\tau) u_\tau, \quad s \geq \tau, t \in R.$$

We take the point $u(s) = u_s$ as an initial point. The solution of (2.1) with the initial condition $u|_{t=s} = u_s$ is

$$u(t) = U_\sigma(t,s) u_s = U_\sigma(t,s) U_\sigma(s,\tau) u_\tau. \tag{2.4}$$

Comparing (2.3) and (2.4), we get

$$U_\sigma(t,s) U_\sigma(s,\tau) = U_\sigma(t,\tau), \forall t \geq s \geq \tau, \tau \in R. \tag{2.5}$$

Further, since the solution of (2.3) satisfies the initial condition, one has

$$U_\sigma(\tau,\tau) = I, \tag{2.6}$$

where I is an identical operator in E.

Two-parametric family of mappings $U_\sigma(t,\tau), t \geq \tau, \tau \in R$ acting in E and satisfying the conditions (2.5), (2.6) is called *a process* associated to the problem (2.1),(2.2) with the time symbol $\sigma(t) \in \Sigma$. We denote by $T(h)$ the translation operator acting on $\sigma(t)$ according to the rule $T(h)\sigma(t) = \sigma(t+h), \quad \forall h \in R$.

It will be assumed that the symbol space Σ is an invariant one with respect to the translation operator, i.e. $T(h)\Sigma = \Sigma, \forall h \in R$. In this case, for the family of processes $U_\sigma(t,\tau), t \geq \tau, \tau \in R$, and the translation operator $T(h)$, the equality holds:

$$U_{T(h)\sigma(t)} = U_{\sigma(t)}(t+h,\tau+h), \quad \forall h \geq 0, \quad t \geq \tau, \quad \tau \in R. \tag{2.7}$$

In fact, the solution of the equation (2.1) with a symbol $\sigma(t+h)$ and initial condition (2.2) has the form:

$$u(t) = U_{\sigma(t+h)}(t,\tau) u_\tau. \tag{2.8}$$

On the other hand, we replace in (2.1) $\sigma(t)$ by $\sigma(t+h)$ and make the change of variables, setting $t+h = s$, where $s \geq t$. This leads to

$$\partial_s u = F(u,\sigma(s)), \quad u|_{s=\tau+h} = u_\tau.$$

The solution of this equation will have the form (after replacement $s \to t$):

$$u(t) = U_{\sigma(t)}(t+h, \tau+h) u_\tau. \tag{2.9}$$

The solution of (2.8) and that of (2.9) coincide for $t = \tau$. Therefore, by virtue of the uniqueness they coincide everywhere, i.e. the equality (2.7) holds.

The relation (2.7) will be called *the translation* property of process-es and the equality (2.5) we shall call *the multiplicative* property of processes. Note that on the symbol space Σ one can consider any semi-group $T(t)$, $t \geq 0$ (it is not necessary for this semigroup to be a trans-lation semigroup), which possesses the property $T(t)\Sigma = \Sigma$, $t \geq 0$ and satisfies the equality (2.7).

The properties (2.5), (2.6) and (2.7) of the solutions of non-autono-mous equations make a base for the construction of the general theory of processes, that are associated to the non-autonomous systems.

If $\sigma_0(t) \in \Sigma$ is some fixed element then the corresponding process $U_{\sigma_0(t)}$ will be called a *fixed* process. It should be noted that the family of processes can be associated to the semigroup $S(t)$, $t \geq 0$ acting on the extended phase space $E \times \Sigma$. The semigroup is defined as follows:

$$S(t)(u,\sigma) = \Big(U_\sigma(t,0)u, \, T(t)\sigma\Big), \quad t \geq 0, \quad (u,\sigma) \in E \times \Sigma.$$

For this semigroup there is proved the existence of a compact in-variant attractor A. Then the attractor A_Σ of a family of processes $U_\sigma(t,\tau)$, $\sigma \in \Sigma$ is connected with the attractor A of the semigroup $S(t)$, $t \geq 0$ thus: $A_\Sigma = \Pi_1 A$, where Π_1 is a projector acting from $E \times \Sigma$ into E according to the rule $\Pi_1(u,\sigma) = u$.

Finally, by analogy with the concept of semigroup we can introduce the concept of the semiprocess. Unlike the process, the semiprocess is defined not for all $t \in R$, but only for $t \geq 0$ [8].

We now give the strong definitions of the concepts of processes, of family of processes and their attractors.

2.2 Processes and Their Attractors. Kernel of Processes, Sections of Kernel

Let us give the general definitions of a process and its attractor without associating them with a concrete system of equations.

Definition 2.1. A two-parametric family of mappings

$$U(t,\tau): E \to E, \quad t \geq \tau, \tau \in R,$$

is said to be a process in a Banach space E if the following properties are valid:

1) $U(\tau,\tau) = I$, $\tau \in R$;
2) $U(t,s)U(s,\tau) = U(t,\tau)$, $\forall t \geq s \geq \tau$, $\tau \in R$.

The process $U(t,\tau)$, $t \geq \tau$, $\tau \in R$ is called continuous if any map-ping $U(t,\tau): E \to E$ is continuous. We shall consider only contin-uous processes in the sequel. To shorten notation $U(t,\tau)$ stand for $U(t,\tau)$, $t \geq \tau$, $\tau \in R$ and it will be used repeatedly below.

Definition 2.2. A set $B_a \subset E$ is said to be an *absorbing* set of process $U(t, \tau)$, $t \geq \tau$, $\tau \in R$, if for any *bounded* set $B \subset E$ there exists $T = T(\tau, B)$ such that

$$U(t, \tau) B \subseteq B_a, \quad t \geq T(\tau, B).$$

A process is said to be a *compact* process if its absorbing set is compact.

Definition 2.3. A set $M \subset E$ is said to be an *attracting* set of the process $U(t, \tau)$, $t \geq \tau$, $\tau \in R$ if for any $\tau \in R$ and any *bounded* set $B \subseteq E$

$$\lim_{t \to \infty} \operatorname{dist}_E (U(t, \tau) B, M) = 0.$$

A process is said to be an *asymptotically compact* process if its attracting set is a *compact* set.

Definition 2.4. *The closed* set $A \subset E$ is called the *attractor of process* $U(t, \tau)$, $t \geq \tau$, $\tau \in R$, if it is an *attracting* set of the process and it is contained in any *closed attracting* set A' of process $U(t, \tau)$, $t \geq \tau$, $\tau \in R$, i.e. the attractor A of process $U(t, \tau)$, $t \geq \tau$, $\tau \in R$, is a *least closed attracting set* of the process $U(t, \tau)$.

Definition 2.5. A curve $u(s) \in E$, $s \in R$ is said to be a complete trajectory of the process $U(t, \tau)$, $t \geq \tau$, $\tau \in R$, if

$$U(t, \tau) u(\tau) = u(t), \quad \forall t \geq \tau, \quad \tau \in R.$$

Definition 2.6. *The kernel* K of the process $U(t, \tau)$ consists of all *bounded complete* trajectories of the process $U(t, \tau)$:

$$K = \{ u(\cdot) : u(t), t \in R \},$$

where $u(\cdot)$ is a complete trajectory of $U(t, \tau)$, $\|u(t)\|_E \leq M_u, \forall t \in R$.

Definition 2.7. The set

$$K(s) = \{ u(s) : u(\cdot) \in K \} \subseteq E$$

is called the *kernel section* K at time $t = s$, $s \in R$. Evidently, the following proposition holds.

Let K be the kernel of the process $U(t, \tau)$ then

$$U(t, \tau) K(\tau) = K(t) \quad \forall t \geq \tau, \quad \tau \in R.$$

If the process $U(t, \tau)$ is a semigroup $S(t)$, $t \geq 0$, i.e.

$$U(t, \tau) = U(t - \tau, 0) = S(t - \tau), \quad t \geq \tau,$$

then the kernel section $K(s)$ does not depend on s : $K(s) = K(0)$.

Let us proceed to the elucidation of the conditions which cause the process $U(t,\tau)$, $t \geq \tau$, $\tau \in R$, acting in E to possess an attractor, i.e. a closed minimal attracting set. Let B be any bounded set in E, $\tau \in R$. We determine the ω-limit set $\omega_\tau(B)$ for the process $U(t,\tau)$:

$$\omega_\tau(B) = \bigcap_{t \geq \tau} \overline{\bigcup_{s \geq t} (U(s,\tau)B)}. \tag{2.10}$$

From now we will be using the following proposition.

Proposition B. If $y \in \omega_\tau(B)$, then there exist some sequences $\{x_n\} \subset B$, $\{t_n\} \subset R_\tau = [\tau, +\infty)$, $t_n \to +\infty$ as $n \to \infty$ such that $U(t_n, \tau)x_n \to y$ as $n \to \infty$. Conversely, if there exist some sequences $\{x_n\} \subset B$, $\{t_n\} \subset R_\tau$, $t_n \to +\infty$ such that the sequence $U(t_n, \tau)x_n$ converges to some y as $n \to \infty$, then $y \in \omega_\tau(B)$. One can prove Proposition **B** by the use of the definition $\omega_\tau(B)$ according to (2.10).

Theorem 2.1. *If the process $\{U(t,\tau)\}$, $t \geq \tau$, $\tau \in R$, possesses a compact attracting set P, then for any $\tau \in R$ and any bounded set $B \subset E$:*

1) $\omega_\tau(B)$ *is nonempty compact set in E and $\omega_\tau(B) \subseteq P$;*
2) $\mathrm{dist}_E(U(t,\tau)B, \omega_\tau(B)) \to 0$, *as $t \to +\infty$;*
3) *if Y is closed and $\mathrm{dist}_E(U(t,\tau)B, Y) \to 0$, as $t \to +\infty$, then $\omega_\tau(B) \subseteq Y$, i.e. $\omega_\tau(B)$ belongs to any closed attracting set.*

Proof. We prove that $\omega_\tau(B) \neq \emptyset$. For any fixed $x \in B$ we consider an arbitrary sequence $\{t_n\} \subset R_\tau$, $t_n \to +\infty$ as $n \to \infty$. According to the attracting property of P

$$\mathrm{dist}_E(U(t_n,\tau)x, P) \to 0, \quad t_n \to +\infty.$$

This means that for some sequence $\{y_n\} \in P$

$$\|U(t_n,\tau)x - y_n\|_E \to 0, \quad \text{as } n \to \infty.$$

From the sequences $y_n \subset P$ by the compactness of P we can extract a convergent subsequence $\{y_{n_j}\}$, $y_{n_j} \to y$, $y \in P$. Hence,

$$U(t_{n_j}, \tau)x \to y, \quad \text{as } n_j \to \infty.$$

By Proposition **B** the point y belongs to $\omega_\tau(B)$, i.e. this set is not empty. Let us show that $\omega_\tau(B) \subseteq P$. (Hence it immediately follows that $\omega_\tau(B)$ is compact set.) Let $y \in \omega_\tau(B)$. Then there exists sequences $\{x_n\} \subset B$, $\{t_n\} \subset R_\tau$, $t_n \to +\infty$ as $n \to \infty$ such that $U(t_n,\tau)x_n \to y$ as $n \to \infty$. Using again the attracting property of P, one gets

$$\mathrm{dist}_E(U(t_n,\tau)x_n, P) \to 0, \quad \text{as } n \to \infty.$$

(It should be noted that the tendency to zero is uniform with respect to the points x_n of B.)

Passing to the limit, we get $\mathrm{dist}_E\,(y, P) = 0$. Here we used the fact that the distance $\rho(x, A)$ between the point x and the set A is continuous function x. Since P is closed, we get $y \in P$. Thus, $\omega_\tau(B) \subseteq P$. Prove the proposition 2) of the theorem. Assume that 2) is false. Then, for some bounded set $B \subset E$ and for $t \geq \tau$ one has:

$$\mathrm{dist}_E\,(U\,(t,\tau)\,B, \quad \omega_\tau(B)) \not\to 0, \quad \text{as } t \to +\infty.$$

This means that for some sequences $\{x_n\} \subset B$ and $\{t_n\} \subset R_\tau$, $t_n \to +\infty$ we have

$$\mathrm{dist}_E\,(U\,(t_n,\tau)\,x_n, \quad \omega_\tau(B)) \geq \delta > 0 \quad \forall\, n \in N. \tag{2.11}$$

However, from the attracting property of P it follows that

$$\mathrm{dist}_E\,(U\,(t_n,\tau)\,x_n, P) \to 0, \quad \text{as } n \to \infty,$$

i.e. there exists a sequence $\{y_n\} \subset P$ such that

$$\|U\,(t_n,\tau)\,x_n \, - \, y_n\|_E \to 0, \quad \text{as } n \to \infty.$$

We can extract from $\{y_n\}$ an convergent subsequence $y_{n_j} \to y$ as $n_j \to \infty$. Consequently, $U\,(t_{n_j},\tau)\,x_{n_j} \to y$. By virtue of Proposition B we have $y \in \omega_\tau(B)$. On the other hand, passing to the limit in (2.11), we find $\mathrm{dist}_E\,(y, \omega_\tau(B)) \geq \delta$, that contradicts the inclusion $y \in \omega_\tau(B)$. Let us prove the proposition 3) of the theorem. Let Y be a closed attracting set of the process $U(t,\tau)$. If $y \subset \omega_\tau(B)$, then there exist sequences $\{x_n\} \subset B, \{t_n\} \subset R_\tau$, $t_n \to +\infty$ as $n \to \infty$ such that $U(t_n,\tau)x_n \to y$ as $n \to \infty$. Since Y is an attracting set, we have

$$\mathrm{dist}_E\,(U\,(t_n,\tau)\,x_n, Y) \to 0, \quad \text{as } n \to \infty,$$

i.e. $\mathrm{dist}_E(y,Y) = 0$ and $y \in Y$, since Y is closed. Thus, $\omega_\tau(B) \subset Y$.

Now we can formulate the theorem on the existence of the attractor of the process $U(t,\tau)$.

Theorem 2.2. *If the process $U(t,\tau)$, $t \geq \tau$, $\tau \in R$, possesses a compact attracting set, then the process possesses a compact attractor A.*

Proof. We denote by B_n the closed balls of radius n:

$$B_n = \{x \in E : \|x\|_E \leq n\}.$$

Let us show that a compact attractor of a process $U(t,\tau)$ is the set

$$A = \overline{\bigcup_{\tau \in R}\bigcup_{n \in N} \omega_\tau(B_n)}. \tag{2.12}$$

Indeed, let $B \subset E$ be any bounded set. Then $B \subseteq B_n$ for some $n \in N$. It is evident that $\omega_\tau(B) \subseteq \omega_\tau(B_n) \subseteq A$. Since $\omega_\tau(B)$ attracts $U(t,\tau)B$, it is obvious that A attracts this set, i.e.

$$\text{dist}_E(U(t,\tau)B, A) \to 0, \quad \text{as } t \to +\infty.$$

Further, the set $\omega_\tau(B_n)$ belongs to any closed attracting set. Consequently, the set A also has this property by the equality (2.12).

The case of uniformly (with respect to τ) processes.

Definition 2.3a. A set $M \subset E$ is said to be uniformly with respect to $\tau \in R$ attracting set for the process $U(t,\tau)$, $t \geq \tau$, $\tau \in R$, if for any bounded set B

$$\sup_{\tau \in R} \text{dist}_E(U(t+\tau,\tau)B, M) \to 0, \quad \text{as } t \to \infty.$$

The process $U(t,\tau)$ is said to be uniformly with respect to $\tau \in R$ asymptotically compact if the attracting set M is compact.

Theorem 2.2a. *Let* $\{U(t,\tau)\}$ *be a continuous uniformly asymptotically compact process with a uniformly attracting set* $P \subset\subset E$. *Then the kernel* K *of the process* $\{U(t,\tau)\}$ *is non-empty, kernel sections* $K(s)$ *are all compact, and* $K(s) \subseteq P$ $\forall s \in R$.

The proof of the theorem is given in [8].

2.3 Families of Processes and Their Attractors

Let Σ be a complete metric space and let $\sigma \in \Sigma$ be an element of this set. We shall call σ the symbol of the family of processes $U_\sigma(t,\tau)$, $t \geq \tau$, $\tau \in R$, $\sigma \in \Sigma$, while Σ itself will be called a symbol space.

Definition 2.8. *The family of processes* acting in Banach space E is a two-parametric family of mappings

$$U_\sigma(t,\tau): E \to E, \quad t \geq \tau, \quad \tau \in R, \quad \sigma \in \Sigma,$$

satisfying the following conditions:

1) $U_\sigma(\tau,\tau) = I, \quad \tau \in R, \quad \sigma \in \Sigma;$
2) $U_\sigma(t,s)U_\sigma(s,\tau) = U_\sigma(t,\tau), \forall t \geq s \geq \tau, \tau \in R, \quad s \in \Sigma.$

Definition 2.9. A family of processes $U_\sigma(t,\tau)$, $\sigma \in \Sigma$ is said to be *uniformly* (with respect to $\sigma \in \Sigma$) *bounded* if for any bounded set B the set

$$\bigcup_{\sigma \in \Sigma} \bigcup_{\tau \in R} \bigcup_{t \geq \tau} (U_\sigma(t,\tau)B)$$

is *bounded.*

Definition 2.10. The set $B_a \subset E$ is said to be *uniformly* (with respect to $\sigma \in \Sigma$) *absorbing* set of processes $U_\sigma(t, \tau)$, $\sigma \in \Sigma$, if for any $\tau \in R$ and any *bounded* set $B \subset E$ there exists $T = T(\tau, B) \geq \tau$ such that

$$\bigcup_{\sigma \in \Sigma} (U_\sigma(t, \tau) B) \subseteq B_a, \quad \forall t \geq T(\tau, B).$$

Definition 2.11. The set $P \subset E$ is said to be *uniformly* (with respect to $\sigma \in \Sigma$) *attracting* set of the family of processes $U_\sigma(t, \tau)$, $\sigma \in \Sigma$, if for any bounded set $B \subset E$ and for any $\tau \in R$

$$\lim_{t \to \infty} \sup_{\sigma \in \Sigma} \text{dist}_E (U_\sigma(t, \tau) B, P) = 0.$$

A family of processes possessing a *compact* uniformly absorbing set is called a *compact* one, while that possessing a compact uniformly attracting set is called an *asymptotically compact* family of processes.

Definition 2.12. The closed set $A_\Sigma \subset E$ is said to be an *uniformly* (with respect to $\sigma \in \Sigma$) *attractor* of the family of processes $U_\sigma(t, \tau)$, $\sigma \in \Sigma$, if it is *uniformly attracting* and it is contained in any closed uniformly attracting set A' of the family of process $U_\sigma(t, \tau)$, $\sigma \in \Sigma$, i.e. $A_\Sigma \subseteq A'$.

Now we shall establish the evidences of the existence the attractor of the family of processes.

Let B be a bounded set, $\tau \in R$ and $\sigma \in \Sigma$. We define the ω-limit set $\omega_{\tau, \Sigma}(B)$:

$$\omega_{\tau, \Sigma}(B) = \bigcap_{t \geq \tau} \overline{\bigcup_{\sigma \in \Sigma} \bigcup_{s \geq t} (U_\sigma(s, \tau) B)}. \tag{2.13}$$

Here the following proposition takes place.

Proposition C. If $y \in \omega_{\tau, \Sigma}(B)$, then there exist sequences $\{\sigma_n\} \subset \Sigma$, $\{x_n\} \subset B$, $\{t_n\} \subset R_\tau$, $t_n \to +\infty$ as $n \to \infty$ such that

$$U_{\sigma_n}(t_n, \tau) x_n \to y, \quad t_n \to +\infty.$$

Conversely, if the sequences $\{\sigma_n\} \subset \Sigma$, $\{x_n\} \subset B$, $\{t_n\} \subset R_\tau$, $t_n \to +\infty$ as $n \to \infty$ such that

$$U_{\sigma_n}(t_n, \tau) x_n \to y, \quad t_n \to +\infty,$$

then $y \in \omega_{\tau, \Sigma}(B)$.

Theorem 2.3. *If the family of processes* $U_\sigma(t, \tau)$, $\sigma \in \Sigma$, *possesses an uniformly (with respect to* $\sigma \in \Sigma$ *) compact attracting set* P, *then for any* $\tau \in R$ *and any bounded set* B *the following properties take place:*

1) $\omega_{\tau, \Sigma}(B)$ *is a nonempty, compact set in* E *and* $\omega_{\tau, \Sigma}(B) \subseteq P$;

2) *if* $\sup_{\sigma \in \Sigma} \mathrm{dist}_E(U_\sigma(t,\tau)B, Y) \to 0$ *as* $t \to +\infty$, *then* $\omega_{\tau,\Sigma}(B)$ $\subseteq Y$, *i.e.* $\omega_{\tau,\Sigma}(B)$ *belongs to any closed uniformly attracting set;*

3) $\sup_{\sigma \in \Sigma} \mathrm{dist}_E(U_\sigma(t,\tau)B, \omega_{\tau,\Sigma}(B)) \to 0$, *as* $t \to +\infty$.

Proof of this theorem is similar to that of Theorem 2.1. Because of this we shall give only a shortened scheme of proving [8]. For any fixed $\sigma \in \Sigma$ and $x \in B$ we take any sequence $\{t_n\} \subset R_\tau$, $t_n \to +\infty$ as $n \to \infty$. From the uniformly attracting property of the set P it follows that

$$\mathrm{dis}_E(U_\sigma(t_n, \tau)x, P) \to 0, \quad \text{as } n \to \infty.$$

This means that there exists a sequence $\{y_n\} \subset P$ such that $\|U_\sigma(t_n,\tau)x - y_n\|_E \to 0$ as $n \to \infty$. By the compactness, from $\{y_n\}$ there can be chosen a convergent subsequence $y_{n_j} \to y \in P$. Then $U_\sigma(t_{n_j},\tau)x \to y$ as $n_j \to \infty$ and by virtue of Proposition **C** we have $y \in \omega_{\tau,\Sigma}(B)$, that is, $\omega_{\tau,\Sigma}(B)$ is nonempty set. Show that $\omega_{\tau,\Sigma}(B) \subseteq P$. (Hence we obtain the compactness of $\omega_{\tau,\Sigma}(B)$.) Let $y \in \omega_{\tau,\Sigma}(B)$. Then, by Proposition **C** there exist sequences $\{\sigma_n\} \subset \Sigma$, $\{x_n\} \subset B$, $\{t_n\} \subset R_\tau$, $t_n \to +\infty$ as $n \to \infty$ such that $U_{\sigma_n}(t_n,\tau)x_n \to y$ as $n \to \infty$. Using the fact that the set P is uniformly attracting set with respect to Σ and $x \in B$, we find

$$\mathrm{dist}_E(U_{\sigma_n}(t_n,\tau)x_n, P) \to 0, \quad \text{as } n \to \infty,$$

i.e. $\mathrm{dist}_E(y, P) = 0$ and $y \in P$ for any $y \in \omega_{\tau,\Sigma}(B)$. The relation

$$\sup_{\sigma \in \Sigma} \mathrm{dist}_E(U_\sigma(t,\tau)B, \omega_{\tau,\Sigma}(B)) \to 0, \quad \text{as } t \to +\infty.$$

is established by reductio ad absurdum.

Let this relation be false for some bounded set B and $t \geq \tau$. Then for some sequences $\{x_n\} \subset B$, $\{\sigma_n\} \subset \Sigma$, $\{t_n\} \subset R_\tau$, $t_n \to +\infty$ one must have

$$\mathrm{dist}_E\left(U_{\sigma_n}(t_n,\tau)x_n, \omega_{\tau,\Sigma}(B)\right) \geq \delta > 0, \quad \forall n \in N. \qquad (2.13')$$

But, it follows from the uniform attraction property of the set P that

$$\mathrm{dist}_E(U_{\sigma_n}(t_n,\tau)x_n, P) \to 0, \; n \to \infty.$$

As before, hence it may be seen that there exists a convergent sequence $y_n \to y$ such that $\|U_{\sigma_n}(t_n,\tau)x_n - y_n\|_E \to 0$ as $n \to \infty$. By virtue of Proposition **C** one has $y \in \omega_{\tau,\Sigma}(B)$. However, from (2.13') it follows that $\mathrm{dist}_E(y, \omega_{\tau,\Sigma}(B)) \geq \delta > 0$. Finally, let Y be a closed uniformly attracting set of the family of processes $U_\sigma(t,\tau)$, $\sigma \in \Sigma$. Hence, if $y \in \omega_{\tau,\Sigma}(B)$, then there exist sequences $\{\sigma_n\} \subset \Sigma$, $\{x_n\} \subset B$, $\{t_n\} \subset R_\tau$, $t_n \to +\infty$ such that $U_{\sigma_n}(t_n,\tau)x_n \to y$.

From the uniform attraction property of the set Y it follows that $\text{dist}_E \left(U_{\sigma_n} \left(t_n, \tau \right) x_n, Y \right) \to 0$, as $n \to \infty$, that is $\text{dist}_E \left(y, Y \right) = 0$ and $y \in Y$. Thus, $\omega_{\tau, \Sigma} \left(B \right) \subseteq Y$. This proves the theorem.

Theorem 2.4. *If the family of processes $U_\sigma(t, \tau)$, $\sigma \in \Sigma$ is uniformly asymptotically compact, then it possesses an uniform attractor A_Σ.*

Proof is similar to that of the Theorem 2.2. As above, we put

$$A_\Sigma = \overline{\bigcup_{\tau \in R} \bigcup_{n \in N} \omega_{\tau, \Sigma} \left(B_n \right)},$$

where $B_n = \{ x \in E : \| x \|_E \leq n \}$.

Evidently, for any bounded set $B \subset E$ there can be found a ball B_n such that $B \subseteq B_n$ and consequently, $\omega_{\tau, \Sigma}(B) \subseteq \omega_{\tau, \Sigma}(B_n) \subseteq A_\Sigma$. By Theorem 2.3, $\omega_{\tau, \Sigma}(B)$ uniformly attracts B, while the set $\omega_{\tau, \Sigma} \left(B_n \right)$ belongs to any uniformly attracting set, then A_Σ is minimal.

Corollary 2.1. A fixed process $U_{\sigma_*} \left(t, \tau \right)$ of the uniformly asymptotically compact family of processes $U_\sigma(t, \tau)$, $\sigma \in \Sigma$ possesses an (ununiform) attractor $A_{\{\sigma_*\}}$ and

$$\bigcup_{\sigma_* \in \Sigma} A_{\{\sigma_*\}} \subseteq A_\Sigma.$$

The inverse inclusion does not hold. If the process has a periodic symbol, then the uniform attractor coincides with ununiform one [8].

2.4 Family of Processes and Semigroups

Let on a symbol space Σ there is given a semigroup $T(t)$, $t \geq 0$ such that $T \left(t \right) \Sigma = \Sigma$, $\forall t \geq 0$.

Condition D. Let the family of processes $U_\sigma(t, \tau)$, $t \geq \tau$, $\tau \in R$, $\sigma \in \Sigma$ and the semigroup $T(t)$, $t \geq 0$ be connected by the following condition

$$U_{T(s)\sigma} \left(t, \tau \right) = U_\sigma \left(t + s, \tau + s \right), \quad s \geq 0.$$

Under Condition **D**, we construct the semigroup $S(t)$ acting on the extended phase space $E \times \Sigma$ which corresponds to the family of processes $U_\sigma(t, \tau)$, $\sigma \in \Sigma$ by the following way

$$S(t)(u, \sigma) = (U_\sigma(t, 0)u, T(t)\sigma), \quad t \geq 0, \ (u, \sigma) \in E \times \Sigma.$$

Show that $S \left(t \right)$, $t \geq 0$ is indeed a semigroup. We have

$$S \left(t_1 + t_2 \right) \left(u, \sigma \right) = \left(U_\sigma \left(t_1 + t_2, 0 \right) u, \ T \left(t_1 + t_2 \right) \sigma \right)$$

$$= \left(U_\sigma \left(t_1 + t_2, t_2 \right) U_\sigma \left(t_2, 0 \right) u, \ T \left(t_1 \right) T \left(t_2 \right) \sigma \right)$$

$$= \left(U_{T(t_2)\sigma} \left(t_1, 0 \right) U_\sigma \left(t_2, 0 \right) u, \ T \left(t_1 \right) T \left(t_2 \right) \sigma \right)$$

$$= S \left(t_1 \right) \left(U_\sigma \left(t_2, 0 \right) u, \ T \left(t_2 \right) \sigma \right) = S \left(t_1 \right) S \left(t_2 \right) (u, \sigma).$$

Definition 2.13. The family of processes $U_\sigma(t,\tau)$, $\sigma \in \Sigma$ acting in E is said to be $(E \times \Sigma, E)$-continuous, if for any fixed t and $\tau, t \geq \tau$, $\tau \in R$, the mapping $(u,\sigma) \to U_\sigma(t,\tau)u$ is continuous from $E \times \Sigma$ into E.

This continuity condition means, in essence, the continuous dependence of the solution of (2»1)-(2.2) on initial data and the right side of the equation. Define two projectors Π_1 and Π_2 from $E \times \Sigma$ into E, putting $\Pi_1(u,\sigma) = u$, $\Pi_2(u,\sigma) = \sigma$.

Theorem 2.5. [8] *Let the following conditions be satisfied:*

1) *the family of processes* $U_\sigma(t,\tau)$, $\sigma \in \Sigma$ *is uniformly asymptotically compact;*

2) *this family is* $(E \times \Sigma, E)$*-continuous;*

3) Σ *is a compact metric space on which there is defined the continuous semigroup* $T(t), t \geq 0$ *such that* $T(t)\Sigma = \Sigma \ t \geq 0$;

4) *the identity holds*

$$U_{T(s)\sigma}(t,\tau) = U_\sigma(t+s,\tau+s), \forall s \geq 0, t \geq \tau, \tau \in R, \sigma \in \Sigma.$$

Then the semigroup $S(t)$, $t \geq 0$ *acting on* $E \times \Sigma$ *and corresponding to the family of processes* $U_\sigma(t,\tau)$, $\sigma \in \Sigma$ *possesses a compact attractor A such that* $S(t)A = A$, $\forall t \geq 0$. *Moreover, the following propositions are valid:*

1) $\Pi_1 A = A_1 = A_\Sigma$ *is a uniform attractor of the family of processes* $U_\sigma(t,\tau)$, $\sigma \in \Sigma$;

2) $\Pi_2 A = A_2 = \Sigma$;

3) $A = \bigcup_{\sigma \in \Sigma} K_\sigma(0) \times \sigma$;

4) $A_1 = A_\Sigma = \bigcup_{\sigma \in \Sigma} K_\sigma(0)$.

Here $K_\sigma(0)$ *is a section at time* $t = 0$ *of the kernel* K_σ *of the process* $U_\sigma(t,\tau)$ *with symbol* $\sigma \in \Sigma$.

Proof. Proving this theorem, we shall use the following evidence of the existence of a compact invariant attractor of semigroup: if $S(t)$, $t \geq 0$ is continuous semigroup acting on the complete metric space X possesses a compact attracting set P, then the set

$$A = \omega(P) = \bigcap_{\tau \geq 0} \overline{\bigcup_{t \geq \tau} S(t)P}$$

is a compact invariant attractor of semigroup $S(t)$, $t \geq 0$. Therefore,

$$A = \{\gamma(0), \gamma \in \Gamma(S)\} = \{\gamma(s), \quad s \in R, \quad \gamma \in \Gamma(S)\},$$

where $\Gamma(S)$ the set of all complete bounded trajectories of semigroup $S(t)$, $t \geq 0$. We recall that the complete bounded trajectory of the semigroup $S(t)$ is the bounded function $\gamma : R \to X$ such that

$$\gamma(t+s) = S(t)\gamma(s), t \geq 0, s \in R.$$

Now we consider the semigroup $S(t)$, $t \geq 0$ acting on the extended phase space $E \times \Sigma$. Since, by the condition of the theorem, P is uniformly attracting compact set of the family of processes $\{U_\sigma(t,\tau)\}$, then the set $P \times \Sigma$ will be a compact attracting set of the semigroup $S(t)$ acting on $E \times \Sigma$. The continuity of semigroup $S(t)$ follows from $(E \times \Sigma, E)$-continuity of the family of processes $U_\sigma(t,\tau)$ and from the continuity of semigroup $T(t)$. Therefore, according to the above general evidence of the existence of the attractor of the semigroup, we conclude that the semigroup $S(t)$ possesses a compact invariant attractor A. Show that $A_1 = \Pi_1 A$ is an uniform attractor A_Σ of the family $U_\sigma(t,\tau)$, $\sigma \in \Sigma$. Let B be a bounded set. Then

$$\mathrm{dist}_{E \times \Sigma}(S(t)(B \times \Sigma), A) \to 0, \quad \text{as } t \to +\infty.$$

Hence,

$$\sup_{\sigma \in \Sigma} \mathrm{dist}_E(U_\sigma(t,0)B, \Pi_1 A) \to 0, \quad \text{as } t \to +\infty. \qquad (2.14)$$

We need to establish the relation

$$\sup_{\sigma \in \Sigma} \mathrm{dist}_E(U_\sigma(t,\tau)B, \Pi_1 A) \to 0, \quad \text{as } t \to +\infty. \qquad (2.15)$$

For that purpose we note that by the condition of the uniform asymptotic compactness the set

$$B_0 = \bigcup_{\sigma \in \Sigma}(U_\sigma(s,\tau)B)$$

is bounded for some $s \geq \tau \geq 0$. Moreover, if $t \geq s$, then

$$U_\sigma(t,\tau)B = U_\sigma(t,s)U_\sigma(s,\tau)B \subseteq U_\sigma(t,s)B_0 = U_{T(s)\sigma}(t-s,0)B_0. (2.16)$$

Taking into account (2.16) and (2.14), we get (2.15). Since attractor of the semigroup consist of the complete bounded trajectories, for any point $u_0 \subset \Pi_1 A$ and for corresponding point $(u_0, \sigma_0) \in A$, there exists a complete bounded trajectory $\gamma(s) = (u(s), \sigma(s))$, $s \in R$, such that $\gamma(0) = (u(0), \sigma(0)) = (u_0, \sigma_0)$. It should be noted that the action of the semigroup $S(t)$ on the trajectory reduces to the argument translation of the semigroup, i.e.

$$S(t)\gamma(s) = S(t)(u(s), \sigma(s)) = (u(t+s), \sigma(t+s)).$$

Changing the notations in this equality $(s \to \tau, \ t \to t - \tau)$, we obtain for $t - \tau \geq 0$

$$S(t-\tau)(u(\tau), \sigma(\tau)) = (u(t), \sigma(t)).$$

On the other hand, taking into account the definition of the semi-group $S(t)$, we find that the following equalities must hold

$$U_{\sigma(\tau)}(t - \tau, 0) u(\tau) = u(t), \quad T(t - \tau) \sigma(\tau) = \sigma(t). \qquad (2.17)$$

We need to show that $u(s)$ is a complete trajectory of the process $U_{\sigma_0}(t, \tau)$ with symbol $\sigma_0 = \sigma(0)$. This means that the equality holds

$$U_{\sigma(0)}(t, \tau) u(\tau) = u(t), t \geq \tau, \quad \tau \in R.$$

This is so indeed. For $\tau \geq 0$ we have with respect to (2.17):

$$U_{\sigma(0)}(t, \tau) u(\tau) = U_{T(\tau)\sigma(0)}(t - \tau, 0) u(\tau) = U_{\sigma(\tau)}(t - \tau, 0) u(\tau) = u(t).$$

Similarly, for $\tau < 0$ we get

$$U_{\sigma(0)}(t, \tau) u(\tau) = U_{T(-\tau)\sigma(\tau)}(t, \tau) u(\tau) = U_{\sigma(t)}(t - \tau, 0) u(\tau) = u(t).$$

It is worthy to note that we used here mainly the relation (2.17), since the semigroup $T(t)$ by the condition of the theorem is not the translation operator. Since $u(s)$ is the complete trajectory of the process $\{U_{\sigma_0}(t, \tau)\}$ with symbol σ_0, one has that $u(0) \in K_{\sigma_0}(0)$ is the kernel section K_{σ_0} at time $t = 0$. Since u_0 was an arbitrary point of the set $\Pi_1 A$, there takes place the inclusion

$$\Pi_1 A \subseteq \bigcup_{\sigma \in \Sigma} K_\sigma(0).$$

We prove the inverse inclusion. Let

$$u_0 \in \bigcup_{\sigma \in \Sigma} K_\sigma(0), \quad u_0 = u(0),$$

where $u(s)$, $s \in R$ is the complete bounded trajectory of the process $\{U_{\sigma_0}(t, \tau)\}$ with symbol $\sigma_0 \in \Sigma$. Since the attractor of the semigroup $S(t)$, $t \geq 0$ consists of the complete bounded trajectories, there exists a complete bounded trajectory $\sigma(s)$, $s \in R$ of the semigroup $T(t)$, $t \geq 0$ such that $\sigma(0) = \sigma_0(t)$. Show that $(u(s), \sigma(s))$ is a complete bounded trajectory of semigroup $S(t)$, $t \geq 0$. For $s \geq 0$ we have

$$S(t)(u(s), \sigma(s)) = \Big(U_{\sigma(s)}(t, 0) u(s), T(t) \sigma(s)\Big)$$

$$= \Big(U_{T(s)\sigma(0)}(t, 0) u(s), \sigma(t + s)\Big)$$

$$= \Big(U_{\sigma(0)}(t + s, s) u(s), \sigma(t + s)\Big) = (u(t + s), \sigma(t + s)).$$

Similarly, for $s < 0$, taking into account that $T(-s)\sigma(s) = \sigma(0)$ and $u(t + s) = U_{\sigma(0)}(T + s, s)u(s) = U_{T(-s)\sigma(s)}(t + s, s) = U_{\sigma(s)}(t, 0)u(s)$, we get

$$S(t)(u(s), \sigma(s)) = \left(U_{\sigma(s)}(t, 0)u(s), T(t)\sigma(s) \right)$$
$$= (u(t + s), \sigma(t + s)), \ t \geq 0, \ s \in R.$$

Since $(u_0, \sigma(0)) \in A$, one has $u_0 \in \Pi_1 A$. Therefore,

$$\bigcup_{\sigma \in \Sigma} K_\sigma(0) \in \Pi_1 A.$$

Finally, we prove that $A_1 = \Pi_1 A$ possesses the minimality property, i.e. $A_1 \subseteq A_\Sigma^*$, where A_Σ^* is any closed uniformly attracting set of the family of processes $U_\sigma(t, \tau)$, $\sigma \in \Sigma$. For that purpose it is sufficient to demonstrate that the point $u(0)$ of any complete bounded trajectory $u(s)$, $s \in R$ of the process under consideration belongs to A_Σ^*.

Let $\sigma(s)$ be the complete bounded trajectory of semigroup $T(t)$ such that $\sigma(0) = \sigma_0(t)$. Consider the set $B_0 = \{u(-n), n \in N\}$. It is evident that, B_0 is a bounded set. Since $u(t) = U_{\sigma(t)}(t, -n)u(-n)$, we have

$$u(0) = U_{\sigma(0)}(0, -n)u(-n) = U_{T(n)\sigma(-n)}(0, -n)u(-n)$$
$$= U_{\sigma(-n)}(n, 0)u(-n).$$

Therefore,

$$\text{dist}_E(u(0)A_\Sigma^*) = \text{dist}_E(U_{\sigma(-n)}(n, 0)u(-n), A_\Sigma^*)$$
$$\subseteq \sup_{\sigma \in \Sigma} \text{dist}_E(U_\sigma(n, 0)B_0, A_\Sigma^*) \to 0, \ n \to +\infty,$$

that is $\text{dist}_E(u(0), A_\Sigma^*) = 0$ and $u(0) \in A_\Sigma^*$. Consequently, $\Pi_1 A \subseteq A_\Sigma^*$. The theorem is proved.

Corollary 2.2. Under conditions of Theorem 2.4 for any $\sigma \in \Sigma$, the kernel of the process $U_\sigma(t, \tau)$ is not empty, i.e. there exists, at least, one complete bounded trajectory of the process $U_\sigma(t, \tau)$ for any $\sigma \in \Sigma$.

The dimension of the attractors of nonautonomous systems. As it follows from Theorem 2.5, the attractor A_Σ of the family of processes has the form:

$$A_\Sigma = \bigcup_{\sigma \in \Sigma} K_\sigma(0).$$

It follows from this relation that the dimension of attractor A_Σ depends in the general case on the dimension of the symbol space Σ. Let, for example, $\Sigma = H(\sigma)$, where $H(\sigma)$ is the hull of almost periodic function $\sigma(t)$. The dimension of the space Σ in this case depends on the number of rationally independent frequencies of almost periodic function $\sigma(t)$. If the number of such independent frequencies is infinite, then the dimension of the attractor A_Σ also may be infinite. In [8] for two-dimensional equations of Navier–Stokes with external quasiperiodic forcing having the form $\sigma(x,t) = \sigma(x,\alpha_1,t,\ldots,\alpha_k,t)$, where α_j, $j = \overline{1,k}$ are rationally independent, there was obtained the following estimate of the dimension of the attractor

$$\dim A_\Sigma \le k + C_1/\nu^2 + C_2\,(k/\nu^2)^{1/3},$$

where ν -is a viscosity coefficient. However, the dimension of the section $K(s)$ of the kernel K at any moment $t = s$ is, in this case, finite and has the upper bound equal to the number

$$\dim K(s) \le C/\nu^2, \quad \forall\, s \in R.$$

The closure of the union of all sections $K(s)$ over s may have Hausdorff dimension which is equal to infinity, i.e.

$$\overline{\bigcup_{s \in R} K(s)} = \infty.$$

2.5 Averaging of Nonlinear Dissipative Systems. Closeness between Attractors of Original and Averaged Systems

The method of averaging for ordinary differential equations is based on the two well known Bogolyubov theorems [6,41,54] on closeness between the solutions of the original and averaged systems on the finite (but asymptotically large) interval and on the infinite time interval.

The difficulty of generalizing these theorems to partial differential equations is that the nonlinear terms of the corresponding equations must satisfy the Lipschitz global condition and this is not possible in many cases. However, if we only consider the class of nonlinear dissipative systems, i.e. systems that have an absorbing set, the Lipschitz condition may be satisfied in the corresponding functional spaces provided the initial data are taken from the interior of the absorbing set. Thus, we can prove not only the closeness between particular solutions but also the closeness between the attractors of the original and averaged systems. Usually one has average systems of standard type. The system can be reduced to an standard one in a number of ways.

It is easy enough to reduce the systems with fast oscillating coefficients or fast oscillating external forcing to standard ones. We consider, for example, the system with fast oscillating forcing

$$\partial_s u + A u + F(u) = f(\omega s), \qquad (2.18)$$

where $\omega \gg 1$. We put $\omega s = t$, $\varepsilon = \omega^{-1}$, and write this equation in the standard form:

$$\partial_t u + \varepsilon A u + \varepsilon F(u) = \varepsilon f(t), \quad u(0) = u_0. \qquad (2.19)$$

By averaging this equation over time, we get

$$\partial_t \bar{u} + \varepsilon A \bar{u} + \varepsilon F(\bar{u}) = \varepsilon \bar{f} \qquad \bar{u}(0) = \bar{u}_0. \qquad (2.20)$$

where

$$\bar{f} = \lim_{t \to \infty} \frac{1}{T} \int_0^T f(t)\, dt. \qquad (2.21)$$

Returning to the previous time s in the averaged equation, we find

$$\partial_s \bar{u} + A \bar{u} + F(\bar{u}) = \bar{f}. \qquad (2.22)$$

Comparing the initial and averaged equations, we see that only right-hand sides of these equations are different: where the forcing f is replaced by its mean $\bar{f}$. We prove that the solutions of (2.19) and (2.20) are close on the asymptotically large time interval $0 \leq t \leq L/\varepsilon$, $L > 0$ for sufficiently small $\varepsilon < \varepsilon_0$ (or, what is the same, the solutions of (2.18) and (2.22) are close on an arbitrary segment $0 \leq s \leq L$ for $\omega > \omega_0(L)$, where ω_0 is large enough). We give the conditions imposed on A, F and f, under which we consider (2.19) and (2.20). Let H denote a Hilbert space with the norm $\| \cdot \|$ and $(,)$ denote the scalar product.

1. Let A be a linear unbounded self-adjoint positive definite operator with the dense domain $D(A) \subset H$ and a compact resolvent. The eigenvectors $\{\phi_j\}$, $j = 1, 2 \ldots$ of the operator A form an orthonormal basis of H. Let $0 < \lambda_1 \leq \lambda_2 \leq \ldots$ be eigenvalues of the operator A and

$$A^\alpha u = \sum_{j=1}^{\infty} \lambda_j^\alpha u_j \phi_j, \quad u_j = (u, \phi_j), \ \alpha \in R,$$

$$D(A^\alpha) = \{ u \in H : \sum_{j=1}^{\infty} \lambda_j^{2\alpha} u_j^2 < \infty \}, + D(A^0) \equiv H.$$

We define the scalar product and the norm of the space $D(A^\alpha)$.

$$(u, v)_{D(A^\alpha)} = (u, v)_\alpha = (A^\alpha u, A^\alpha v),$$

$$\|u\|_{D(A^\alpha)} = \|u\|_\alpha = \|A^\alpha u\| = (\sum_{j=1}^{\infty} \lambda_j^{2\alpha} u_j^2)^{1/2}.$$

It should be noted that the embedding $D(A^\alpha) \subset D(A^{\alpha-\gamma})$ is compact for any $\alpha \in R$ and $\gamma > 0$. The estimates $(0 < a \leq \lambda_1)$ hold:

$$\|e^{-tA}u\| \leq e^{-at}\|u\|, \quad t \geq 0, \|A^\alpha e^{-tA}u\|$$
$$\leq c_\alpha t^{-\alpha} e^{-at}\|u\|, \ t > 0, \ \alpha \in [0,1).$$

2. The nonlinear operator $F : D(A) \to H$ is

$$F(u) = F_0 + F_1(u) + F_2(u) + \ldots + F_N(u),$$

where $F_k(u)$ is k-linear continuous operator from $D(A)^k$ to H. Here $D(A)^k$ is a direct product of k samples of $D(A)$ and

$$F_k(u) = F_k(u_1, u_2, \ldots, u_k) \big|_{u_1=u_2=\ldots=u_k=u}.$$

We suppose that for $0 < \beta < 1$ the operators $F_k(u)$ are continuous from $D(A)^k$ to $D(A^\beta)$, and $F_0 \in D(A)$. For simplicity we restrict ourselves to the case in which $n = 2$. The above assumptions lead to

$$\|F_1(u)\|_\beta \leq c_1\|u\|_1, \quad \|F_2(u,v)\|_\beta \leq c_2\|u\|_1\|v\|_1, \quad \|F_0\|_1 \leq c_0.$$

We will also omit the terms F_0 and F_1 because they may be easily taken into consideration.

3. We suppose that the function F is defined on the whole axis R and $f \in D(A^\gamma)$, $0 < \gamma < 1$, and the condition is satisfied:

$$\|\frac{1}{T}\int_t^{t+T} f(s)\,ds - \bar{f}\|_\gamma \to 0, \quad T \to +\infty,$$

uniformly with respect to $t \in R$. We need, in fact, to estimate this integral when $T = \tau/\varepsilon$, $0 \leq \tau \leq L$. In this case we suppose that

$$\|\frac{1}{T}\int_t^{t+T} f(s)\,ds - \bar{f}\|_\gamma \leq \sup_{0\leq\tau\leq L} \|\varepsilon/\tau \int_t^{t+\tau/\varepsilon} f(s)\,ds - \bar{f}\|_\gamma = K(\varepsilon),$$
$$K(\varepsilon) \to 0, \quad \varepsilon \to 0.$$

4. We require that the problem (2.19) (for $\varepsilon = 1$), if $u_0 \in H$, has the unique solution

$$u \in C([0, +\infty) H) \cap L_2(0, T, D(A^{\frac{1}{2}})).$$

(If u_0 is smoother, say, $u_0 \in D(A^{1/2})$, then the solution also becomes smoother depending on the smoothness of f.)

5. The semigroup $S(t)$, $t \geq 0$ generated by averaged equation (2.22), for $\varepsilon = 1$ has an absorbing set in the spaces H and $D(A)$.

These sets may be considered to be balls $B_H(r)$ and $B_{D(A)}(r)$ of radius r centered at zero. We suppose that the semigroup $S(t)$, $t \geq 0$ is bounded uniformly in these spaces, i.e. there exists a ball of radius R such that

$$S(t)B_H(r) \subset B_H(R),\ t > 0,\ S(t)B_{D(A)}(r) \subset B_{D(A)}(R),\ t > 0. \quad (2.23)$$

Obviously, we may consider the ball of radius r to be contained in the ball R with a ρ–neighborhood, i.e. in (2.23) R may be replaced by $R - \rho$, $\rho > 0$. Let $v(t) = u(t) - \bar{u}(t)$ be difference between the solutions of (2.19) and (2.22).

We estimate this difference on the segment $0 \leq \tau \leq L/\varepsilon$, where $L > 0$ is any given number. We suppose that

1) $u(0) = \bar{u}(0) \in B_{D(A)}(r)$;

2) the solution $u(\tau)$ with the initial condition $u(0) \in B_{D(A)}(r)$ remains in the ball $B_{D(A)}(R)$ on the whole segment $0 \leq t \leq L/\varepsilon$. (We later show that this is so indeed.) Thus, let the conditions 1–5 be satisfied. Writing (2.19) and (2.22) in the integral form and subtracting one equation from the other, we obtain

$$v(t) = -\varepsilon \int_0^t e^{-\varepsilon(t-s)A}\left(F_2(v(s),\,u(s)) + F_2(\bar{u}(s),\,v(s))\right)ds$$

$$+\varepsilon \int_0^t e^{-\varepsilon(t-s)A}\left(f(s) - \bar{f}\right)ds.$$

We estimate the terms in the right-hand side of this equality in the norm $D(A)$. The estimate of the first term is

$$\left\|\varepsilon \int_0^t e^{-\varepsilon(t-s)A}\left(F_2(v(s),\,u(s)) + F_2(\bar{u}(s),\,v(s))\right)ds\right\|_1$$

$$= \left\|\varepsilon \int_0^t A^{1-\beta}e^{-\varepsilon(t-s)A}\,A^\beta\left(F_2(v(s),\,u(s)) + F_2(\bar{u}(s),\,v(s))\right)ds\right\|$$

$$\leq c_2\,\varepsilon^\beta\,C_{1-\beta}\int_0^t (t-s)^{\beta-1}\,e^{-\varepsilon a(t-s)}\left(\|u(s)\|_1 + \|\bar{u}(s)\|_1\right)\|v(s)\|_1\,ds$$

$$\leq c_2\,2\,R\,\varepsilon^\beta\,C_{1-\beta}\int_0^t (t-s)^{\beta-1}\,e^{-\varepsilon a(t-s)}\|v(s)\|_1\,ds.$$

Before estimating the second term, we must integrate it by parts. We have

$$\varepsilon \int_0^t e^{-\varepsilon(t-s)A}\left(f(s) - \bar{f}\right)ds = \varepsilon\,e^{-\varepsilon(t-s)A}\int_t^s \left(f(\tau) - \bar{f}\right)d\tau\ \Big|_{s=0}^{s=t}$$

$$+\varepsilon^2 \int_0^t \left(A\,e^{-\varepsilon(t-s)A}\int_s^t \left(f(\tau) - \bar{f}\right)d\tau\right)ds.$$

Hence,

$$\|\varepsilon \int_0^t e^{-\varepsilon\,(t-s)\,A}(f(s) - \bar{f})\,ds\|_1$$

$$\leq \|\varepsilon\,t\,A^{1-\gamma}\,e^{-\varepsilon t A}\,A^\gamma t^{-1} \int_0^t (f(\tau) - \bar{f})\,d\tau\|$$

$$+\|\varepsilon^2 \int_0^t (A^{2-\gamma}\,(t\,-\,s)\,e^{-\varepsilon\,(t-s)\,A}\,A^\gamma\,(t\,-\,s)^{-1} \int_s^t (f(\tau)\,-\,\bar{f})\,d\tau)\,ds\|$$

$$\leq C_{1-\gamma}\,(\varepsilon\,t)^\gamma\,e^{-\varepsilon a t}\,K\,(\varepsilon) + C_{2-\gamma}\,K\,(\varepsilon)\,\varepsilon^\gamma \int_0^t (t\,-\,s)^{\gamma-1}\,e^{-\varepsilon a(t-s)}\,ds$$

$$\leq C\,L^\gamma\,K\,(\varepsilon),\quad C = C_{1-\gamma} + \gamma^{-1}\,C_{2-\gamma}.$$

Thus,

$$\|v(t)\|_1 \leq C\,L^\gamma\,K\,(\varepsilon) + b\,\varepsilon^\beta \int_0^t (t\,-\,s)^{\beta-1}\,e^{-\varepsilon\,a(t-s)}\,\|v(s)\|_1\,ds,$$

$$b = 2\,R\,C_{1-\beta}\,c_2.$$

We now use the known inequality [73]. If

$$u\,(t) \leq a\,+\,b \int_0^t (t\,-\,s)^{\beta-1}\,u\,(s)\,ds,\quad 0 < \beta \leq 1,$$

then

$$u\,(t) \leq a\,G_\beta\,([b\,\Gamma\,(\beta)]^{\frac{1}{\beta}}\,t),$$

where $G_\alpha\,(x)$ is a monotonic function, while $\Gamma\,(\beta)$ is a gamma-function. In our case we have

$$\|v(t)\|_1 \leq C\,L^\gamma\,K\,(\varepsilon)\,G_\beta\,([b\,\varepsilon^\beta\,\Gamma\,(\beta)]^{\frac{1}{\beta}}\,t)$$
$$\leq C\,L^\gamma\,K\,(\varepsilon)\,G_\beta\,(L\,[b\,\Gamma\,(\beta)]^{\frac{1}{\beta}}) = \delta_L\,(\varepsilon) \to 0,\quad \varepsilon \to 0. \qquad (2.24)$$

The estimate derived is true, when the solution $u\,(t)$ remains in the ball $B_{D(A)}$ on the segment $0 \leq t \leq L/\varepsilon$.

Let $t = t^* < L/\varepsilon$ be a moment, when the solution $u\,(t)$ leaves this ball, i.e. let $u\,(t^*) = R$. We choose ε small enough for the inequality to hold

$$\|u\,(t)\,-\,\bar{u}\,(t)\|_1 \leq \rho/2,\quad 0 \leq t \leq t^*.$$

Since we assume that the solution $\bar{u}\,(t)$ of the averaged equation is always in the above ball with its ρ-neighborhood, one get $\|\bar{u}\,(t^*)\|_1 \leq R\,-\,\rho$. We have

$$\|u\,(t^*)\|_1 \leq \|u\,(t^*)\,-\,\bar{u}\,(t^*)\| + \|\bar{u}\,(t^*)\| \leq R\,-\,\rho/2.$$

Thus, the solution $u(t)$ remains in the ball $B_{D(A)}(R)$ for $t \in [0, L/\varepsilon]$, $L > 0$. Now we proved the following

Theorem 2.6. *Let the above conditions 1–5 be satisfied. Then the inequality*

$$\|u(t) - \bar{u}(t)\|_1 \le \delta_L(\varepsilon), \quad \delta_L(\varepsilon) \to 0, \quad \varepsilon \to 0 \qquad (2.25)$$

holds on the segment $0 \le t \le L/\varepsilon$, $L > 0$, where $u(t)$ and $\bar{u}(t)$ are the solutions of (2.19) and (2.20), respectively, starting at the moment $t = 0$ from the same point of the absorbing set $B_{D(A)}(r)$.

Notice that if the initial conditions are given for $t = t_0$ rather than for $t = 0$, then the estimate (2.24) (or (2.25)) does not change, i.e. this estimate is uniform with respect to t_0.

Applications. The main requirement of the above theorem on averaging consists in the conditions imposed on nonlinearity. We have used the assumption that the inequality holds

$$\|F_2(u,v)\|_\beta \le c_2 \|u\|_1 \|v\|_1, \quad 0 < \beta < 1,$$

and there exists an absorbing set in the space $D(A)$. One can consider other conditions imposed on nonlinearity. For example, let

$$\|F_2(u,v)\|_{-\beta} \le c_2 \|u\|_{1/2} \|v\|_{1/2}, \quad 0 < \beta < 1/2,$$

and let the system have an absorbing set in $D(A^{1/2})$. The theorem on averaging, in this case, is proved in a similar manner. This is the case, say, for the barotropic vorticity equation on the rotating sphere S, that is [28]:

$$\partial_t \omega + J(\Delta^{-1}\omega, \omega + l) = \nu \Delta \omega + f(\theta t). \qquad (2.26)$$

where $J(a,b)$ is the Jacobian. Let $A = -\Delta$ be the Laplace–Beltrami operator on the sphere with domain

$$D(A) = \{f \in L_2^0(S) : Af \in L_2^0(S)\},$$

$$L_2^0(S) = \{f \in L_2(S) : \int_S f \, ds = 0\}, \quad \|\cdot\|_{L_2^0(S)} = \|\cdot\|.$$

We determine the degrees of the operator in the usual way:

$$A^\alpha f = \sum_{n=1}^{\infty} \sum_{|m| \le n} \lambda_n^\alpha f_{mn} Y_{mn}.$$

Here Y_{mn} are spherical harmonics. It is known that if $0 < \beta < 1/2$, then

$$\|f\|_{L_{2/1-2\beta}^0} \le c \|f\|_{(D(A^\beta))} = c \|f\|_\beta, \quad \|f\|_{L_{1/\beta}^0} \le c \|f\|_{D(A^{1/2})}.$$

Using these estimates, we obtain

$$| (J (\Delta^1 \omega, \omega), h) | \leq \|\nabla (\Delta^{-1} \omega)\|_{L^0_{1/\beta}} \|\nabla \omega\|_{L^0_2} \|h\|_{L^0_{2/1-2\beta}}$$
$$\leq c \|\nabla (\Delta^{-1}\omega)\|_{D(A^{1/2})} \|\omega\|_{D(A^{1/2})} \|h\|_{D(A^\beta)}$$
$$\leq c \|\omega\|_{D(A^{1/2})} \|\omega\|_{(D(A^{1/2})} \|h\|_{D(A^\beta)}.$$

Hence,

$$\|J (\Delta^{-1}\omega, \omega)\|_\beta \leq c\|\omega\|^2_{1/2}.$$

Similarly, we have

$$\|J (\Delta^{-1} \omega, l)\|_{-\beta} \leq c\|l\|_{1/2} \|\omega\|_{1/2}.$$

Now the closeness between the solution of (2.26) and that of the averaged equation

$$\partial_t \bar\omega + J (\Delta^{-1} \bar\omega, \bar\omega + l) = \nu \Delta \bar\omega + \bar f$$

with the initial data $\omega(0) = \bar\omega(0)$ that belong to the absorbing set in $D(A^{1/2})$ may be proved in the same way as in Theorem 2.6 as $\theta \to +\infty$ ($\varepsilon = 1/\theta \to 0$), i.e.

$$\|\omega(t) - \bar\omega(t)\|_{1/2} \leq \delta(\varepsilon), \quad \delta(\varepsilon) \to 0, \quad \varepsilon \to 0 \, (\theta \to +\infty).$$

As the second example we consider the Navier–Stokes 2-D equations in the bounded domain $\Omega \subset R^2$ with the smooth boundary $\partial\Omega \in C^2$ and the fast oscillating external forcing $\phi(\omega t)$, $\omega \gg 1$, that is

$$\partial_t u + (u\nabla) u = \nu \Delta u - \nabla p + \phi(\omega t),$$
$$\text{div } u = 0, \quad u \,|_{\partial\Omega} = 0, \quad u \,|_{t=0} = u_0. \tag{2.27}$$

One usually considers (2.27) in the spaces H and V, that are the closures of the set of functions $\{u \in C_0^\infty(\Omega)^2, \text{ div } u = 0\}$ in the norms $L_2(\Omega)^2$. and $H_0^1(\Omega)^2$. Putting

$$A = -\Pi\Delta, B(u,v) = \Pi(u\nabla)v = \Pi \left(\sum_{i=1}^2 u_i\partial_i v \right), \quad \Pi\phi = f,$$

where Π is the orthogonal projection in $L^2(\Omega)^2$ on H, we write (2.27) as follows

$$\partial_t u + \nu A u + B(u,u) = f(\omega t).$$

Setting $\tau = \omega t$, $\varepsilon = \omega^{-1}$, we find $\partial_\tau u + \varepsilon \nu A u + \varepsilon B(u,u) = \varepsilon f(\tau)$. Averaging this equation, we get

$$\partial_\tau \bar u + \varepsilon \nu A \bar u + \varepsilon B(\bar u, \bar u) = \varepsilon \bar f. \tag{2.28}$$

Let us return to the previous argument in (2.28):

$$\partial_t \bar{u} + \varepsilon \nu A \bar{u} + \varepsilon B (\bar{u}, \bar{u}) = \varepsilon \bar{f}. \tag{2.29}$$

Here A is a positive self-adjoint operator on H with the dense domain $D(A) = V \cap H^2(\Omega)^2$. The operator A^{-1} is compact. The spaces $D(A^\alpha)$, $\alpha \in (0, 1]$ are defined in the usual way [125].

Equation (2.29) (to be more exact, the semigroup generated by the problem (2.27)) has absorbing set in the spaces $D(A^{\frac{1}{2}})$, $D(A)$ and H. Therefore, we must choose the spaces in which the nonlinear operator $B(u, u)$ is bounded, as the operator from these spaces into spaces $D(A^{\frac{1}{2}}) = V$ and $D(A)$. It may be shown that as such spaces one can take the spaces $D(A^{-\beta})$ and $D(A^{\frac{1}{2}})$, $0 < \beta < 1/2$. Therefore, for $\tau \in [0, L/\varepsilon]$, $L > 0$ the following propositions take place:

1) if $u(0) = \bar{u}(0) \in B_{D(A)}(r)$, then

$$\|u(\tau) - \bar{u}(\tau)\|_1 \le \delta_{1L}(\varepsilon), \quad \delta_{1L}(\varepsilon) \to 0, \quad \varepsilon \to 0, \gamma \in (0, 1);$$

2) if $u(0) = \bar{u}(0) \in B_{D(A^{1/2})}(r)$, then

$$\|u(t) - \bar{u}(t)\|_{1/2} \le \delta_{1/2L}(\varepsilon), \quad \delta_{1/2L}(\varepsilon) \to 0, \quad \varepsilon \to 0, \gamma \in (0, 1).$$

We now establish the closeness between the attractor $A_\Sigma(\varepsilon)$ of the original system (2.19) and the attractor $\bar{A}$ of the averaged system (2.20). The closeness, that is the upper semicontinuity, will be established in the norm, in which the closeness of the solution of the original system and the solution of the averaged one was proved. Thus, we must prove that

$$\mathrm{dist}_{D(A)}(A_\Sigma(\varepsilon), \bar{A}) \to 0, \quad \varepsilon \to 0.$$

This, in turn, implies that however small the ρ-neighborhood of the attractor $\bar{A}$ is, there can be found ε_0 such that $A_\Sigma(\varepsilon) \subset O_\rho(\bar{A})$ for $\varepsilon < \varepsilon_0$.

If the ρ-neighborhood of the attractor $\bar{A}$ is an uniformly absorbing set of the family of processes of the initial equation, then the closeness of the attractors $A_\Sigma(\varepsilon)$ and $\bar{A}$ in the above sense exists, since the attractor $A_\Sigma(\varepsilon)$ represents by the definition an uniformly attracting set of the above family of processes.

We make the following assumptions (in addition to those made in the proof of Theorem 2.6).

6. Let f be an almost periodic function taking values in $D(A^\gamma)$, $0 < \gamma < 1$. We denote by $H(f)$ its hull. For any function $g \in H(f)$, the solution of (2.19) with initial condition $u(\tau) = u_\tau$ will have the form

$$u(t) = U_g(t, \tau) u_\tau, \quad t \ge \tau, \quad \tau \in R,$$

where $\{U_g(t,\tau)\}$, $g \in H(f) = \Sigma$ is the corresponding family of processes [8]. We suppose that this family of processes possesses an uniform attractor $A_\Sigma(\varepsilon)$, which is uniformly bounded as $\varepsilon \to 0$.

7. The semigroup $S(t)$ generated by the averaged equation possesses a global attractor $\bar{A}$. Let us prove the closeness between the attractors $A_\Sigma(\varepsilon)$ and $\bar{A}$. We use the fact that any neighborhood of $\bar{A}$ is an absorbing set, i.e. if a solution $\bar{u}(t)$ enters this neighborhood, then it never leaves this neighborhood.

Let $\rho > 0$ be an arbitrary small number. We have already proved the theorem on averaging is true for any segment $[\tau, \tau + L/\varepsilon]$, $L > 0$, if $u(\tau) = \bar{u}(\tau) \in B_{D(A)}(r) \equiv B_a$. Since L is arbitrary number, we choose it according to condition

$$S(t) B_a \subset O_{\rho/3}(\bar{A}), \quad \forall t \geq L(\rho/3).$$

This condition implies that all the points of the absorbing set B_a for $t \geq L(\rho/3)$ enter the $\rho/3$-neighborhood of $\bar{A}$ and remain there. Let $u_0 \in B_a$. Let the solutions $\bar{u}(t) = S(t) u_0$ and $u(t) = U_g(t,0) u_0$ start from this point. We choose $\varepsilon_0 = \varepsilon_0(L, \rho/3)$ so that the difference between these solutions on the segment $[0, \frac{L}{\varepsilon_0}]$ is less than $\rho/3$. Let $t_1 = L/\varepsilon_0 > L$. Then we have $\bar{u}(t_1) \in O_{\rho/3}(\bar{A})$ and $u(t_1) \in O_{\frac{2}{3}\rho}(\bar{A})$ at $t = t_1$. We now take the point $u(t_1) = u_1$ as the initial point and start from it the solutions $\bar{u}(t)$ and $u(t)$. We consider these solutions on the segment $t_1 \leq t + t_1 \leq t_1 + L/\varepsilon_0 = 2t_1$, $0 \leq t \leq t_1$

$$\bar{u}(t) = S(t) u_1, \quad u(t) = U_g(t + t_1, t_1) u_1$$
$$= U_g(t + t_1, t_1) U_g(t_1, 0) u_0 = U_g(t + t_1, 0) u_0, \ 0 \leq t \leq t_1.$$

(Notice that the point u_1 does not lie on the trajectory $S(t) u_0$, but it lies on the trajectory of the solution $u(t) = U_g(t,0) u_0$.) On this segment the solutions will diverge by the value less than $\rho/3$. Since $S(t_1) u_1 \in O_{\rho/3}(\bar{A})$, we get $u(2t_1) \in O_{2\rho/3}(\bar{A})$.

Therefore, on the segment $t_1 \leq t + t_1 \leq 2t_1$ the solution $u(t)$ does not leave the ρ-neighborhood of the attractor $\bar{A}$ and at $t = 2t_1$ we have $S(2t_1) u_1 \in O_{\rho/3}(\bar{A})$, $u(2t_1) \in O_{(2/3)\rho}(\bar{A})$.

If we take the point $u_2 = u(2t_2)$ as the initial one, then we see that the situation is similar to that occurred at the previous step.

Repeating this process, we arrive at a conclusion that the solution $u(t) = U_g(t,0) u_0$ for all $t \geq t_1$ is in the ρ-neighborhood of the attractor $\bar{A}$. For any bounded set $B \subset B_a$, there can be chosen L (uniformly with respect to all the point of the set B) so that $S(t) B \subset O_{\rho/3}(\bar{A})$, $\forall t \geq L$.

In other words, the choice of L and ε_0 does not depend on the initial point u_0. It is obvious that our reasonings are also independent of the choice of $g \in H(f)$. Therefore, $U_g(t, 0) B \subset O_\rho(\bar{A})$ for $t \geq t_1$.

Finally, the solution $U_g(t, 0) u_0$, $t > 0$ may be replaced by $U_g(t, \tau) u_\tau$, $t \geq \tau$, $t \in R$. Thus, we can state the following

Theorem 2.7. *Let the conditions* 1–7 *be satisfied. Then*

$$\text{dist}_{D(A)}(A_\Sigma(\varepsilon), \bar{A}) \to 0, \quad \varepsilon \to 0,$$

where $A_\Sigma(\varepsilon)$ *is the attractor of the original system* (2.19), *while* $\bar{A}$ *is the attractor of the averaged system* (2.20).

The theorem implies, in particular, the closeness between the attractor of the initial equation and that of the averaged one both for the barotropic vorticity equation on the sphere (2.26) and for the two-dimensional Navier–Stokes equations (2.27).

More exactly, let f in (2.26) be an almost periodic function taking values in $D(A^\gamma)$, $0 < \gamma < 1$ and let $A_f(\theta)$ be an uniform attractor of the family of processes associated with the equation (2.26). Then

$$\text{dist}_{D(A^{1/2})}(A_f(\theta), \bar{A}) \to 0, \quad \theta \to \infty,$$

where $\bar{A}$ is an attractor of the corresponding averaged equation.

The analogous proposition takes place for the Navier–Stokes equations (2.27). Let ϕ in (2.27) be an almost periodic function taking values in $D(A^\gamma)$, $0 < \gamma < 1$ and let $A_\phi(\omega)$ be an uniform attractor of the corresponding family of processes, while $\bar{A}$ is the attractor of the averaged system. Then

$$\text{dist}_{D(A^{1/2})}(A_\phi(\omega), \bar{A}) \to 0, \quad \omega \to \infty.$$

2.6   On Closeness of Solutions of Original and Averaged Nonlinear Dissipative Systems on Infinite Time Interval

In this section we consider some class of nonlinear dissipative systems with high oscillation forcing. Such systems can be found in the ocean dynamics, where as the high oscillating forcing there can be taken a wind forcing.

These systems can be written in standard form and then they can be averaged over explicitly containing time. The task is to prove the closeness of the solutions of original and averaged systems on infinite time interval. We restrict ourselves with the case, when the system of 2D Navier–Stokes equations is taken as the nonlinear dissipative system.

Problem. Let Ω be a bounded domain in R^2 with smooth boundary $\partial\Omega \subset C^2$. We consider in Ω the following problem

$$\frac{\partial u}{\partial t} + (u \cdot \nabla)u = \nu \Delta u - \nabla p + f(x, \omega t), \tag{2.30}$$

$$\operatorname{div} u = 0, \tag{2.31}$$

$$u\big|_{\partial\Omega} = 0, \tag{2.32}$$

$$u\big|_{t=0} = u_0, \tag{2.33}$$

where p is the pressure, $u = (u_1, u_2)$ the velocity vector, $u_j = u_j(x,t)$, $j = 1,2$, $x = (x_1, x_2) \in R^2$. We shall assume that $\omega \gg 1$, so that $\varepsilon = 1/\omega$ is the small parameter.

As usually, we denote by H and V the spaces are the closures of the function set $\{u : u \in C_0^\infty(\Omega)^2, \operatorname{div} u = 0\}$ over the norm of spaces $L_2(\Omega)^2$ and $H_0^1(\Omega)^2$. Let P be the projector $L_2(\Omega)^2$ onto H. Introduce the notations

$$A = -P\Delta, \quad B(u,v) = P(u \cdot \nabla)v = P\left(\sum_{j=1}^2 u_j \partial_j v\right), Pf = F,$$

$$b(u,v,w) = \; < B(u,v), w >_{V \times V'} = \int_\Omega u_j \partial_j v_k w_k dx.$$

We write the problem (2.30)-(2.33) as follows

$$\partial_t u + B(u,v) = -\nu Au + F(\omega t), \quad u\big|_{t=0} = u_0. \tag{2.34}$$

Setting in (2.34) $\omega t = s$, $\varepsilon = 1/\omega$, we find

$$\partial_s u + \varepsilon B(u,u) = -\nu\varepsilon Au + \varepsilon F(s). \tag{2.35}$$

Let there exist a mean

$$\lim_{T \to \infty} \frac{1}{T} \int_0^T F(s)ds = F_0 \tag{2.36}$$

More precisely, we assume that

$$\left\|\left(\frac{1}{T}\int_t^{t+T} F(s)ds - F_0\right)\right\|_{D(A^\gamma)} \le \rho_\gamma(T,t), \tag{2.37}$$

where $\rho_\gamma(T,t) \to \infty$ for $T \to \infty$ is uniform with respect to $t \in R,$. The value of γ will be given below.

We shall consider alongside with (2.35) the averaged equation

$$\partial_s w + \varepsilon B(w,w) = -\nu\varepsilon Aw + \varepsilon F_0 \tag{2.38}$$

or (since $s = t/\varepsilon$)

$$\partial_t w + B(w,w) = -\nu Aw + F, \quad w\big|_{t=0} = w_0. \tag{2.39}$$

The semigroup $S(t), t \geq 0$ generated by the problem (2.39) possesses the absorbing sets in the spaces H, V and $D(A)$. As such absorbing sets there can be taken the balls $B(r)$ in these spaces with zero center and of radius r. The semigroup $S(t)$ is bounded in the above spaces, i.e. for the ball $B(r)$ there exists a ball $B(R)$ such that $S_t B(r) \subset B(R),\ t > 0$.

Let $F(t)$ be almost-periodic function defined for all $t \in R$ taking their values in $D(A^\gamma)$. We write these equations as follows

$$\frac{\partial u}{\partial t} + \varepsilon B(u, u) = -\nu \varepsilon A u + \varepsilon F(t). \tag{2.40}$$

$$\frac{\partial w}{\partial t} + \varepsilon B(w, w) = -\nu \varepsilon A w + F_0. \tag{2.41}$$

The theorems proved in the previous section can be reformulated.

Theorem 2.8 *Let :*

1) $u(t_0) = w(t_0) \in B_{D(A^{1/2})}(r), t_0 \geq 0$ *i.e. initial data of the original and averaged systems belong to the absorbing set* $B_{D(A^{1/2})}(r)$ *of the space* V; 2) $\gamma > -1/2$. *Then on the interval* $t_0 \leq t \leq t_0 + L/\varepsilon$, *where* $L > 0$ *is arbitrarily given number, the inequality*

$$\|u(t) - w(t)\|_{D(A^{1/2})} \leq \tau_L(\varepsilon)$$

holds, where $\tau_L(\varepsilon) \to 0$ *for* $\varepsilon \to 0$.

Theorem 2.9. *Let:*

1) $u(t_0) = w(t_0) \in B_{D(A)}, t_0 \geq 0$, *i.e. the initial data of the original and averaged systems belong to the absorbing set* $B_{D(A)}(r)$ *of the space* $D(A)$; 2) $\gamma > 0$. *Then on the interval* $t_0 \leq t \leq t_0 + L/\varepsilon$, *where* L *is the arbitrarily given number, the inequality*

$$\|u(t) - w(t)\|_{D(A)} \leq \delta_L(\varepsilon)$$

holds, where $\delta_L(\varepsilon) \to 0$ *for* $\varepsilon \to 0$.

We shall use the above theorems to obtain the closeness of the solutions of the original and averaged systems on infinite time interval. For this purpose we require that the solution of the averaged system must be uniformly asymptotically stable. We shall prove the theorem.

Theorem 2.10. *Let:*

1) *the conditions of Theorem 2.8 be satisfied ;*

2) *the solution* $w = w(t)$ *of the averaged equation* (2.41) *is defined for all* $t \geq 0$ *and lies in the ball* $B_V(r)$ *with some ρ-neighborhood;*

3) *the solution* $w = w(t)$ *is uniformly asymptotically stable.*

Then for any $\eta > 0$ *one can find* ε_0 *such that for* $\varepsilon < \varepsilon_0$ *the inequality holds*

$$\|u(t) - w(t)\|_{D(A^{1/2})} < \eta \text{ for all } t \geq 0.$$

Proof. Let $w(t)$ be uniformly asymptotically stable solution of (2.41). We note that *uniform* stability of the solution $w_0(t)$ means the following. Let $\eta/2 > 0$ and $\hat{t} \geq 0$ are every given numbers. Then there can be found the number $\rho = \rho(\eta/2) < \eta$ such that

$$\|w(t) - w_0(t)\|_{D(A^{1/2})} < \eta/2, \quad \forall t > \hat{t}, \tag{2.42}$$

if only

$$\|w(\hat{t}) - w_0(\hat{t})\|_{D(A^{1/2})} \leq \rho, \tag{2.43}$$

where $w(t)$ is any solution of (2.41). It should be emphasized that ρ depends only on $\eta/2$, but it does not depend on $\hat{t}$. Further, by virtue of the uniform *asymptotic* stability of the solution $w_0(t)$, it follows that one can choose $\hat{L}$ such that for $t > \hat{t} + \hat{L}/\varepsilon$ the inequality fulfills

$$\|w(t) - w_0(t)\|_{D(A^{1/2})} < \rho/2, \quad \forall t > \hat{t} + \hat{L}/\varepsilon, \tag{2.44}$$

where ρ is defined from the conditions (2.42) and (2.43). Note that here $\hat{L} = \hat{L}(\rho/2)$, but $\hat{L}$ does not depend on $\hat{t}$. Now for fixed ρ and $\hat{L}$ we choose $\varepsilon_0 = \varepsilon_0(\rho, \hat{L})$ so that for $\varepsilon < \varepsilon_0$ on the interval $[0, \hat{L}/\varepsilon]$ the inequality holds

$$\|u(t) - w_0(t)\|_{D(A^{1/2})} < \rho, \quad t \in [0, \hat{L}/\varepsilon], \varepsilon < \varepsilon, \tag{2.45}$$

where $u(t)$ is the solution of original system (2.40) and $u(0) = w_0(0)$. The inequality (2.45) takes place according to Theorem 2.8.

We assume now that the inequality

$$\|u(t) - w_0(t)\|_{D(A^{1/2})} < \eta \tag{2.46}$$

does not hold for all $t \geq 0$. Then there can be found a moment $t^* > \hat{L}/\varepsilon$, such that the equality holds

$$\|u(t^*) - w_0(t^*)\|_{D(A^{1/2})} = \eta. \tag{2.47}$$

If $t < t^*$, then (2.46) holds. Let t^* be the first moment, when the difference

$$\|u(t) - w_0(t)\|_{D(A^{1/2})}$$

approaches to the value η. Then there can be found t_q such that $\|u(t_q) - w_0(t_q)\|_{D(A^{1/2})} = \rho$. There can exist several points as t_q. We take the largest of them $t^{**} = \max_q t_q$. Then

$$\|u(t^{**}) - w_0(t^{**})\|_{D(A^{1/2})} = \rho \tag{2.48}$$

and for $t > t^{**}$ one has

$$\|u(t) - w_0(t)\|_{D(A^{1/2})} > \rho, \quad t > t^{**}. \tag{2.49}$$

Let $\hat{w}(t)$ be any solution of the averaged system (2.41) such that $\hat{w}(t^{**}) = u(t^{**})$. Then, evidently,

$$\rho = \|u(t^{**}) - w_0(t^{**})\|_{D(A^{1/2})} = \|\hat{w}(t^{**}) - w_0(t^{**})\|_{D(A^{1/2})}. \quad (2.50)$$

If there will be taken t^{**} as $\hat{t}$ then one has (see (2.42) and (2.43))

$$\|w_0(t) - \hat{w}(t)\|_{D(A^{1/2})} < \eta/2, \quad t > t^{**}$$
$$\|w_0(t) - \hat{w}(t)\|_{D(A^{1/2})} < \rho/2, \quad t \geq t^{**} + \hat{L}/\varepsilon.$$

Using the averaging theorem and the condition $u(t^{**}) = \hat{w}(t^{**})$ for $\varepsilon_1 < \varepsilon_0$ on the interval $[t^{**}, t^{**} + \hat{L}/\varepsilon]$ we have

$$\|u(t) - \hat{w}(t)\|_{D(A^{1/2})} < \rho/2.$$

On the other hand, if $t \in [t^{**}, t^{**} + \hat{L}/\varepsilon_1]$, then

$$\|u(t) - w_0(t)\|_{D(A^{1/2})} \leq \|u(t) - \hat{w}(t)\|_{D(A^{1/2})} + \|\hat{w}(t) - w_0(t)\|_{D(A^{1/2})}$$
$$< \rho/2 + \eta/2 < \eta \quad \text{i.e. } t^{**} + \hat{L}/\varepsilon_1 < t^*.$$

However, for $t = t^{**} + \hat{L}/\varepsilon > t^{**}$ one has

$$\|u(t^{**} + \hat{L}/\varepsilon_1) - w_0(t^{**} + \hat{L}/\varepsilon_1)\|_{D(A^{1/2})}$$
$$\leq \|u(t^{**} + \hat{L}/\varepsilon_1) - \hat{w}(t^{**} + \hat{L}/\varepsilon_1)\|_{D(A^{1/2})}$$
$$+ \|\hat{w}(t^{**} + \hat{L}/\varepsilon_1) - w_0(t^{**} + \hat{L}/\varepsilon_1)\|_{D(A^{1/2})} < \rho/2 + \rho/2 = \rho.$$

The inequality obtained contradicts the inequality (2.49). The theorem is proved.

Remark. Similar theorem on the closeness of solutions on the infinite time interval can be proved if we replace Theorem 2.8 by Theorem 2.9. In this case the closeness of the solutions will take place in the norm $D(A)$.

The second theorem on averaging on infinite time interval. There can be proved another theorem on closeness of solutions of original and averaged systems on infinite time interval. This theorem is similar to the classic Bogolyubov theorem for the finite-dimensional case. Thus, let us consider the original equation

$$\frac{\partial u}{\partial t} + \varepsilon B(u, u) = -\varepsilon \nu Au + \varepsilon \nu Au + \varepsilon f(t) \quad (2.51)$$

and corresponding averaged equation

$$\frac{\partial w}{\partial t} + \varepsilon(w, w) = -\varepsilon \nu Aw + \varepsilon f_0. \quad (2.52)$$

It is known that there exists at least one stationary solution $w_0 \in D(A^{1/2})$ to (3.2):

$$B(w_0, w_0) + \nu A w_0 = f_0. \tag{2.53}$$

We make in (2.51) the replacement

$$u = w_0 + z. \tag{2.54}$$

Then (2.51) has the form

$$\frac{\partial z}{\partial t} = -\varepsilon \left(\nu A z + B(z, w_0) + B(w_0, z) + B(z, z) + f_0 - f \right). \tag{2.54}$$

We pass to a new variable h in this equation, setting

$$z = h - \varepsilon v(t, \varepsilon), \tag{2.55}$$

where the function $v(t, \varepsilon)$ satisfies the equation

$$\frac{\partial v}{\partial t} = -\varepsilon \nu A v - (f - f_0). \tag{2.56}$$

After simple transformations, we get the following equation for the definition of function h:

$$\frac{\partial h}{\partial t} = -\varepsilon \left(\nu A h + B(h, w_0) + B(w_0, h) \right) + \varepsilon F(h, \varepsilon, t), \tag{2.57}$$

where

$$F(h, \varepsilon, t) = F_1 + F_2 - B(h, h)$$
$$F_1 = B(\varepsilon v, w_0) + B(w_0, \varepsilon v) - B(\varepsilon v, \, \varepsilon v),$$
$$F_2 = B(h, \varepsilon v) + B(\varepsilon v, h). \tag{2.58}$$

It follows from (2.57) that F contains: 1) the known term F_1 depending on the stationary solution and the known function v and having the order ε; 2) the term F_2 linear with respect to h of the order ε; 3) the term $B(h, h)$ which is non-linear with respect to h. In other words, εF contains: 1) the linear terms of the order $O(\varepsilon^2)$, 2) the nonlinear term of the order $O(\varepsilon)$, 3) the known term of the order $O(\varepsilon^2)$. Before we pass to the study of the equation (2.57), we which to give some properties of the function v satisfying (2.56).

Lemma 2.1. *Let*

$$\left\| \frac{1}{T} \int_t^{t+T} (f(s) - f_0) ds \right\|_{D(A^\gamma)} \leq \rho_\gamma(T, t),$$

where $\rho_\gamma(T,t) \to 0$, for $T \to +\infty$, is uniform relatively to $t \in R$. Then for $\alpha < 1 + \gamma$ (2.56) has in $D(A^\alpha)$ the unique uniform over $t \in R$ bounded solution $v(t,\varepsilon)$, which satisfies the condition

$$\|\varepsilon v(t,\varepsilon)\|_{D(A^\alpha)} \to 0, \quad \varepsilon \to 0.$$

If $f(t)$ is the almost-periodic taking values in $D(A^\alpha)$, then v is also almost-periodic function in $D(A^\alpha)$ with frequency basis contained in the frequency basis f.

The proof of this proposition is similar to the finite-dimensional case. The bounded solution of (2.56) has the form

$$v(t,\varepsilon) = \int_t^{-\infty} e^{\varepsilon \nu A s}(f(s)) - f_0)ds.$$

The estimate of the norm $\|\varepsilon v(t,\varepsilon)\|_{D(A^{1/2})}$ is obtained by the simple calculation. If $\|h\|_{D(A^{1/2})} \le \sigma$ then the mapping $F(h,\varepsilon,t)$ (as the mapping from $D(A^{1/2})$ into $D(A^{-\beta})$, $\beta > 0$) satisfies the following equations:

$$\|F(h_1,\varepsilon,t) - F(h_2,\varepsilon,t)\|_{D(A^{-\beta})} \le \lambda(\varepsilon,\sigma)\|h_1 - h_2\|_{D(A^{1/2})}$$
$$\|F(0,\varepsilon,t)\|_{D(A^{-\beta})} \le K(\varepsilon),$$

where $\Lambda(\varepsilon,\sigma)$ and $K(\varepsilon)$ tends to zero for $\varepsilon \to 0$ and $\sigma \to 0$.

We return now to the equation (2.57). Setting in this equation $t = \tau/\varepsilon$, we write it as follows [6]

$$\frac{\partial h}{\partial \tau} + Lh = Q(h,\varepsilon,\tau), \tag{2.59}$$

where

$$Q(h,\varepsilon,\tau) = F(h,\varepsilon,\tau/\varepsilon). \tag{2.60}$$

We note that according to the above said the function Q satisfies the following conditions:

$$\|Q(h_1,\varepsilon,\tau) - Q(h_2,\varepsilon,\tau)\|_{D(A^{-\beta})} \le \lambda(\varepsilon,\sigma)\|h_1 - h_2\|_{D(A^{1/2})},$$
$$\|Q(0,\varepsilon,\tau/\varepsilon)\|_{D(A^{-\beta})} \le K(\varepsilon), \quad \|h_k\|_{D(A^{1/2})} \le \sigma, k = 1,2,$$
$$\lambda(\varepsilon,\sigma) \to 0, \quad K(\varepsilon) \to 0, \varepsilon \to 0. \tag{2.61}$$

It can be shown that if the function Q satisfies the conditions (2.61) and $\gamma > -1/2$ then for $\varepsilon < \varepsilon_0$ the equation (2.59) has unique bounded solution $h^*(t)$ such that

$$\|h^*\|_{c_b(R,D(A^{1/2}))} \le \delta(\varepsilon), \quad \delta(\varepsilon) \to 0\, \varepsilon \to 0.$$

Chapter 3

Analysis of Barotropic Model

3.1 Existence of Global Attractor

In this chapter we shall consider the following equations of the barotropic atmosphere on rotating sphere $S^2 \subset R^3$ with radius a:

$$\frac{\partial}{\partial t}\triangle\psi + J(\psi,\triangle\psi + l) = -\sigma\triangle\psi + \nu\triangle^2\psi + f, \quad \psi|_{t=0} = \psi_0, \quad (3.1)$$

and

$$\frac{\partial}{\partial t}\triangle\psi + J(\psi,\triangle\psi + l) = -\sigma\triangle\psi + \nu(-\triangle)^{s+1}\psi + f, \quad \psi|_{t=0} = \psi_0, \quad (3.2)$$

where ψ is the stream-function, $J(a,b)$ is the Jacobian, $l = 2\Omega\sin\phi$, $s \geq 1$, while the parameters $\Omega > 0$ and $\sigma \geq 0$ have the dimensions $[\Omega] = T^{-1}$, $[\sigma] = T^{-1}$, respectively.

The parameter $\nu > 0$ in (3.1) has a dimension $[\nu] = L^2T^{-1}$, while in (3.2) the dimension $\nu = \nu(s)$ must be associated with the parameter s, i.e. $[\nu(s)] = L^{2s}T^{-1}$.

In what follows we shall use other forms of the equations (3.1) and (3.2). Setting $\triangle\psi = \omega$, $\psi = \triangle^{-1}\omega$, we write (3.1) and (3.2) as follows:

$$\frac{\partial}{\partial t}\omega + J(\triangle^{-1}\omega,\omega + l) = -\sigma\omega + \nu\triangle\omega + f, \quad \omega|_{t=0} = \omega_0, \quad (3.3)$$

$$\frac{\partial\omega}{\partial t} + J(\triangle^{-1}\omega,\omega + l) = -\sigma\omega - \nu(-\triangle)^s\omega + f, \quad \omega|_{t=0} = \omega_0. \quad (3.4)$$

Let us introduce the necessary for the later use spaces of functions on S^2.

We denote by $L_0^2(S^2)$ the subspace of functions of $L^2(S^2)$, which are orthogonal to the constant (S^2 is two-dimensional sphere embedded in R^3):

$$L_0^2(S^2) = \left\{ f : f \in L^2(S^2), \quad \int_{S^2} f ds = 0 \right\},$$

$$(f,g) = \int_{S^2} f \bar{g} ds, \quad ||f||_{L_0^2(S)} \equiv ||f||.$$

Spherical harmonics

$$Y_{mn}(\lambda, \varphi) = P_{mn}(\sin \varphi) e^{im\lambda}, \quad |m| \leq n,$$

$$(-\Delta) Y_{mn} = \lambda_n Y_{mn}, \quad (Y_{mn}, Y_{m'm'}) = \delta_{mm'} \delta_{nn'},$$

where P_{mn} are the associated Legendre polynomials, $0 \leq \lambda \leq 2\pi$, $|\varphi| \leq \pi/2$ form the basis in $L_0^2(S^2)$ and, for $\forall f \in L_0^2(S^2)$, one has the expansion

$$f = \sum_{n=1}^{\infty} \sum_{|m| \leq n} f_{mn} Y_{mn} = \sum_{n=1}^{\infty} F_n, \quad F_n = \sum_{m=-n}^{n} f_{mn} Y_{mn}.$$

For $\forall \alpha \in R$ one has the equality

$$(-\Delta)^\alpha f = \sum_{n=1}^{\infty} \sum_{|m| \leq n} \lambda_n^\alpha f_{mn} Y_{mn} = \sum_{n=1}^{\infty} \lambda_n^\alpha F_n.$$

We associate the operator $(-\Delta)$ to the scale of Hilbert spaces $H_0^\alpha(S^2)$, $\alpha \in R$, setting

$$(f,g)_\alpha = (f,g)_{H_0^\alpha(S^2)} = ((-\Delta)^{\alpha/2} f, g(-\Delta)^{\alpha/2} \psi),$$

$$||f||_\alpha = ||f||_{H_0^\alpha(S^2)} = ||(-\Delta)^{\alpha/2} f|| = (\sum_{n=1}^{\infty} \sum_{|m| \leq n} \lambda_n^\alpha f_{mn}^2)^{1/2}.$$

Here $H_0^\alpha(S^2)'$ denotes the dual space of $H_0^\alpha(S^2)$. Notice that

$$(H_0^\alpha(S^2))' = H_0^{-\alpha}(S^2), \quad \alpha \geq 0.$$

In the sequel, the following inequalities will be needed:

1) $||f||_\alpha \geq \lambda_1^{\frac{\alpha-\beta}{2}} ||f||_\beta, \quad \alpha > \beta, \quad f \in H_0^\alpha(S^2)$;

2) $\lambda_1 ||f||^2 \leq ||\nabla f||^2 = ||(-\Delta)^{1/2} f||^2, \quad f \in H_0^1(S^2)$;

3) $||f||_{L^4(S^2)} \leq 2^{1/4} ||f||_{L^2(S^2)}^{1/2} ||\nabla f||_{L^2(S^2)}^{1/2}$.

We introduce now the spaces V_α, $\alpha \geq 0$, setting

$$V_\alpha = H_0^\alpha(S^2)^2, \quad V_\alpha' = (H_0^\alpha(S^2)^2)' = H_2^{-\alpha}(S^2)^2 = V_{-\alpha}$$
$$u = (u_1, u_2) \in V_\alpha.$$

Evidently,

$$\|u\|_{V_\alpha} \geq \lambda_1^{\frac{\alpha - \beta}{2}} \|u\|_{V_\beta}, \quad \alpha > \beta.$$

If $V_0 = V_0'$, then we obtain the embeddings

$$\ldots \subset V_2 \subset V_1 \subset V_0 \subset V_1' \subset V_2' \subset \ldots,$$

Here the embeddings are continuous and compact. We introduce bilinear forms $a_1(u, v)$, $a_2(u, v)$, $b_1(u, v)$ and a trilinear form $b_2(u, v, w)$, setting

$$a_1(u, v) = \int_{S^2} \nabla u \cdot \nabla v ds, \quad a_2(u, v) = \int_{S^2} \Delta u \cdot \Delta v ds,$$

$$b_1(u, v) = \int_{S^2} J(l, u) v ds$$

$$b_2(u, v, \omega) = \int_{S^2} J(\Delta u, v) \Delta \omega ds = \int_{S^2} J(\omega, v) \Delta u ds.$$

The following propositions are evident.

1. Bilinear form $a_1 : H_0^1 \times H_0^1 \to R$ is continuous and coercive, that is

$$|a_1(u, v)| \leq \|u\|_1 \|v\|_1, \quad \forall u, v \in H_0^1(S^2),$$
$$a_2(u, u) = \|u\|_1^2, \quad \forall u \in H_0^1(S^2). \tag{3.5}$$

2. Bilinear form $a_2 : H_0^2 \times H_0^2 \to R$ is continuous and coercive, that is

$$|a_2(u, v)| \leq \|u\|_2 \|v\|_2, \quad \forall u, v \in H_0^2(S^2),$$
$$a_2(u, u) = \|u\|_2^2, \quad \forall u \in H_0^2(S^2). \tag{3.6}$$

3. Bilinear form $b_1 : H_0^1 \times L_0^2 \to R$ is continuous and skew-symmetric, that is

$$b_1(u, v) \leq \begin{cases} \max_{S^2} |\nabla l| \|u\|_1 \|v\|, & \forall u \in H_0^1(S^2), \\ & \forall v \in L_0^2(S^2), \\ \max_{S^2} |l| \|u\|_1 \|v\|_1, & \forall u, v \in H_0^1(S^2); \end{cases}$$

$$b_1(u,v) = -b_1(v,u), \quad b_1(u,u) = 0. \tag{3.7}$$

4. Trilinear form $b_2 : H_0^2 \times H_0^2 \times H_0^2 \to R$ is continuous and allows the next estimates:

$$b_2(u,v,\omega) \leq \begin{cases} \left(\dfrac{2}{\lambda_1}\right)^{1/2} \|u\|_2 \|v\|_2 \|\omega\|_2, \\ 2^{1/2} \|u\|_3^{1/2} \|u\|_2^{1/2} \|v\|_2^{1/2} \|v\|_1^{1/2} \|\omega\|_1, \\ 2^{1/2} \|u\|_4^{1/2} \|u\|_3^{1/2} \|v\|_2^{1/2} \|v\|_1^{1/2} \|\omega\|, \end{cases}$$

$$b_2(u,v,\omega) = -b_2(u,\omega,v), \quad b_2(u,v,v) = 0. \tag{3.8}$$

Above propositions allow us to introduce in the usual way the operators $A_1, A_2, B_1(u), B_2(u,v)$, putting

$$\begin{aligned} a_1(u,v) &= \langle A_1 u, v \rangle, & A_1 &: H_0^1 \to H_0^{-1}, \\ a_2(u,v) &= \langle A_2 u, v \rangle, & A_2 &: H_0^2 \to H_0^{-2}; \\ b_1(u,v) &= \langle B_1(u), v \rangle, & B_1(u) &: H_0^1 \to L_0^2, \\ b_2(u,v,\omega) &= \langle B_2(u,v), \omega \rangle, & B_2(u,v) &: H_0^2 \times H_0^2 \to H_0^{-2}. \end{aligned} \tag{3.9}$$

Definition 3.1. Let $\psi_0 \in H_0^1(S^2)$ and $f \in L^2(0,T; L_0^2)$ for $T > 0$. The generalized solution of the problem (3.1) is said to be the function $\psi(t)$ satisfying the conditions:

$$\psi \in L^2(0,T; H_0^2) \cap L^\infty(0,T; H_0^1), \quad \frac{\partial}{\partial t} a_1(\psi,v) + \sigma a_1(\psi,v)$$
$$+\nu a_2(\psi,v) + b_1(t,v) + b_2(\psi,\psi,v) = -(f,v). \tag{3.10}$$

One may give another (equivalent) formulation of the generalized solution of (3.1).

Definition 3.2. Let $\psi_0 \in H_0^1(S^2)$ and $f \in L^2(0,T; H_0^{-1})$ for $T > 0$. The generalized solution of (3.1) in this case is the function $\psi(t)$ satisfying conditions:

$$\psi \in L^2(0,T; H_0^2) \cap L^\infty(0,T; H_0^1), \quad \psi' \in L^2(0,T; H_0^{-1}),$$
$$A_1 \frac{\partial \psi}{\partial t} + \sigma A_1 \psi + \nu A_2 \psi + B_1(\psi) + B_2(\psi,\psi) = f \; \psi|_{t=0} = \psi_0. \tag{3.11}$$

Proposition 3.1. If $\psi_0 \in H_0^2(S^2)$, then there exists a unique generalized solution of (3.1) such that

$$\psi \in L^2(0,T; H_0^3) \cap C([0,T], H_0^2), \quad \psi' \in L^2(0,T; H_0^1).$$

Then the mapping $\psi_0 \to \psi(t)$ from H_0^2 into H_0^2 is continuous for any $t \in [0,T]$. If f does not depend on t, then the problem (3.1) generates the continuous semigroup $S(t)$, $t \geq 0$ acting on $H_0^2 : S(t)\psi_0 = \psi(t)$.

Let us multiply the equation (3.1) by ψ. We find

$$\frac{1}{2}\frac{\partial}{\partial t}\|\psi\|_1^2 + \sigma\|\psi\|_1^2 + \nu\|\psi\|_2^2 = -(f,\psi)$$

$$\leq \|f\|_{-1}\|\psi\|_1 \leq \frac{\nu}{2}\|\psi\|_2^2 + \frac{\|f\|_{-1}^2}{2\nu\lambda_1}. \tag{3.12}$$

Hence we get the following inequality

$$\|\psi\|_1^2 \leq \|\psi_0\|_1^2 e^{-\gamma t} + \frac{\|f\|_{-1}^2}{\nu\lambda_1\gamma}(1 - e^{-\gamma t}), \quad \gamma = 2\sigma + \lambda_1\nu. \tag{3.13}$$

From (3.12) one can obtain the inequalities:

$$\|\psi\|_1^2 \leq \|\psi_0\|_1^2 e^{-\gamma_1 t} + \frac{\|f\|_{-1}^2}{\sigma\gamma_1}(1 - e^{-\gamma_1 t}),$$

$$\|\psi\|_1^2 \leq \|\psi_0\|_1^2 e^{-\gamma_2 t} + \frac{\|f\|_{-1}^2}{\gamma_2^2}(1 - e^{-\gamma_2 t}),$$

$$\gamma_1 = \sigma + 2\nu\lambda_1, \quad \gamma_2 = \sigma + \nu\lambda_1. \tag{3.14}$$

In the sequel, we shall use the inequality (3.13). Let us introduce the dimensionless number

$$G = \frac{\|f\|_{-1}}{\nu\gamma}. \tag{3.15}$$

Let $\|\psi_0\|_1 \leq R$ and $\delta > 0$. Then from (3.13) we get the inequality

$$\|\psi\|_1^2 \leq \frac{1+\delta}{\lambda_1}G, \quad \forall\, t \geq t_1(R,\delta),$$

$$t_1(R,\delta) = \frac{1}{\gamma}\ln\frac{\lambda_1 R_1^2}{\delta\|f\|_{-1}G}. \tag{3.16}$$

We introduce the notation

$$\mu_1^2 = \frac{1+\delta}{\lambda_1}\|f\|_{-1}G.$$

Then (3.16) becomes

$$\|\psi\|_1^2 \leq \mu_1^2, \quad \forall\, t \geq t_1(R_1\delta).$$

From (3.12) we also get

$$\frac{\partial}{\partial t}\|\psi\|_1^2 + \nu\|\psi\|_2^2 \leq \frac{\|f\|_{-1}^2}{\nu\lambda_1}.$$

Consequently, for $t \geq t_1$

$$\int\limits_t^{t+1} \|\psi(\tau)\|_2^2 d\tau \leq \frac{\mu_1^2}{\nu} + \frac{\|f\|_{-1}^2}{\lambda_1 \nu^2} \equiv \rho_1^2. \tag{3.17}$$

Multiplying (3.1) by $\triangle\psi$, we obtain

$$\frac{1}{2}\frac{\partial}{\partial t}\|\psi\|_2^2 + \sigma\|\psi\|_2^2 + \nu\|\psi\|_3^2 = -(f, \triangle\psi) \leq \frac{\nu}{2}\|\psi\|_3^2 + \frac{\|f\|_{-1}^2}{2\nu}. \tag{3.18}$$

From (3.18) we get

$$\|\psi\|_2^2 \leq \|\psi_0\|_2^2 e^{-\gamma t} + \frac{\|f\|_{-1}^2}{\nu\gamma}(1 - e^{-\gamma t}). \tag{3.19}$$

Notice that (3.18) gives the inequalities

$$\|\psi\|_2^2 \leq \|\psi_0\|_2^2 e^{-\gamma_1 t} + \frac{\|f\|^2}{\sigma\gamma_1}(1 - e^{-\gamma_1 t}),$$

$$\|\psi\|_2^2 \leq \|\psi_0\|_2^2 l^{-\gamma_2 t} + \frac{\|f\|^2}{\gamma_2^2}(1 - l^{-\gamma_2 t}). \tag{3.20}$$

Let $\|\psi_0\|_2 \leq R_2$ and $\delta > 0$. Hence,

$$\|\psi\|_2^2 \leq (1+\delta)\frac{\|f\|_{-1}^2}{\nu\gamma} = (1+\delta)\|f\|_{-1}G \equiv M_2^2, \quad t \geq t_2(R_2,\delta),$$

$$t_2 = \frac{1}{\gamma}\ln\frac{R_2^2}{\delta\|f\|_{-1}G}. \tag{3.21}$$

Setting $\lambda R_1^2 \leq R_2^2$, we see that for $t \geq t_2$ both inequalities (3.16) and (3.21) are satisfied and (3.17) also holds.

It follows from (3.21) that the ball $B_{M_2}(0) \subset H_0^2(S^2)$ centered at zero with radius M_2 will be an absorbing set for the semigroup $S(t)$ of $H_0^2(S^2)$. Notice that from (3.18) one can obtain also the inequality

$$\int\limits_t^{t+1} \|\psi\|_3^2 d\tau \leq \frac{M_2^2}{\nu} + \frac{\|f\|_{-1}^2}{\nu^2} = \rho_2^2, \quad \forall t \geq t_2. \tag{3.22}$$

Multiplying (3.1) by $\triangle^2\psi$, we get

$$\frac{1}{2}\frac{\partial}{\partial t}\|\psi\|_3^2 + \sigma\|\psi\|_3^2 + \nu\|\psi\|_4^2 - (J(\psi, \triangle\psi), \triangle^2\psi)$$

$$= (f, \triangle^2\psi) \leq \|f\|\|\psi\|_4 \leq \frac{\nu}{2}\|\psi\|_4^2 + \frac{1}{2\nu}\|f\|^2. \tag{3.23}$$

Notice that

$$|(J(\psi, \triangle\psi), \triangle^2\psi)| \leq 2^{1/2}\|\psi\|_1^{1/2}\|\psi\|_2^{1/2}\|\psi\|_3^{1/2}\|\psi\|_4^{3/2}$$

$$\leq \frac{\nu}{2}\|\psi\|_4^2 + \left(\frac{3}{2\nu}\right)^3 \|f\|_1^2\|\psi\|_2^2\|\psi\|_3^2.$$

Therefore,

$$\frac{\partial}{\partial t}\|\psi\|_3^2 \leq \left[2\left(\frac{3}{2\nu}\right)^3 \|\psi\|_1^2\|\psi\|_2^2\right]\|\psi\|_3^2 + \frac{\|f\|^2}{\nu}.$$

Using the uniform Gronwall Lemma for $t \geq t_2$, we find

$$\|\psi(t+1)\|_3^2 \leq \left(\frac{1}{\nu}\|f\|^2 + \rho_2^2\right)\exp\left[2\left(\frac{3}{2\nu}\right)^3 \mu_1^2\mu_2^2\right] = \rho_3^2. \qquad (3.24)$$

From (3.24), it follows that an absorbing set for the semigroup $S(t)$ is the ball $B_{\rho_3}(0) \subset H_0^3(S^2)$ centered at zero with radius ρ_3, i.e. the semigroup $S(t)$ is uniformly compact. Consequently, the following proposition takes place.

Proposition 3.2. The semigroup $S(t)$ possesses a global attractor A in the space $H_0^2(S^2)$. The attractor A is a compact (connected) set of $H_0^2(S^2)$ which attracts all bounded sets of the space $H_0^2(S^2)$ and has as its basin of attraction the whole space $H_0^2(S^2)$.

Remark. In terms of vorticity, i.e. for the equality (3.3), the Proposition 3.2 means that the semigroup $S(t)$ generated by (3.3) ($\omega_0 \in L_0^2(S^2)$) possesses a global attractor in $L_0^2(S^2)$.

3.2 Estimate of Dimension of Attractor

To derive an estimate of the attractor dimension A we write down the variation equation for (3.1). The equation has the form

$$\frac{\partial}{\partial t}\triangle\tilde{\phi} = -\sigma\triangle\tilde{\phi} + \nu\triangle^2\tilde{\phi} + J(\tilde{\phi}, \triangle\psi) + J(\psi, \triangle\tilde{\phi}) + J(\tilde{\phi}, l), \qquad (3.25)$$

$$\tilde{\phi}(0) = \xi.$$

Proposition 3.3. Let $\xi \in H_0^2(S^2)$. Then there exists a unique solution $\tilde{\phi}(t)$ of the problem (3.25) such that

$$\tilde{\phi} \in L^2(0, T; H_0^3(S^2)) \cap C([0, T], H_0^2(S^2)).$$

Hence, $\tilde{\phi}(t) = L(t, \psi_0)\xi$, $t > 0$, where $L(t, \psi_0)$ is the Freshet derivative of the mapping $\psi_0 \to S(t)\psi_0$ at the point ψ_0 of $H_0^2(S^2)$.

This proposition is proved in the usual way. Setting in (3.25) $\Delta\tilde{\phi} = \phi$, $\Delta\phi = \omega$, $\Delta\phi_0 = \omega_0$, we write (3.25) in the following form:

$$\frac{\partial}{\partial t}\phi = -\sigma\phi + \nu\Delta\phi + J(\Delta^{-1}\phi,\omega)$$

$$+J(\Delta^{-1}\omega,\phi) + J(\Delta^{-1}\phi,l) \equiv L(t,\omega_0)\phi. \qquad (3.26)$$

where $\omega = \omega(t,\omega_0)$. The equality (3.26) is the first variation equation corresponding to the equation (3.3). Hereafter it will be convenient to use the first variation equation of the form (3.26). Notice that, multiplying the equation (3.3) by ω, we get

$$\int_0^t \|\omega(\tau)\|_1^2 d\tau \leq \frac{t\|f\|_{-1}^2}{\nu^2} + \frac{\|\omega_0\|_1^2}{\nu}$$

and

$$\limsup_{t\to\infty} \sup_{\omega_0\in A} \frac{1}{t}\int_0^t \|\omega(\tau)\|_1^2 d\tau \leq \frac{\|f\|_{-1}^2}{\nu^2}. \qquad (3.27)$$

Let $\{\phi_j(t)\}$, $j = \overline{1,N}$, be the set of linearly independent solutions of (3.26) which correspond to the initial conditions $\phi_j(0) = \phi_{j0}$, $j = \overline{1,N}$. We denote by $Q_N(t)$ the orthogonal projector in $L_0^2(S^2)$ onto $\text{Span}\{\phi_1(t),\phi_2(t),\ldots,\phi_N(t)\}$. Then the Hausdorff dimension of the attractor A will be bounded above

$$\dim_H A < N^*,$$

where N^* is chosen from the condition

$$\limsup_{t\to\infty} \sup_{\omega_0\in A} \frac{1}{t}\int_0^t tr(L(\tau,\omega_0)Q_N(\tau))d\tau \leq c < 0 \qquad (3.28)$$

for $\forall N \geq N^*$. The problem consists in the estimate of the spur of the operator $L(t,\omega_0)Q_N(t)$. If $\{\varphi_j\}$, $j = \overline{1,N}$, is the orthonormal basis of $Q_N(t)L_0^2(S^2)$, the above spur is calculated by the formula

$$tr(L(t,\omega_0)Q_N(t)) = \sum_{j=1}^N (L(t,\omega_1)\varphi_j,\varphi_j)$$

$$= -\nu\sum_{j=1}^N \|\nabla\varphi_j\|^2 - \sigma\sum_{j=1}^N \|\varphi_j\|^2 - \sum_{j=1}^N \int_{S^2} J(\Delta^{-1}\varphi_j,\omega)\varphi_j ds.$$

Following [21, 75], one can show that

$$\sum_{j=1}^{N} \int_{S^2} J(\Delta^{-1}\varphi_j, \omega)\varphi_j \, dS \leq N^{1/2} \|\nabla\omega\| \|\rho\|_{\infty}^{1/2},$$

where

$$\rho = \sum_{j=1}^{N} |\nabla(\Delta^{-1}\varphi_j)|^2.$$

Since the estimate for $\|\nabla\omega\| = \|\omega\|_1$ on the attractor A has the form of (3.27), the problem reduces to the estimate $\|\rho\|_{\infty}^{1/2}$.

In the work [75] was proved the next inequality: for the integer $k \geq 0$ and $\forall \varphi \in H_0^3(S^2)$

$$2\sqrt{\pi}\|\nabla\varphi\|_{\infty}$$
$$\leq \|\Delta\varphi\| \left[(2\ln(k+1)+1)^{1/2} + (k+1)^{-1}\left(\frac{2}{\lambda_1}\right)^{1/2}\|\nabla(\Delta\varphi)\|\right].$$

Using this inequality, one can obtain the estimate for $\|\rho\|_{\infty}^{1/2}$. After making the corresponding calculations for $\sigma = 0$, we get the next Hausdorff dimension of the attractor A [75]:

$$\dim_H A \leq \left(\frac{12}{\sqrt{\pi}}\right)^{2/3} G^{2/3} \left(\frac{1}{2} + \ln\frac{3\sqrt{2}}{\sqrt{\pi}} + \ln G\right)^{1/3}, \qquad (3.29)$$

where $G = \dfrac{\|f\|_{-1}}{\lambda_1 \nu^2}$.

3.3 Statistical Solutions and Invariant Measures on Attractor

As is known, the observed fields which are used by the numerical calculations as the initial fields are known with some error. Therefore, it is natural to consider the problem, where instead of the precisely given initial conditions $\psi_0 \in H_0^2(S^2)$ there are given only the probability distributions of the initial data. In the present section we shall consider the equation (3.1) in the form of (3.3), the later we shall write in the new form, replacing ω by u:

$$\frac{\partial u}{\partial t} + J(\Delta^{-1}u, u+l) = -\sigma u + \nu\Delta u + f, \quad u|_{t=0} = u_0. \qquad (3.3')$$

We notice that if $f \in H_0^{-1}(S^2)$, then by multiplying this equality by u one can get the inequality

$$\frac{\partial}{\partial t}\|u\|^2 + 2\sigma\|u\|^2 + \nu\|u\|_1^2 \leq \frac{\|f\|_{-1}^2}{\nu} \leq \frac{\|f\|^2}{\lambda_1 \nu},$$

and by integrating it from 0 to t, we obtain

$$\|u(t)\|^2 + \nu \int_0^t \|u\|_1^2 d\tau \leq \|u_0\|^2 + \frac{t}{\nu}\|f\|_{-1}^2. \tag{3.30}$$

This energetic inequality plays the significant role in the construction of an invariant measure.

We introduce on the space of the initial data $u_0 \in L_0^2(S^2)$ the probability measure, i.e. there is chosen $\sigma_B(L_0^2)$-algebra of Borel sets and on it there is given σ-additive nonnegative function $\mu_0(\omega)$ of the set $\omega \in \sigma_B$ such that $\mu_0(L_0^2) = 1$ (for brevity, we shall write σ_B instead of $\sigma_B(L_0^2)$). Using the probability measure $\mu_0(\omega)$, one can introduce the concept of the statistical solution of the equation (3.3) as the family of measures $\mu(t, \omega)$, $\omega \in \sigma_B$ which satisfy the condition [130]:

$$\mu(t, \omega) = \mu_0(S(t)^{-1}\omega), \quad \forall \omega \in \sigma_B,$$
$$S(t)^{-1}\omega = \{u \in L_0^2 : S(t)u \in \omega\}. \tag{3.31}$$

It should be noticed that since the operator $S(t)$ is continuous, one has $S(t)^{-1}\omega \in \sigma_B$, i.e. $\mu(t, \omega)$ is the probability measure on σ_B. From (3.31), it follows that

$$\int F(S(t)u)\mu_0(du) = \int F(u)\mu(t, du) \tag{3.32}$$

for any continuous functional $F : L_0^2 \to R$, for which there exists one of the integrals in (3.32). For the statistical solution $\mu(t, \omega)$, it is easy to establish the analog of the energetic inequality (3.30). For that purpose we replace in (3.30) $u(t)$ by $S(t)u_0$ and integrate (3.30) over measure $\mu_0(du_0)$. Then we get

$$\int \|S(t)u_0\|^2 \mu_0(du_0) + \nu \int_0^t d\tau \int \|S(\tau)u_0\|_1^2 \mu_0(du_0)$$
$$\leq \frac{t}{\nu}\|f\|_{-1}^2 + \int \|u_0\|^2 \mu_0(du_0),$$

or

$$\int \|u_0\|^2 \mu(t, du_0) + \nu \int_0^t d\tau \int \|u_0\|_1^2 \mu(\tau, du_0)$$
$$\leq \frac{t}{\nu}\|f\|_{-1}^2 + \int \|u_0\|^2 \mu_0(du_0).$$

Replacing u_0 by u, we find

$$\int \|u\|^2 \mu(t, du) + \nu \int_0^t d\tau \int \|u\|_1^2 \mu(\tau, du)$$

$$\leq \frac{t}{\nu}\|f\|_{-1}^2 + \int \|u\|^2 \mu_0(du). \tag{3.33}$$

The last is the analog of the energetic inequality (3.30) for the statistical case. Taking into account that

$$\|f\|_{-1}^2 \leq \frac{1}{\lambda_1}\|f\|^2,$$

the inequality (3.33) may be written:

$$\int \|u\|^2 \mu(t, du) + \int_0^t d\tau \int \|u\|_1^2 \mu(\tau, du)$$

$$\leq \frac{t}{\lambda_1 \nu}\|f\|^2 + \int \|u\|^2 \mu_0(du). \tag{3.34}$$

Here we have assumed that there exists an integral

$$\int \|u\|^2 \mu_0(du) < \infty.$$

The statistical solution $\mu(t, \omega)$ of the equality (3.3') is called *stationary solution*, if the family of measures $\mu(t, \omega)$ does not depend on t, i.e.

$$\mu(t, \omega) \equiv \mu(\omega), \quad \forall\, t \geq 0.$$

Proposition 3.5. The measure $\mu(\omega)$, $\omega \in \sigma_B$, is *invariant* with respect to the semigroup $S(t)$ of (3.3') if and only if $\mu(\omega)$ is a *stationary statistical solution* of the equation (3.3').

Proposition 3.6. [130]. Let $\mu(\omega)$ is the probability measure given on $\sigma_B(L_0^2(S^2))$. If for some $\gamma > 0$

$$\int \|u\|_k^\gamma \mu(du) < \infty, \quad k = 1, 2,$$

then the measure $\mu(\omega)$ is supported on the space $H_0^k(S^2)$.

The sequence of measures $\nu_k(\omega)$, $\omega \in \sigma_B$ is said to be weakly convergent to $\nu(\omega)$, $\omega \in \sigma_B$, if for any continuous (bounded) function $f(u)$, $u \in L_0^2(S^2)$ one has

$$\int f(u)\nu_k(du) \to \int f(u)\nu(du), \quad k \to \infty.$$

Using the statistical solution $\mu(t,\omega)$, we construct by the averaging the probability measures $\mu_T(\omega)$ [25]:

$$\mu_T(\omega) = \frac{1}{T} \int_0^T \mu(t,\omega)dt, \quad \forall \in \sigma_B.$$

Using the results obtained by A.V.Fursikov [64, 65] for the two-dimensional Navier–Stokes equality one can prove the following

Proposition 3.7. There exists a weak limit $\mu_T(\omega) \to \hat{\mu}(\omega)$, $T \to \infty$, i.e.

$$\int f(u)\mu_T(du) \to \int f(u)\hat{\mu}(du), \quad T \to \infty,$$

or

$$\lim_{T \to \infty} \mu_T(\omega) = \hat{\mu}(\omega),$$

thereby $\hat{\mu}(\omega)$ is the probability measure on $L_0^2(S^2)$ and it is invariant with respect to the semigroup $S(t)$. The measure $\hat{\mu}(\omega)$ is supported on the attractor $A \subset L_0^2(S^2)$ of the semigroup $S(t)$.

The proof of this proposition is similar to that of the proposition for two-dimensional Navier–Stokes equalities [64]. The weak convergence follows from the well-known Theorem of Prokhorov [130]. Indeed, according to (3.34) one has

$$\frac{1}{T} \int\limits_0^T d\tau \int \|u\|_1^2 \mu(\tau,du) = \int \|u\|_1^2 \mu_T(du)$$

$$\leq \frac{\|f\|^2}{\lambda_1 \nu^2} + \frac{1}{T} \int \|u\|^2 \mu_0(du) \leq \text{const}$$

if $T > T_0 > 0$, that is,

$$\int \|u\|_1^2 \mu_T(du) \leq C, \quad \forall T > T_0.$$

Hence, from the sequence of probability measures $\mu_T(\omega)$, $\omega \in \sigma_B$ there can be chosen a sequence $\mu_{T_k}(\omega)$ weakly convergent on $L_0^2(S^2)$ as $T_k \to \infty$ such that $\mu_{T_k}(\omega) \to \hat{\mu}(\omega)$.

To prove the invariance of the measure $\hat{\mu}$ with respect to the semigroup $S(t)$ it is sufficient to show that

$$\int F(u)\hat{\mu}(du) = \int F(S(\tau)u)\hat{\mu}(du)$$

for any continuous bounded on $L_0^2(S^2)$ functional $F(u)$ and $\forall \tau > 0$.

This equality is established as for the case of two-dimensional Navier–Stokes equations. Proof of the fact that $\hat{\mu}(A) = 1$ is given by reductio ad absurdum.

Thus, any initial probability measure $\mu_0(\omega)$ generates the invariant measure $\hat{\mu}(\omega)$ on the attractor A, as "forgetting" its initial state. If $S(t)$ is ergodic then $\hat{\mu}(\omega)$ is unique. However the problem of ergodicity is still open.

3.4 Estimate of Attractor Dimension with Respect to Orography

We consider with respect to the orography the equation (3.3) [68]

$$\frac{\partial}{\partial t}\Delta\omega + J(\Delta^{-1}\omega, \omega + l + h) = -\sigma\omega + \nu\Delta\omega + f, \quad \omega|_{t=0} = \omega_0, \quad (3.3'')$$

where $h = h(\lambda, \varphi)$ is a given function.

We now give some estimates. Multiplying (3.3'') by $-\Delta^{-1}\omega$, we find

$$\|\nabla(\Delta^{-1}\omega)\|^2 \leq \|\nabla(\Delta^{-1}\omega_0)\|^2 e^{-\gamma t} + \frac{\|\nabla(\Delta^{-1}f)\|^2}{\gamma^2}(1 - e^{-\gamma t}), \quad (3.35)$$

$$\gamma = \sigma + \lambda_1\nu.$$

From (3.31) it follows that on the attractor A an estimate is valid

$$\|\nabla(\Delta^{-1}\omega)\|^2 \leq \frac{\|\nabla(\Delta^{-1}f)\|}{(\sigma + \lambda_1\nu)^2}. \quad (3.36)$$

Notice that in the estimate the effect of orography is lacking. Multiplying (3.3") by ω, we obtain

$$\frac{1}{2}\frac{\partial}{\partial t}\|\omega\|^2 + \sigma\|\omega\|^2 + \nu\|\nabla\omega\|^2$$
$$\leq (\|f\| + \|\nabla h\|_\infty\|\nabla(\Delta^{-1}\omega)\|)\,\|\omega\|, \quad (3.37)$$

or

$$\frac{\partial}{\partial t}\|\omega\|^2 + (\sigma + \lambda_1\nu)\|\omega\|^2$$
$$\leq \frac{1}{\sigma + \lambda_1\nu}\left(\|f\| + \|\nabla h\|_\infty\|\nabla(\Delta^{-1}\omega)\|\right)^2. \quad (3.38)$$

Hence we find that on the attractor A the estimate holds

$$\|\omega\|^2 \leq \frac{1}{(\sigma + \lambda_1\nu)^2}\left(\|f\| + \|\nabla h\|_\infty\frac{\|\nabla(\Delta^{-1}f)\|}{\sigma + \lambda_1\nu}\right)^2$$
$$\equiv \frac{f_h^2}{(\sigma + \lambda_1\nu)^2}, \quad f_h = \|f\| + \|\nabla h\|\frac{\|\nabla(\Delta^{-1}f)\|^2}{\sigma + \lambda_1\nu}. \quad (3.39)$$

We find also from (3.33) that

$$\frac{\partial}{\partial t}\|\omega\|^2 + 2\nu\|\nabla\omega\|^2 \leq \frac{1}{2\sigma}\left(\|f\| + \|\nabla h\|_\infty \|\nabla(\Delta^{-1}\omega)\|\right)^2.$$

Consequently,

$$\lim_{T\to\infty} \frac{1}{T} \int_0^T \|\nabla\omega\|^2 dt \leq \frac{f_h^2}{4\sigma\nu}. \tag{3.40}$$

The first variation equation for (3.3") has the form

$$\frac{\partial}{\partial t}U = -J(\Delta^{-1}U, \omega + l + h) - J(\Delta^{-1}\omega, U) - \sigma U + \nu\Delta U \equiv \mathcal{L}(t,\omega_0)U,$$

$$U|_{t=0} = u_0.$$

Let $U(t)$ is the solution of the first variation equation with the initial condition $U_j(0) = U_{0j}$, $j = \overline{1,n}$. We denote by Q_n the orthoprojector $L_0^2(S^2)$ on $\mathrm{Span}\{U_1(t),\ldots,U_n(t)\}$. As it was noted above, it is necessary to find the number n^* such that for $n > n^*$ the inequality holds

$$\lim_{t\to\infty} \sup \sup_{\omega_0\in A} \frac{1}{t} \int_0^t tr(\mathcal{L}(\tau,\omega_0)\cdot Q_n(\tau))d\tau \leq C < 0.$$

Then $\dim_H A < n^*$. Let $\varphi_j(t)$, $j = \overline{1,n}$ be orthonormal basis in $Q_n(t)L_0^2(S^2)$. We have

$$tr(\mathcal{L}\cdot Q_n) = \sum_{g=1}^n (\mathcal{L}\varphi_j,\varphi_j) = -\sigma n - \nu\sum_{j=1}^n \|\nabla\varphi_j\|^2$$

$$-\sum_{j=1}^n (J(\Delta^{-1}\varphi_j, \omega + h), \varphi_j)$$

$$\leq -\sigma n - \nu\sum_{j=1}^n \|\nabla\varphi_j\|^2 + \|\rho_n\|_\infty^{1/2} \cdot n^{1/2}\|\nabla(\omega + h)\|,$$

where

$$\rho_n = \sum_{j=1}^n |\nabla(\Delta^{-1}\varphi_j)|^2.$$

We use now the estimate obtained in [74] (k is an integer):

$$\|\varphi\|_\infty^{1/2} \leq \frac{1}{\sqrt{\pi}}\left\{\left(\ln(k+1) + \frac{1}{2}\right)^{1/2}\right.$$

$$\left. +(k+1)^{-1}\left(\lambda_1^{-1}\sum_{j=1}^n \|\nabla\varphi_j\|^2\right)^{1/2}\right\}^{1/2}, k \geq 0.$$

Setting here $k = \left[\lambda_1^{-1} \sum_{j=1}^{n} \| \nabla \varphi_j \|^2 \right]^{-1}$, we find

$$\| \rho_n \|_{\infty}^{1/2} \leq \frac{1}{\sqrt{\pi}} \left(\ln \lambda_1^{-1} \sum_{j=1}^{n} \| \nabla \varphi_j \|^2 + \frac{3}{2} \right)^{1/2}.$$

Now, using the Jensen's inequality, we obtain

$$\frac{1}{t} \int_0^t tr(\mathcal{L} Q_n) d\tau \leq -\sigma n - \nu \lambda_1 T_n$$

$$+ \frac{1}{\sqrt{\pi}} \left(\ln T_k + \frac{3}{2} \right)^{1/2} n^{1/2} \left(\frac{f_h}{2\sqrt{\sigma \nu}} + \| \nabla h \|_{\infty} \right),$$

where

$$T_n = \frac{1}{t} \lambda_1^{-1} \int_0^t \sum_{j=1}^{n} \| \nabla \varphi_j \|^2 d\tau \geq \frac{n^2}{4}.$$

To obtain the estimate of the attractor dimension A we need to find the number n for which the inequality holds

$$-\sigma n - \nu \lambda_1 T_n + \frac{a}{\sqrt{\pi}} \left(\ln T_n + \frac{3}{2} \right)^{1/2} n^{1/2} < 0, \qquad (3.41)$$

where

$$a = \frac{f_h}{2\sqrt{\sigma \nu}} + \| \nabla h \|_{\infty},$$

$$f_h = \| f \| + \| \nabla h \|_{\infty} \frac{\| \nabla (\Delta^{-1} f) \|}{\sigma + \lambda_1 \nu} \leq \| f \| \left(1 + \frac{\| \nabla h \|_{\infty}}{\lambda_1^{1/2} (\sigma + \lambda_1 \nu)} \right).$$

Taking into account that

$$T_n \geq \frac{n^2}{4}, \qquad (3.42)$$

we need to solve two inequalities (3.41) and (3.42) for $n = 1, 2, \ldots$, i.e. one must find the number n^* such that for $n \geq n^*$ these two inequalities will hold.

Numerical calculation of such a number n^*, obviously, is not complicated one. The existence of such a number n^* follows from the analysis of the inequalities (3.41) and (3.42), if we write them in the form

$$\begin{cases} \frac{a}{\sqrt{\pi}} \left(\ln T_n + \frac{3}{2} \right)^{1/2} < \sigma n^{1/2} + \nu \lambda_1 \frac{T_n}{n^{1/2}}, \\ \frac{T_n}{n^{1/2}} > \frac{n^{3/2}}{4}. \end{cases}$$

One can also obtain the analytical solutions of the inequalities (3.41) and (3.42). In fact, taking into account the growth of T_n according to (3.42), one can neglect in (3.41), for large n, the contribution of the term σn. Then (3.41) has the form

$$\frac{a}{\sqrt{\pi}} \left(\ln T_n + \frac{3}{2} \right)^{1/2} n^{1/2} < \nu \lambda_1 T_n.$$

If the inequality does not hold, then one must have

$$\frac{a}{\sqrt{\pi}} \left(\ln T_n + \frac{3}{2} \right)^{1/2} \sqrt{2} T^{1/4} > \nu \lambda_1 T_n,$$

that is

$$\ln \left(T_n + \frac{3}{2} \right) > \frac{\pi}{2} \left(\frac{\nu \lambda_1}{a} \right)^2 T_n^{3/2}. \tag{3.43}$$

Now it is convenient to use the inequalities

$$\ln x + c \leq 2 e^{c-2/2} x^{1/2}, \quad \forall \, x > 0.$$

Then, from (3.43) we find

$$T_n \leq \frac{4}{\pi} e^{-1/4} \left(\frac{a}{\nu \lambda_1} \right), \; \ln T_n \leq \ln \left(\frac{a}{\nu \lambda_1} \right) + \ln \left(\frac{4}{\pi} e^{-1/4} \right). \tag{3.44}$$

Combining (3.44) with (3.43), we finally obtain

$$\frac{n^2}{4} \leq T_N < \left(\frac{2}{\pi} \right)^{2/3} \left(\frac{a}{\nu \lambda_1} \right)^{4/3} \left(\ln \left(\frac{a}{\nu \lambda_1} \right) + \frac{3}{2} + \ln \left(\frac{4}{\pi} e^{-1/4} \right) \right)^{2/3}.$$

Hence

$$\dim_H A \leq n^* < 2 \left(\frac{2}{\pi} \right)^{1/3} \left(\frac{a}{\nu \lambda_1} \right)^{2/3} \left(\ln \left(\frac{a}{\nu \lambda_1} \right) + \frac{3}{2} + \ln \left(\frac{4}{\pi} e^{-1/4} \right) \right)^{1/3}.$$

One can obtain from (3.41) still another estimate of the attractor dimension, if ν is small and n is not too large. In this sense, the term $\nu \lambda_1 T_n$ in (3.41) can be neglected and (3.41) can be replaced by the inequality

$$\sigma n < \frac{a}{\sqrt{\pi}} n^{1/2} \left(\ln T_n + \frac{3}{2} \right)^{1/2}.$$

After making simple calculations, we get

$$\dim_H A < \frac{a}{\pi \sigma^2} \left(\ln \left(\frac{a}{\nu \lambda_1} \right) + \frac{3}{2} + \ln \left(\frac{4}{\pi} e^{-1/4} \right) \right).$$

3.5   Galerkin Approximations

Let $\omega = \sum\limits_{l=1}^{\infty} \sum\limits_{m=-l}^{l} \omega_{ml} Y_{ml}$. We introduce the operator P_N, setting

$$P_N \omega \equiv \omega_N = \sum_{l=1}^{N} \sum_{m=-l}^{l} \omega_{ml} Y_{ml}.$$

By applying the operator P_N to the equation (3.3"), we find

$$\frac{\partial}{\partial t} \omega_N + P_N J(\Delta^{-1} \omega_N, \omega_N + l + h) = -\sigma \omega_N + \nu \Delta \omega_N + P_N f.$$

The system obtained is the system $N^2 - 1$ of ordinary differential equations. For every $N \geq 1$ this system has an attractor A_N. However, the attractor A_N is in some sense the projection of the attractor A of the initial system on the finite-dimensional space H_N. It is natural that $\dim_H A_N \leq N^2 - 1$. Indeed, the attractor dimension may be less that the space dimension $H_N (\dim_H H_N = N^2 - 1)$. We try to get the estimate of the attractor dimension A_N depending on the number N. Let

$$\Phi_N(\lambda, \varphi) = \sum_{l=1}^{N} \sum_{m=-l}^{l} \Phi_{ml} Y_{ml}.$$

In [46] the following identity was proved

$$\sum_{m=-l}^{l} |\nabla Y_{ml}(\lambda, \varphi)|^2 \equiv \frac{\lambda_1}{2} \frac{l(l+1)(2l+1)}{4\pi},$$

by the use of which one can obtain the inequality

$$\|\nabla \Phi_N\| \leq (2\ln(N+1) + 1)^{1/2} \|\Delta \Phi_N\|.$$

As in the infinite-dimensional case, we find that

$$\|\rho_n\|_{\infty}^{1/2} \leq \frac{1}{\sqrt{\pi}} \left(\ln(N+1) + \frac{1}{2} \right)^{1/2}.$$

Now the inequality (3.41) becomes

$$\sigma n + \nu \lambda_1 T_n > \frac{a}{\sqrt{\pi}} \left(\ln(N+1) + \frac{1}{2} \right)^{1/2} n^{1/2}, \quad 1 \leq n \leq N^2 - 1,$$

or

$$\sigma n + \nu \lambda_1 \frac{n^2}{4} > \frac{a}{\sqrt{\pi}} \left(\ln(N+1) + \frac{1}{2} \right)^{1/2} n^{1/2}.$$

Thus, we need to find the number n^* such that the last inequality holds for $n \geq n^*$. Putting $x_n = n^{1/2}$, we reduce it to

$$\nu \lambda_1 x_n^3 + \sigma x_n > \frac{a}{\sqrt{\pi}} \left(\ln(N+1) + \frac{1}{2} \right)^{1/2}.$$

A numerical solution of the above inequality may be easily obtained. We notice that if $\dfrac{\nu \lambda_1 n}{4} \ll \sigma$, then from (3.48) we find

$$\dim A_N < \frac{a}{\pi \sigma^2} \left(\ln(N+1) + \frac{1}{2} \right).$$

3.6   Existence of Inertial Manifold

Let us consider on the rotating sphere the equation

$$\frac{\partial u}{\partial t} + J\left(\Delta^{-1} u, \, u + 2\mu\right) = -\sigma u - \nu(-\Delta)^s u + f, \qquad (3.45)$$

$$u|_{t=0} = u_0.$$

It should be noted that this equation coincides with (3.4), for which there were taken the notations $\omega \equiv u$, $\mu \equiv \sin \varphi$.

Following the works [45,46], we establish the existence of the inertial manifold for this equation, if $s > 1$. We show that for the above equation the conditions of Theorem 2.1 are valid for $s > 1$.

We shall assume that (3.45) is written in dimensionless variables so that the radius of the sphere is $a = 1$, the angular velocity of sphere is $\Omega = 1$. Let us introduce the notations

$$A = \nu(-\Delta)^s, \quad B(u, v) = J(\Delta^{-1} u, v),$$

$$C u = J(\Delta^{-1} u, 2\mu) + \sigma u = 2 \frac{\partial}{\partial \lambda} \Delta^{-1} u, + \sigma u. \qquad (3.46)$$

Then (3.45) has the form

$$\frac{\partial u}{\partial t} + Au + B(u, u) + Cu = f. \qquad (3.47)$$

We associate the operator $(-\Delta)$ to the scale of Hilbert spaces $H_0^\alpha(S^2)$ with the norm $\|u\|_\alpha = \|(-\Delta)^{\alpha/2} u\|$, $\alpha \in R$.

In these spaces we shall consider the equation (3.47). Here $(,)$ and $\|\cdot\|$ denote the scalar product and the norm in $L_0^2(S^2) \equiv H \equiv H_0^0(S^2)$, respectively.

Notice that $J(u, v) = \nabla_s u \cdot \nabla'_s v$, where ∇u is the gradient of the scalar function on S representing the tangent vector field to the sphere, while $\nabla'_s = -l_n \times \nabla_s$, l_n is a normal vector to the surface of sphere S. The operator J satisfies the following relations:

$$J(u, v) = -J(v, u), \quad (J(u, \mu), (-\Delta)^\beta u) = 0, \quad \beta \in R,$$
$$(J(u,v), h) = (J(v,h), u) = (J(h, u), v), \quad (J(u,v), F(u)) = 0, \quad (3.48)$$

where $F(u)$ is any differentiable function. Let us give some estimates for the operator J.

Lemma 3.1. *Trilinear form* $(J(u, v), h)$ *is continuous on the space* $H_0^2 \times H \times H_0^p$, $p > 1$, *i.e.*

$$\mid (J(u, v), h) \mid \leq c^* \|u\|_2 \|v\| \|h\|_p.$$

Proof. Let $p > 2$. Then

$$\mid (J(u,v), h) \mid = \mid (J(u, h), v) \mid = \mid (\nabla u \nabla' h, v) \mid \leq \|\nabla h\|_{C^0} \|\nabla u\| \|v\|$$
$$= \|\nabla h\|_{C^1} \|u\|_1 \|v\| \leq 2^{-1/2} c(p) \|h\|_p \|u\|_2 \|v\|.$$

Here we used the embedding of the space H_0^p into C^1 for $p > 2$ and the estimates

$$\|u\|_{C^1} \leq c(p) \|u\|_p, \quad 2^{1/2} \|u\|_1 \leq \|u\|_2.$$

It should be noted that the constant $c(p)$ may be explicitly calculated [46]. If $p = 2$, then

$$\mid (J(u, v), h) \mid \leq \|\nabla u\|_{L_4} \|\nabla h\|_{L_4} \|v\|$$
$$\leq c^2(4; 1) \|\nabla u\|_1 \|\nabla h\|_1 \|v\| = c^2(4; 1) \|u\|_2 \|h\|_2 \|v\|.$$

Here we used the embedding of the space H_0^β into L_α for $\beta + 2/\alpha \geq 1$, $\alpha \geq 1$ and the estimate $\|u\|_{L_\alpha} \leq c(\alpha; \beta) \|u\|_\beta$ for $\alpha = 4$, $\beta = 1$. If $1 < p < 2$, then

$$\mid (J(u, v), h) \mid \leq \|\nabla u\|_{L_{2/p-1}} \|\nabla h\|_{L_{2/2-p}} \|v\|$$
$$\leq c(\frac{2}{2-p}, p-1) c(\frac{2}{p-1}, 1) \|u\|_2 \|h\|_p \|v\|.$$

Setting

$$c^*(p) = \begin{cases} 2^{-1/2} c(p), & p > 0, \\ c^2(4; 1), & p = 2, \\ c(\frac{2}{2-p}; p-1) c(\frac{2}{p-1}, 1), & 1 < p < 2, \end{cases} \quad (3.49)$$

we complete the proof of the lemma.

Lemma 3.2. *Bilinear operator* $B(u, v)$ *is continuous from* $H \times H$ *into* H_0^{-p} *for* $p > 1$.
Proof. Let $u, v, h \in L_2^0(S^2)$. Then

$$| ((-\Delta)^{-p/2} J (\Delta^{-1} u, v), h) | = | (J (\Delta^{-1} u, v), (-\Delta)^{-p/2} h) |$$
$$\leq c^* (p) \|\Delta^{-1} u\|_2 \|v\| \|(-\Delta)^{-p/2} h\|_p \leq c^* (p) \|u\| \|v\| \|h\|.$$

Consequently,

$$\|B (u, v)\|_{-p} = \|J (\Delta^{-1} u, v)\|_{-p} \leq c^* (p) \|u\| \|v\|.$$

Lemma 3.3. *Bilinear operator* $B(u, v)$ *is continuous from* $H_0^p \times H_0^p$ *into* H *for* $p \geq 1$.
The proof mimics that of Lemma 3.2.
We are coming now to the proof of the existence of the inertial manifold. Let us introduce the notation

$$R(u) = B(u, u) + C u - f.$$

Then (3.47) is

$$\frac{\partial u}{\partial t} + A u + R(u) = 0. \qquad (3.50)$$

We show that for this equation the conditions of Theorem (1.54) are valid.
Lemma 3.4. *If* $\alpha = 0$, $s > 1$ *and* $f \in H_0^{-s}$, *then the nonlinear operator* $R(u)$ *satisfies the inequalities*

$$\|R(u)\|_{D(A^{\alpha-1/2})} \equiv \|R(u)\|_{D(A^{-1/2})} \leq c_1, \quad \|u\| \leq r,$$
$$\|R(u) - R(v)\|_{D(A^{-1/2})} \leq c_2 \|u - v\|, \quad \|u\|, \|v\| \leq r.$$

Proof. We have

$$\|R(u)\|_{D(A^{-1/2})} = \|\nu^{-1/2} (-\Delta)^{-s/2} R(u)\|$$
$$\leq \nu^{-1/2}[\|J(\Delta^{-1}u, u)\|_{-s} + \|J(\Delta^{-1}u, 2\mu)\|_{-s} + \sigma\|u\|_{-s} + \|f\|_{-s}]$$
$$\leq \nu^{-1/2}[c^*(s)(\|u\|^2 + 2\|\mu\|\|u\|) + 2^{-s/2}\sigma\|u\| + \|f\|_{-s}]$$
$$\leq \nu^{-1/2}[c^*(s)(r^2 + 2r\sqrt{4\pi/3}) + 2^{-s/2}\sigma r + \|f\|_{-s}] \equiv c_1.$$

Similarly,

$$\|R(u) - R(v)\|_{D(A^{-1/2})}$$
$$\leq \nu^{-1/2}[c^*(s)\|u-v\|(\|u\|+\|v\|+2\|mu\|)+2^{-s/2}\sigma\|u-v\|] \leq c_2\|u-v\|,$$

where

$$c_2 = \nu^{-1/2} (2 c^* (s)(r + \sqrt{4\pi/3} + 2^{-s/2} \sigma).$$

Thus, the conditions imposed on nonlinearity in the Theorem 1.54 are satisfied for $\alpha = 0$ and c_1, c_2 from the present lemma. We verify the fulfillment of the spectral gap condition for $\lambda_n = \nu\,(n\,(n+1))^s$, $n = 1, 2, \ldots$.

We take into account that the multiplicity of any eigennumber λ_n is equal to $2\,n + 1$. Therefore $\lambda_{N+1} \neq \lambda_N$ if $N = k^2 - 1$, where $k \geq 2$ is an integer. We have

$$\lambda_{N+1}^{1/2} - \lambda_N^{1/2} > \nu^{1/2}\,s\,(k-1)^{s-1}.$$

Let $s > 1$. Then as k increases the right side will increase too, and consequently for some $N > N_0$ the spectral gap condition will be fulfilled. Thus we have proved

Theorem 3.1. *If $s > 1$ and $f \in H = L_0^2\,(s)$, then the equation (3.45) possesses an inertial manifold in the space H.*

Let us estimate the dimension of the inertial manifold. We take as the absorbing set B_a the ball of radius $\rho_\varepsilon = 1 + \varepsilon/\sigma + \lambda_1\|f\|$, $\varepsilon > 0$ centered at zero of the space H. The constant C_3 from Theorem 1.54 for our case will have the form

$$C_3 = 6\,\nu^{-1/2}\left[2\,c^*\,(\rho_\varepsilon + \sqrt{\pi/3}) + 2^{-s/2}\,\sigma + \frac{\|f\|_{-s}}{3\,\rho_\varepsilon}\right].$$

After making the necessary calculations, we find [11]

$$\bar{C}_3 = \inf C_3 = \nu^{-1/2}\,C_0,$$

$$\rho_\varepsilon > \rho_0,$$

where $\rho_0 = \|f\|/(\sigma + \lambda)$ and

$$C_0 = \begin{cases} 6\,(b + 2\,\sqrt{2/3\,c^*\,\|f\|\,2^{-s/2}}), & \|f\| \leq a, \\ 6\,(b + 2\,c^*\,\rho_0 + \|f\|/3 \cdot 2^{s/2}\,\rho_0), & \|f\| > a, \end{cases} \qquad (3.51)$$

hence $a = 6 \cdot 2^{s/2} c^* \rho_0^2$, $b = 2c^*\sqrt{\frac{\pi}{2}} + 2^{-s}\sigma$. The spectral gap condition is

$$\nu\,(k\,(k+1))^s > \nu^{-1}\,C_0^2\,(20 + 8\,e^{-1/2})^{1/2},$$

$$\nu^{1/2}\,k^{s/2}\,[(k+1)^{s/2} - (k-1)^{s/2}] > 18\,\nu^{-1/2}\,C_0.$$

Hence,

$$k > \left(2.24\,\frac{C_0}{\nu}\right)^{1/s}, \qquad (3.52)$$

$$k > 1 + (18\,C_0/\nu\,s)^{1/(s-1)}. \qquad (3.53)$$

Notice that in the numerical experiments ν is usually a small parameter, while the addition of the term $\nu(-\Delta)^s u$ in (3.45) allow us to overcome the nonstability of the numerical algorithm (this term suppresses the short waves). Therefore one can obtain the estimate of the parameter ν by the index L of the spherical harmonics, beginning with this index the corresponding harmonics are strongly suppressed. This harmonic is such a solution of the equations

$$\frac{\partial \varphi}{\partial t} = -\nu(-\Delta)^s \varphi,$$

that dampes in e times per time unit, i.e. $|\varphi(1)| = |\varphi(0)|e^{-1}$.

Therefore, it may be assumed that $\nu = (L(L+1))^{-s} \approx L^{-2s}$. Since $N = k^2 - 1 \approx k^2$, we find from (3.51)

$$N > \left(\frac{18\,C_0}{s}\right)^{1/(s-1)} L^{2s/(s-1)} > \left(\frac{220\,C_0}{s}\right)^{1/(s-1)} L^{2s/(s-1)}.$$

Hence it is seen that the attractor dimension of the inertial manifold for $s > 1$ increases with the growth of L. Since the dimension D of the phase space of not very strong suppressed harmonics has order L^2, one has

$$N > \left(\frac{220\,C_0}{s}\right)^{1/(s-1)} D^{1s/(s-1)}, \quad s > 1.$$

For conclusion we give the estimate of the time of the attraction of the solution to the inertial manifold. Let $u_0 \in H$. Then the solution $u(t) = S(t)u_0$ is attracted to the inertial manifold M according to the following low [125]:

$$\mathrm{dist}_H\left(S(t)\,u_0,\,M\right) \le a\,e^{-t/t_0},$$

where

$$t_0 = 1/2\,C_3^2, \quad a = 2\,\mathrm{dist}_H\left(B_a,\,M\right)e^{T(u_0)/t},$$

while $T(u_0)$ is a time in which the solution starting from the point u_0 enters into the absorbing set B_a. In our case

$$T = \frac{\ln\left(\|u_0\|^2 - \rho_0^2\right) - \ln\left(\rho_\varepsilon^2 - \rho_0^2\right)}{\sigma + 2^s\,\nu},$$

if $\|u_0\|^2 > \rho_0^2$, and $T = 0$ in the opposite case. Further, one may consider that $\mathrm{dist}_H\left(Ba,\,M\right) \le 2\,\rho_\varepsilon$. Therefore, $a \le 4\,\rho_\varepsilon \exp\left(2\,C_3^2\,T\right)$.

As it was noted above, here one need to replace C_3 by $\bar{C}_3 = \nu^{-1/2}\,C_0$. Then $t_0 = \nu/2\,C_0^2$, where C_0 is determined by (3.51). Since $C_0 > 12\,c^*$, where c^* is determined from (3.49), one has

$$t_0 < \nu/300\,c^{*2} = \left(300\,c^{*2}\,L^2\right)^{-1}.$$

Hence it follows that $t_0 \ll 1$ and, thus, the dynamics of the equation (3.45) in time comparable with the unit becomes in fact the finite-dimensional dynamics on the inertial manifold.

Chapter 4

Discretization of Systems Possessing Attractors

For the numerical realization of the systems possessing attractors, it is common practice to use the time and space finite difference schemes and other approximations of such systems. In essence, one deals with the replacing of an initial infinite-dimensional system by different finite dimensional systems. In connection with this the question emerges: what properties of the initial system are kept under such approximation. Since we consider the systems possessing attractors, the question must be answered first: what happens with attractors of such systems under above time-space discretization.

4.1 Discretization of Systems Possessing Inertial Manifolds

We note at once that the systems possessing an inertial manifold (attractor in this case lies in inertial manifold) are best suited to approximating. We now give the results that are presented in [18].

Time discretization. Let us choose some time difference (stable in some sense) approximation, for example, the semiimplicit one and let τ be the time step. As a rule, if the initial nonlinear dissipative system of partial differential equations has an inertial manifold M of finite dimension N, then the corresponding system obtained from it by the time difference approximation with the time step τ for sufficiently small τ possesses an inertial manifold M_τ of the same dimension N, thereby $M_\tau \to M$ as $\tau \to 0$, i.e. the inertial manifold M_τ approximates the manifold of the initial system M.

Thus, we deal here with the time discretization and the initial infinite-dimensional system is replaced by another infinite-dimensional system.

Let us give some theorem of this class [18]. We consider in a Hilbert space H the equation

$$\frac{du}{dt} + Au + Cu + F(u) = 0, \quad u(0) = u_0, \tag{4.1}$$

where $A : H \to H$ is an unbounded self-adjoint positively definite operator and A^{-1} is compact, while C is a bounded skew-symmetric operator and F is a nonlinear operator such that

$$\|A^\gamma(F(u) - F(v))\|_\alpha \le L\|u - v\|_\alpha, \quad \forall \, u, v \in D(A^\alpha),$$

$$\gamma \in (0, 1/2], \quad \alpha \in R.$$

Let N be such that

$$\lambda_{N+1} \ge 3L^2\lambda_1^{2\gamma-1/2}, \quad \lambda_{N+1} - \lambda_N \ge 30L(\lambda_{N+1}^\gamma - \lambda_N^\gamma), \tag{4.2}$$

then (4.1) has an inertial manifold of finite dimension. (Here λ_j are eigenvalues of the operator A.) We carry out the following time approximation of (4.1) (see [18]):

$$\frac{u^{n+1/2} - u^n}{\tau} + Au^{n+1/2} + F(u^n) = 0,$$

$$\frac{u^{n+1} - u^{n+1/2}}{\tau} + C\frac{u^{n+1} + u^{n+1/2}}{2} = 0. \tag{4.3}$$

Proposition 4.1. [18]. For every τ such that $\tau N < 1$, the system (4.3) has an inertial manifold M_τ of the same dimension as the manifold M of the initial equation (4.3).

If Φ is the graph of the manifold M, while Φ_τ is the graph of the manifold M_τ, then, for every $\tau \in (0, \lambda_{N+1}^{-1})$, we have the estimate

$$\|\Phi - \Phi\|_\alpha \le C\tau^\beta \ln \tau, \quad \beta = 1 - \gamma, \tag{4.4}$$

where N is defined by the condition (4.2).

Space discretization. Usually, for the spatial variables, the Galerkin approximation procedure is applied. If one replaces the initial system possessing an inertial manifold by the system of ordinary differential equations of order $m \ge N$, where N is defined by (4.2), the results are:

1) from $m \ge N$ on, all Galerkin approximations will have the inertial manifolds M_m of the same dimension as the inertial manifold M of the initial system;

2) $M_m \to M$ as $m \to \infty$, i.e. the manifolds M_m approximate the manifolds M as precisely as we need.

Thus, beginning with some sufficiently large number m, Galerkin approximations adequately describe the behavior of the initial system, i.e. the space discretization in this sense is justified.

It is noteworthy that, physically, the equation of the inertial manifold $q = \Phi(p)$ represents the parametrization of the small eddies (scales) q by the large eddies p. Because of this, the number m of necessary Galerkin approximations, when such a parametrization may be carried out, is theoretically sufficiently large: $q = \Phi(p)$, $p = (p_1, \ldots, p_m)$, $q = (q_{m+1}, q_{m+2}, \ldots)$, but in any particular case one can obtain the number m that will be sufficiently less than the theoretical one.

If one takes the number of Galerkin approximations smaller than the theory requires, then these approximations do not assure the existence of inertial manifolds (with the exception of the trivial case – the phase space itself for Galerkin system).

To summarize, we can say that *the time and space discretization of systems possessing an inertial manifold is possible if the time step and the number of Galerkin approximations are chosen appropriately.* In this case, the time and space discretization describes adequately the initial system. Roughly speaking there is occurred the continuous dependence of the inertial manifold on the time and space discretization.

4.2 Time-Space Discretization of Systems Possessing Attractor

If a system does not have an inertial manifold, but it possesses an attractor, then the situation with the time and space discretization (i.e. the situation with the continuous dependence of an attractor on parameters) is something specific.

Spatial approximation. Once again, we consider Galerkin approximations. In the general case, if the initial dissipative system of partial differential equations possesses a global attractor A, then any m-Galerkin approximation will possess its own global attractor A_m.

The dimension and the properties of attractors A_m of Galerkin approximation may differ from the dimension and properties of the attractor A of the initial system, if the number m of Galerkin approximations is less than the dimension of the attractor of the initial system. However, even if the number of Galerkin approximations is sufficiently large(it is about the dimension of the attractor of the initial system), then the attractor of such a Galerkin approximation may differ from the attractor of the initial system.

Moreover, if the number m of Galerkin approximation tends to the infinity, then A_m, in the general case, will not tend to coincide with A, i.e. $A_m \not\to A$, as $m \to \infty$. It is known only that the limit attractor A_∞ is embedded in the attractor A of the initial system [4,76]:

$$\lim_{m\to\infty} A_m = A_\infty \subset A.$$

Thus, analyzing some Galerkin approximation of the initial system, we analyze the attractor of this approximation only and nothing more. The next approximation will also possess an attractor, but it is another attractor.

Time discretization. In this case the situation is something different from the previous one. We can construct the difference schemes that approximate the uniformly asymptotically stable sets of the initial system, in particular, attractors. To do this, we need to construct a Lyapunov function in this case. We shall consider the time difference schemes for the systems that possess the uniformly asymptotically stable sets, in particular, attractor.

Time difference schemes and Lyapunov functions. We consider the semigroup $S(t)$, $t \geq 0$ acting on a Hilbert space H, which is generated by the evolution equation

$$\partial_t u = F(u), \quad u|_{t=0} = u_0. \tag{4.5}$$

Let the system (4.5) be nonlinear, dissipative one possessing an attractor A (or it possess the uniformly asymptotically stable set M). We proceed now from (4.5) to the corresponding finite difference equation

$$u_{n+1} = u_n + \tau F_\tau(u_n, u_{n+1}), \quad n = 0, 1, \ldots \tag{4.6}$$

In the general case, we know the estimate on the segment $0 \leq t \leq T$:

$$\|u(n\tau) - u_n\| \leq c\tau^k, \quad c = O(e^T), \tag{4.7}$$

where $k \geq 1$ is the order of the difference scheme.

As is seen from the estimate (4.7), if $T \to \infty$, then this estimate does not hold. Naturally, the question emerges: when the difference scheme fits for the approximating of the solution of (4.5) on the infinitely large time interval? The necessity of the consideration of the case in which $T \to \infty$ follows from the fact that the systems (4.5) having the attractor should be investigated for very large T, when the system enters the small neighborhood of the attractor.

It is known that if the system (4.5) is finite-dimensional and possesses the simplest attractors as the asymptotically stable stationary states and the stable limit cycle, then the estimate similar to the estimate (4.7) takes place on the infinite (time) interval.

It turns out that, in the general case also, when the system (4.5) is finite-dimensional and possesses an attractor, the similar proposition is valid [80]. We give now the relevant results.

Finite-dimensional case. We consider the system of ordinary differential equations

$$\dot{x} = F(x), \quad x \in R^N, \quad x(0) = x_0. \tag{4.8}$$

Let this system possess an attractor A. Hence one can assume that the solutions of the system (4.8) belong to some ball $B_R(0)$ of the space R^N containing A. We shall assume that $F(x)$ satisfies the Lipschitz condition in the ball $B_R(0)$ with a Lipschitz constant L. We denote by $x(t, x_0)$ the solution of (4.8).

Since the global attractor A of the system (4.8) is uniformly asymptotically stable set of this system, in accordance with [136] the system (4.8) possesses Lyapunov function $V(x)$ which is defined in some neighborhood $S(A, \rho)$ of the attractor A, where

$$S(A, \rho) = \{x \in R^N : \text{dist}_{R^N}(x, A) < \rho\}.$$

More precisely, the following theorem is valid [80].

Theorem 4.1. *Let $M \subset R^N$ be a nonempty compact uniformly asymptotically stable set for the system (4.8), while $F(x)$ satisfies the Lipschitz condition (with the Lipschitz constant L) in a neighborhood of the set M. Then there exists Lyapunov function $V(x)$ defined in a neighborhood of the set M such that*

1) $V : S(M, \rho) \to R_+, \ \rho > 0;$

2) $|V(x) - V(x')| \le L\|x - x'\|, \ \forall x, x' \in S(M, \rho);$

3)   *there exist continuous positive strictly increasing functions $a(r), b(r)$ such that $\alpha(r) < \beta(r), \ r > 0, \ \alpha(0) = \beta(0),$ $a(\text{dist}_{R^N}(x, M)) \le V(x) \le b(\text{dist}_{R^N}(x, M)), \ \forall x \in S(M, \rho);$*

4) *there exists a constant $C > 0$ such that*

$$D^+V(x) \le -CV(x), \quad \forall x \in S(M, \rho),$$

where

$$D^+V(x) = \overline{\lim_{h \to 0}} \frac{V(x + hF(x)) - V(x)}{h}.$$

We note that if $x(t, x_0) \in S(M, \rho), \ \forall t \ge 0$, then from the conditions of the theorem it follows that

$$V(x(t, x_0)) \le e^{-ct}V(x_0),$$

i.e. Lyapunov function decreases exponentially on every solution of the system (4.8). However, if M is an invariant set, then setting $x_0 \in M$, we find $x(t, x_0) \in M$ for $\forall t \ge 0$.

Therefore,

$$a\left(\text{dist}_{R^N}\left(x\left(t,x_0\right),M\right)\right) = a(0) \leq V\left(x(t,x_0)\right)$$
$$\leq b(\text{dist}_{R^N}(x(t,x_0)M)) = b(0),$$

i.e. $V(x(t,x_0)) = 0$, $\forall\, t \geq 0$. Thus, the function $V(x) = 0$, if $x \in M$ and M is invariant.

It is shown in [80], that if the system (4.8) has an uniformly asymptotically stable set M, then the difference system (4.6) associated with it will also possess an uniformly asymptotically stable set M_τ for small τ and M_τ tend to M as $\tau \to 0$ in the Hausdorff metric. In other words, the continuous dependence M_τ on τ holds. We give now the strong formulation of the corresponding theorem [80].

Theorem 4.2. *Let F and its first k-derivatives be uniformly bounded in R^N and the system (4.5) possess the compact uniformly asymptotically stable set M. Then there exists τ^* such that for every $0 < \tau < \tau^*$ the system (4.6) possesses the compact uniformly asymptotically stable set M_τ which contains M and converges to M in the Hausdorff metric as $\tau \to 0$.*

Moreover, there exists a bounded open set U_0 which does not depend on τ and contains M_τ and there is the time $T_0(\tau) = C_1 + C_2 k \ln \tau^{-1}$ such that $x_n \in M_\tau$ for all $n\tau \geq T_0(\tau)$, $x_0 \in U$, $0 < \tau < \tau^*$. In proving this theorem one uses essentially the condition

$$\|x(\tau) - x_1\| = \|x(\tau) - x_0 - \tau F_\tau(x_0)\| \leq C_k \tau^k, \ k \geq 2, \ C_k = \text{const.} \quad (4.9)$$

Construction of the Lyapunov function in infinite-dimensional case and time difference schemes. Let H be a Hilbert space and let $S(t)$, $t \geq 0$ be the semigroup in H, while A is the global attractor of the semigroup $S(t)$. We denote by $B_R(0)$ the absorbing set of the semigroup $S(t)$. Clearly, $A \subset B_R(0)$. Let $O_\delta(M)$ denote δ-neighborhood of the set $M \subset H$, i.e.

$$O_\delta(M) = \bigcup_{x \in M} O_\delta(x), \quad O_\delta(x) = \{y : \|x - y\| < \delta\}.$$

We shall assume that $\|S(t)u - S(t)v\| \leq Le^{\mu t}\|u - v\|$, $L > 0$, $\mu > 0$; and for all $u \in B_R(0)$, there exists $T(\delta)$ such that

$$\text{dist}_H(S(t)u, A) < \delta, \quad \forall\, t \geq T(\delta), \quad \forall\, u \in B_R(0).$$

This is the condition of the uniform asymptotic stability of the attractor A. In other words,

$$S(t)u \in O_\delta(A), \quad \forall\, t \geq T(\delta), \quad \forall\, u \in B_R(0).$$

We put $\delta = 1/k$, $k = 1, 2, \ldots$ and introduce the function

$$g\left(\frac{1}{k}\right) = e^{-(c+\mu)T(\frac{1}{k})}, \quad c > 0, \ k = 1, 2, \ldots,$$

where $T\left(\frac{1}{k}\right)$ is chosen so that

$$S(t)u \in O_{\frac{1}{k}}(A), \quad \forall\, t \geq T\left(\frac{1}{k}\right), \quad \forall\, u \in B_R(0),$$

that is

$$\mathrm{dist}_H(S(t)u, A) < \frac{1}{k}, \quad \forall\, t \geq T\left(\frac{1}{k}\right), \quad \forall\, u \in B_R(0).$$

We denote by $V_k(u)$ the function of the form

$$V_k(u) = e^{-(C+\mu)T(\frac{1}{k})} \sup_{t \geq 0} e^{ct}\mathrm{dist}_H(S(t)u, O_{\frac{1}{k}}(A))$$

$$= g\left(\frac{1}{k}\right) \sup_{0 \leq t \leq T(\frac{1}{k})} e^{ct}\mathrm{dist}_H\left(S(t)u, O_{\frac{1}{k}}(A)\right), \forall\, u \in B_R(0).$$

Properties of functions $V_k(u)$, $k = 1, 2, \ldots$:
1) $V_k(u) = 0$, $\forall\, k \geq 1$, if $u \in A$; $V_k(u) = 0$, if $u \in O_{\frac{1}{k}}(A)$;
2) $V_k(u)$ is bounded, i.e. $0 \leq V_k(u) \leq R$, $\forall\, u \in B_R(0)$:

$$V_k(u) \leq g\left(\frac{1}{k}\right) e^{cT}(\frac{1}{k}) \sup_{0 \leq t \leq T(\frac{1}{k})} \mathrm{dist}_H\left(S(t)u, O_{\frac{1}{k}}(A)\right)$$

$$\leq e^{-\mu T(\frac{1}{k})} \sup_{t \geq 0}(S(t)u, A) \leq R;$$

3) $|V_k(u) - V_k(v)| \leq L\|u - v\|$.
Indeed, taking into account that in H the inequality holds

$$|\varphi(u) - \varphi(v)| \leq \|u - v\|, \quad \varphi(u) = \mathrm{dist}_H(u, M), \quad M \subset H,$$

we find

$$|V_k(u) - V_k(v)| = e^{-(c+\mu)T(\frac{1}{k})} \sup_{0 \leq t \leq T(\frac{1}{k})} e^{ct}\mathrm{dist}\left(S(t)u, O_{\frac{1}{k}}(A)\right)$$

$$-\mathrm{dist}_H\left(S(t)v, O_{\frac{1}{k}}(A)\right) \leq e^{-(c+\mu)T(\frac{1}{k})}e^{cT(\frac{1}{k})} \sup_{0 \leq t \leq T(\frac{1}{k})} \|S(t)u - S(t)v\|$$

$$\leq e^{-\mu T(\frac{1}{k})}Le^{\mu T(\frac{1}{k})}\|u - v\| \leq L\|u - v\|;$$

4) $\dot{V}_k(u) \leq -cV_k(u)$, $\forall\, u \in B_R(0)$, $c > 0$, i.e.

$$\frac{d}{dt}V_k(S(t)u) \leq -cV_k(S(t)u).$$

Indeed,

$$V_k(S(h)u) = e^{-(c+\mu)T(\frac{1}{k})}\sup_{t\geq 0} e^{ct}\mathrm{dist}_H\left(S(t+h)u, O_{\frac{1}{k}}(A)\right)$$

$$\leq e^{-(c+\mu)T(\frac{1}{k})}e^{-ch}\sup_{\tau\geq h} e^{c\tau}\mathrm{dist}_H\left(S(\tau)u, O_{\frac{1}{k}}(A)\right) \leq e^{-ch}V_k(u).$$

Hence,

$$\frac{d}{dt}V(S(t)u)|_{t=0} = \lim_{h\to 0+}\frac{V_k(S(h)u) - V_k(S(0)u)}{h}$$

$$\leq \lim_{h\to 0+}\frac{e^{-ch} - 1}{h}V_k(u) = -cV_k(u).$$

We define now Lyapunov function $V(u)$ in the following way

$$V(x) = \sum_{k=1}^{\infty}\frac{1}{2^k}V_k(x), \quad \forall u \in B_R(0).$$

In accordance with the above said the following properties of Lyapunov function are apparent:

1. $|V(u) - V(v)| \leq L\|u - v\|_H$.
2. $\dot{V}(u) \leq -CV(u)$, $\forall u \in B_R(0)$.
3. $V(u) = 0$, if $u \in A$.
4. $a\left(\mathrm{dist}_H(u, A)\right) \leq V(u) \leq b\left(\mathrm{dist}_H(u, A)\right)$, $a(0) = b(0) = 0$, $a(r) < b(r)$, $r > 0$, where $a(r)$ and $b(r)$ are non-negative monotonically increasing functions r.

We omit the proof of the property 4, since the attempts to construct the functions $a(r)$ and $b(r)$ have not met with success. Using Lyapunov function $V(x)$ constructed above, one can choose the classes of time-difference schemes in the infinite-dimensional case, that will be globally stable.

Theorem 4.3. *Let the time-approximation scheme* (4.6) *for the evolution equation* (4.5) *be such that*

$$\|u(\tau) - u_1\|_H \leq c(k)\tau^k, \quad k \geq 2, \tag{4.10}$$

and let (4.5) *possess the attractor A. Then there exists τ_0 such that for $0 < \tau < \tau_0$ the system* (4.6) *has the uniformly asymptotically stable set M_τ, which contains the attractor A, i.e. $A \subset M_\tau$ and M_τ tends to A as $\tau \to 0$ in the Hausdorff metric, i.e.*

$$\mathrm{dist}(M_\tau, A) \to 0, \quad \tau \to 0.$$

Remark 1. In general, it is difficult to verify the condition (4.10) in infinite-dimensional case. If the system under consideration possesses the "good" Lyapunov function, then the condition (4.10) does not require the verification.

By a "good" Lyapunov function is meant the function $V(u)$, which is defined on the invariant (with respect to the semigroup $S(t)$) set $M \subset H$ and is such that the function $V(S(t)u)$ is a nonincreasing function t for $\forall u \in M$, and if $V(u) = V(S(t)u)$ for some $t > 0$, then $u = S(t)u = Z$ is a fixed point of the semigroup $S(t)$ (i.e. $S(t)Z = Z$). In this case one can establish the continuous dependence of the attractor of the initial system on the parameter τ [76].

Remark 2. To solve the problem (see Chapter 3)

$$\frac{\partial}{\partial t}\omega + J(\Delta^{-1}\omega, \omega + l) = -\sigma\omega + \nu\Delta\omega + f, \quad \omega|_{t=0} = \omega_0,$$

one can propose some different semi-implicit time-difference schemes. We consider in detail one of such possible schemes. Namely, let

$$\frac{\omega_{n+1} - \omega_n}{\tau} + \sigma\omega_{n+1} - \nu\Delta\omega_{n+1} + J(\Delta^{-1}\omega_n, \omega_{n+1}) = f - J(\Delta^{-1}\omega_n, l),$$

$$n = 0, 1, 2, \ldots, \tag{4.11}$$

or

$$(1 + \sigma\tau - \nu\tau\Delta)\omega_{n+1} + \tau J(\Delta^{-1}\omega_n, \omega_{n+1}) = \omega_n + \tau f - \tau J(\Delta^{-1}\omega_n, l),$$

$$n = 0, 1, 2, \ldots \tag{4.12}$$

Solvability of finite difference problems. All finite difference equations (4.12) have the form

$$L_\tau u + \tau J(\Delta^{-1}w, u) = F(\tau, w), \tag{4.13}$$

where $\tau > 0$ is fixed, w is the known function:

$$L_\tau = (1 + \tau\sigma - \tau\nu\Delta)u, \quad F(\tau, w) = w + \tau f - \tau J(\Delta^{-1}w, l). \tag{4.14}$$

We denote by $a_\tau(u, v)$ the bilinear form

$$a_\tau(u, v) = \int_{S^2}(uv + \sigma\tau uv\tau + \nu\tau\nabla u\nabla v)ds$$

$$+ \tau\int_{S^2} J(\Delta^{-1}w, u)vds. \tag{4.15}$$

Let $f \in L_0^2(S^2)$, $w \in L_0^2(S^2)$. The generalized solution of the problem (4.13) is the function $u \in H_0^1(S^2)$ such that

$$a_\tau(u, v) = (F(\tau, w), v), \quad \forall v \in H_0^1. \tag{4.16}$$

Lemma 4.1. *The problem* (4.13) *has a unique generalized solution.*

This assertion follows from the Lax–Milgramm theorem, since

$$1)\, |a_{(\tau)}(u,v)| \leq (1+\sigma\tau)\|u\| \cdot \|v\| + \nu\tau\|\nabla u\| \cdot \|\nabla v\|$$

$$\leq \left(\frac{1+\sigma\tau}{\lambda_1} + \nu\tau + c\tau\right)\|u\|_{H_0^1}\|v\|_{H_0^1}; \tag{4.17}$$

$$2)\, c = \|\Delta^{-1}w\|_{L^\infty(S^2)}, \quad a_\tau(u,u) \geq \nu\tau\|u\|_{H_0^1}^2. \tag{4.18}$$

Estimates of solutions of iteration problems. Absorbing set. Attractor. To shorten the calculations we restrict ourselves to the case, when $l = 0$. We have for $n = 0$

$$L_\tau\omega_1 + \tau J(\Delta^{-1}\omega_0, \omega_1) = \omega_0 + \tau f. \tag{4.19}$$

By taking the scalar product of (4.19) with ω_1 in $L_0^2(S^2)$, we obtain

$$(1+\sigma\tau)\|\omega_1\|^2 + \nu\tau\|\nabla\omega_1\|^2 = (\omega_0 + \tau f, \omega_1),$$

or

$$(1+\sigma\tau+\nu\tau\lambda_1)\|\omega_1\|^2 \leq (\|\omega_0\| + \tau\|f\|)\|\omega_1\|,$$

that is

$$\|\omega_1\| \leq \frac{\tau\|f\| + \|\omega_0\|}{1+\tau\sigma+\tau\nu\lambda_1} \equiv \alpha(\tau\|f\| + \|\omega_0\|), \tag{4.20}$$

where

$$\alpha = \frac{1}{1+\tau\sigma+\tau\nu\lambda_1} < 1.$$

In the similar way, for $n = 1$, we get

$$\|\omega_2\| \leq \alpha(\tau\|f\| + \|\omega_1\|) \leq (\alpha + \alpha^2)\tau\|f\| + \alpha^2\|\omega_0\|.$$

Continuing the analogous calculations, we obtain on n iteration

$$\|\omega_n\| \leq \alpha(\tau\|f\| + \|\omega_{n-1}\|) \leq \tau\|f\| \sum_{k=1}^{n} \alpha^k + \alpha^n\|\omega_0\|. \tag{4.21}$$

Lemma 4.2. *The difference scheme* (4.12) *possesses an absorbing set in* $L_0^2(S^2)$, *which is the ball* $B_R(0)$ *in* $L_0^2(S^2)$ *centered at zero with radius*

$$R = \frac{\|f\|}{\sigma + \nu\lambda_1}(1+\delta), \quad \delta > 0.$$

This absorbing set coincides with the absorbing set of the initial problem (3.3) (see Chapter 3).

Proof. Setting in (4.21) $n \to \infty$, we find that the right side of (4.21) has a limit equal to

$$\tau \|f\| \sum_{k=1}^{\infty} \alpha^k = \frac{\|f\|}{\sigma + \nu \lambda_1}.$$

Therefore, for any $\delta > 0$ one can find $N(\delta)$ such that for $n \geq N(\delta)$ one has

$$\|\omega_n\| \leq \frac{\|f\|}{\sigma + \nu \lambda_1}(1 + \delta) = R,$$

i.e. all ω_n will enter the ball $B_R(0)$, from the number N on, consequently, $B_R(0)$ is the absorbing set. We multiply now (3.3) (see Chapter 3) by ω in $L_0^2(S^2)$. This gives

$$\frac{1}{2}\frac{\partial}{\partial t}\|\omega\|^2 + \sigma\|\omega\| + \nu\|\nabla\omega\|^2 = (f, \omega).$$

Taking into account that $\lambda_1\|\omega\|^2 \leq \|\nabla\omega\|^2$, we get

$$\frac{1}{2}\frac{\partial}{\partial t}\|\omega\|^2 + (\sigma + \nu\lambda_1)\|\omega\|^2 \leq (f, \omega) \leq \|f\| \cdot \|\omega\| \leq \frac{\sigma + \nu\lambda_1}{2}\|\omega\|^2$$

$$+ \frac{1}{2(\sigma + \nu\lambda_1)}\|f\|^2,$$

that is

$$\frac{\partial}{\partial t}\|\omega\|^2 + (\sigma + \nu\lambda_1)\|\omega\|^2 \leq \frac{\|f\|^2}{\sigma + \nu\lambda_1}.$$

Consequently,

$$\|\omega(t)\|^2 \leq \|\omega_0\|e^{(\sigma + \nu\lambda_1)t} + \frac{\|f\|^2}{(\sigma + \nu\lambda_1)^2}(1 - e^{(\sigma + \nu\lambda_1)t}).$$

This means that for any $\delta > 0$ there can be found $T(\delta)$ such that as $t \geq T(\delta)$

$$\|\omega(t)\| \leq \frac{\|f\|}{\sigma + \nu\lambda_1}(1 + \delta).$$

Thus, the absorbing set of the continuous problem and that of the discrete problem coincide in $L^2(S^2)$.

Lemma 4.3. *The difference scheme* (4.12) *possesses the global attractor* A_τ.

Proof. In accordance with Lemmas 4.1 and 4.2 the difference scheme (4.12) generates the discrete semigroup $S_\tau^n : \omega_0 \to \omega_n$, $n = 0, 1, 2, \ldots$, acting in $L_0^2(S^2)$.

This semigroup, as it shown above, possesses an absorbing set in $L_0^2(S^2)$. The compactness of the semigroup to be proved follows from the inequality (the detailed analysis of the difference schemes and their attractors will be given in the next sections)

$$\|s_\tau^1\omega_0\| = \|\omega_1\| \leq \frac{1}{\nu\tau\lambda_1^{1/2}}(\|\omega_0\| + \tau\|f\|), \quad \tau > 0.$$

Similar, one can consider the difference scheme of the form

$$\frac{\omega_{n+1} - \omega_n}{\tau} + \sigma\omega_{n+1} - \nu\Delta\omega_{n+1} + J(\Delta^{-1}\omega_{n+1}, \omega_{n+1} + l) = f.$$

4.3 Globally Stable Difference Schemes for Barotropic Vorticity Equation

Let us consider the barotropic vorticity equation on the sphere with radius a rotating with the constant angular velocity Ω:

$$\frac{\partial}{\partial t}\Delta\psi + J(\psi, \Delta\psi + l) + \sigma\Delta\psi - \nu\Delta^2\psi = f, \qquad (4.22)$$

where $\psi = \psi(t, \lambda, \mu)$ is the stream function, λ is the longitude $(0 \leq \lambda \leq 2\pi)$, $\mu = \sin\vartheta$, ϑ is the latitude, $(-\frac{\pi}{2} \leq \vartheta \leq \frac{\pi}{2})$, $l = 2\Omega\mu$. Parameters entering into the equation (4.22) have the dimensions

$$[\psi] = L^2T^{-1}, \quad [\sigma] = T^{-1}, \quad [\nu] = L^2T^{-1}, \quad [\Omega] = T^{-1}, \quad [f] = T^{-2}.$$

Passing in (4.22) to dimensionless values

$$\psi' = \frac{\psi}{a^2\Omega}, \quad t' = \Omega t, \quad \sigma' = \frac{\sigma}{\Omega}, \quad \nu' = \frac{\nu}{a^2\Omega}, \quad f' = \frac{f}{\Omega^2}$$

and omitting primes, we get

$$\frac{\partial}{\partial t}\Delta\psi + J(\psi, \Delta\psi + 2\mu) + \sigma\Delta\psi - \nu\Delta^2\psi = f. \qquad (4.23)$$

We combine this equation with the initial condition

$$\psi|_{t=0} = \psi_0(\lambda, \mu). \qquad (4.24)$$

We shall consider the problem (4.23)-(4.24) in the spaces H_0^α, where α is any real number. We give the following definition of these spaces.

Let $Y_{mn}(\lambda, \mu)$ be spherical harmonic and let Δ be the operator of Laplace–Beltrami on the unit sphere S. Let $L_0^2(S)$ denote the space

$$L_0^2(S) = \{f : f \in L^2(S), \int_S f \, ds = 0\}$$

with a scalar product and a norm

$$(f, g) = \int_S f\bar{g} \, ds, \quad \|f\|^2 = (f, f) = \sum_{n=1}^{\infty} \sum_{|m| \le n} |f_{mn}|^2,$$

where

$$f_{mn} = \int_S f\bar{Y}_{mn} \, ds.$$

We associate the operator $(-\Delta)$ to the scale of Hilbert spaces $H_0^{\alpha}(S)$, $\alpha \in \mathbf{R}$, setting

$$(f, g)_{\alpha} = (f, g)_{H_0^{\alpha}(S)} = ((-\Delta)^{\alpha/2} f, (-\Delta)^{\alpha/2} g),$$

$$\|f\|_{\alpha} = \|f\|_{H_0^{\alpha}(S)} = \|(-\Delta)^{\alpha/2} f\| = (\sum_{n=1}^{\infty} \sum_{|m| \le n} \lambda_n^{\alpha} |f_{mn}|^2)^{1/2},$$

where $\lambda_n = n(n + 1)$ are eigenvalues of the operator of Laplace–Beltrami on the unit sphere (with multiplicity $2n + 1$), which are associated with the eigenfunctions Y_{mn}, $|m| \le n$. Notice that

$$\|f\|_{\alpha}^2 \ge \lambda_1^{\alpha - \beta} \|f\|_{\beta}^2, \quad \alpha \ge \beta. \tag{4.25}$$

In the sequel, we shall use aside with this inequality the inequalities of the form [5]:

$$\|u\|_{L^4(S)} \le 2^{1/4} \|u\|^{1/2} \|u\|_1^{1/2}, \, \forall u \in H_0^1(S),$$

$$\|\nabla u\|_{L^4(S)} \le 2^{1/4} \|u\|_1^{1/2} \|u\|_2^{1/2}, \, \forall u \in H_0^2(S). \tag{4.26}$$

We shall assume that the right-hand side f of the equation (4.55) does not depend on time and suppose that $f = f(\lambda, \mu) \in L_0^2(S)$. Above it was shown that the problem (4.23)-(4.24) possesses an attractor $\mathcal{A}$ which is a global attractor in the spaces $H_0^1(S)$ and $H_0^2(S)$.

Let X be a complete metric space and let $\mathcal{T}$ be nontrivial subgroup of the additive group $\mathbf{R}$ of real numbers. Let $\mathcal{T}_+ = \mathcal{T} \cap [0, +\infty)$ be a semigroup of non-negative elements of $\mathcal{T}$ and let $W : \mathcal{T}_+ \times X \to X$ be the continuous in the second argument mapping possessing the semigroup property

$$W(t_1, W(t_2, x)) = W(t_1 + t_2, x), \quad \forall t_1, t_2 \in \mathcal{T}_+, \quad \forall x \in X.$$

The triplet $\{W, \mathcal{T}_+, X\}$ is called a semidynamical system (SDS), hence X is the phase space of the states of the system, the time is measured by elements $\mathcal{T}_+$, while the evolution of the system is defined by continuous operators $W(t, \cdot)$. As $\mathcal{T}_+$ usually there is taken $\mathcal{T}_+ = \mathbf{R}_+ = [0, +\infty)$ or in the case in which the system has a discrete time, $\mathcal{T}_+ = \mathbf{Z}_+ = \{0, 1, 2, \ldots\}$ and $\mathcal{T}_+ = \mathbf{Z}_+\tau = \{0, \tau, 2\tau, \ldots\}$. Then the following theorem is valid [76].

Theorem 4.4. *Let for SDS $\{W, \mathcal{T}_+, X\}$ the following conditions are satisfied:*

1) there exists a bounded set B_0 such that for all $x \in X$ there can be found some $T = T(x) \in \mathcal{T}_+$, for which $W(T, x) \in B_0$;

2) for any bounded sequence $x_k \in X$, $k = 1, 2, \ldots$ and any increasing sequence $t_k \in \mathcal{T}_+$, $t_k \to \infty$ the sequence $W(t_k, x_k)$ is precompact.

Then SDS $\{W, \mathcal{T}_+ X\}$ possesses a nonempty global attractor $\mathcal{A}$, i.e. 1) $\mathcal{A}$ is a compact attracting set; 2) $\mathcal{A}$ is an invariant set, i.e. $W(t, \mathcal{A}) = \mathcal{A}, \forall t \in \mathcal{T}_+$.

Theorem 4.5. *Let the parameter d runs through some complete metric space D with metric $\rho(\cdot, \cdot)$ and let d_0 be nonisolated point D. Let the family of SDS $\{W_d, \mathcal{T}_+, X\}$ depending on the parameter be such that*

1) for any d SDS $\{W_d, \mathcal{T}_+ X\}$ possesses a compact attractor $\mathcal{A}_d$;

2) for any sequence convergent to d_0 $d_k \in D$, $d_k \neq d_0$ (convergent to d_0), the closure in X of the union for all $k = 1, 2, \ldots$ of the attractors $\mathcal{A}_{d_k}$ is compact;

3) if $d_k \to d_0$, $x_k \in \mathcal{A}_{d_k}$ and $x_k \to x_0$, then $W_{d_k}(t, x_k) \to W_{d_0}(t, x_0)$ for some $t > 0$.

Then for any neighborhood $\mathcal{O}$ of the attractor $\mathcal{A}_{d_0}$ there can be found δ such that $\mathcal{A}_d \subset \mathcal{O}$ for all $d \in D$, that are apart from d_0 on the value not large than δ, i.e. $\rho(d, d_0) \leq \delta$.

In the present section we investigate the global stability of discrete time difference scheme for the barotropic vorticity equation (4.22). We give the a priori estimates on the solutions of the equation (4.22), in particular, for the case in which the initial data are taken on the attractor $\mathcal{A}$. We also consider the semi-implicit time difference scheme with the time step τ. For this scheme we prove the existence of the global attractor $\mathcal{A}_\tau$. Further, the convergence of this scheme is considered and the estimates of the rate of its convergence are derived.

We then prove the global stability of the difference scheme under consideration. The global stability means that for the sufficiently small time step τ the attractor $\mathcal{A}_\tau$ of the difference scheme enters any given neighborhood of the attractor $\mathcal{A}$ of the initial system.

Further, we shall also prove that the difference scheme possesses the uniformly asymptotically stable sets M_τ, that contain the attractor $\mathcal{A}$ and converge to $\mathcal{A}$ as τ tends to zero.

A priori estimates. We denote by $V(t, \cdot)$ the solving operator of the problem (4.23)-(4.24) with the right side $f = f(\lambda, \mu) \in L_0^2(S)$. We show that for the attractor $\mathcal{A}$ of the problem the estimate holds

$$\|\psi_0\|_2 \leq R_0 = \frac{\|f\|}{\delta}, \quad \forall \psi_0 \in \mathcal{A}, \tag{4.27}$$

where $\delta = \sigma + \nu\lambda_1 = \sigma + 2\nu$.

Indeed, by taking the scalar product of (4.23) with $\Delta\psi$ in $L_0^2(S)$, we obtain

$$1/2\frac{\partial}{\partial t}\|\psi\|_2^2 + \sigma\|\psi\|_2^2 + \nu\|\psi\|_3^2 = (f, \Delta\psi) \leq \|f\|\|\psi\|_2$$

$$\leq \frac{\delta\|\psi\|_2^2}{2} + \frac{\|f\|^2}{2\delta}. \tag{4.28}$$

Since $\|\psi\|_3^2 \geq \lambda_1\|\psi\|_2^2$, we find from (4.28)

$$\frac{\partial}{\partial t}\|\psi\|_2^2 + \delta\|\psi\|_2^2 \leq \frac{\|f\|^2}{\delta}.$$

Consequently,

$$\|\psi\|_2^2 \leq \|\psi_0\|_2^2 e^{-\delta t} + \frac{\|f\|^2}{\delta^2}(1 - e^{-\delta t}). \tag{4.29}$$

Thus, the ball $B_{R_0}(0) = \{u \in H_0^2(S) : \|u\|_2 \leq R_0\}$ is the attracting set for the solving operator $V(t, \psi_0)$. Since $\mathcal{A} \subset B_{R_0}$, (4.27) holds. In the sequel we shall denote by R_i, $i = 1, 2, \ldots$ positive constants which depend on σ, ν and $\|f\|$.

Lemma 4.4. *The attractor $\mathcal{A}$ is bounded in $H_0^3(S)$, that is,*

$$\|\psi_0\|_3 \leq R_1, \quad \forall\, \psi_0 \in \mathcal{A}. \tag{4.30}$$

Proof. From (4.28) the inequality follows

$$\nu \int_0^1 \|\psi(t)\|_3^2 dt \leq \|\psi_0\|_2^2 + \frac{1}{\delta}\|f\|^2.$$

Taking into account (4.27), we get

$$\int_0^1 \|V(t, \psi_0)\|_3^2 \, dt \leq \frac{R_0^2}{\nu} + \frac{1}{\nu\delta}\|f\|^2 \equiv R_2^2, \ \forall\, \psi_0 \in \mathcal{A}. \tag{4.31}$$

By taking now the scalar product of (4.23) with $t\, \Delta^2\psi$ in $L^2(S)$, we obtain

$$\frac{1}{2}t\frac{d}{dt}\|\psi\|_3^2 + \sigma t\|\psi\|_3^2 + \nu t\|\psi\|_4^2 = (J(\psi, \Delta\psi) - f, t\, \Delta^2\psi).$$

Let us estimate the expression

$$| (J(\psi, \Delta\psi), \Delta^2\psi) | \le \|\nabla\psi\|_{L^4(S)} \|\nabla\Delta\psi\|_{L^4(S)} \|\psi\|_4.$$

Using the inequalities (4.26), we find

$$| (J(\psi, \Delta\psi), \Delta^2\psi) | \le \sqrt{2} \|\psi\|_1^{1/2} \|\psi\|_2^{1/2} \|\psi\|_3^{1/2} \|\psi\|_4^{3/2}$$

$$\le 2^{1/4} \|\psi\|_2 \|\psi\|_3^{1/2} \|\psi\|_4^{3/2}.$$

By applying the Young inequality, we obtain

$$| (J(\psi, \Delta\psi), \Delta^2\psi) | \le \frac{\nu}{2} \|\psi\|_4^2 + \frac{27}{16\nu^3} \|\psi\|_2^4 \|\psi\|_3^2.$$

Further,

$$| (f, \Delta^2\psi) | \le \frac{\nu}{2} \|\psi\|_4^2 + \frac{1}{2\nu} \|f\|^2.$$

Thus,

$$\frac{d}{dt} (t\|\psi\|_3^2) + 2\sigma\, t\|\psi\|_3^2 \le \|\psi\|_3^2 + \frac{27}{8\nu^3} t\|\psi\|_2^4 \|\psi\|_3^2 + \frac{t}{\nu} \|f\|^2.$$

Integrating the inequality over time between 0 and 1 and presenting result in terms of the solving operator $V(t, \psi_0)$, we have

$$\|V(1, \psi_0)\|_3^2 \le [1 + \frac{27}{8\nu^3} \max_{0 \le t \le 1} \|V(t, \psi_0)\|_2^4] \int_0^1 \|V(t, \psi_0)\|_3^2 dt + \frac{\|f\|^2}{2\nu}.$$

If $\psi_0 \in \mathcal{A}$, then with respect to (4.27) and (4.31), we find

$$\|V(1, \psi_0)\|_3^2 \le \left(1 + \frac{27}{8\nu^3} R_0^4\right) R_2^2$$

$$+ \frac{\|f\|^2}{2\nu} \equiv R_1^2, \quad \forall\, \psi_0 \in \mathcal{A}. \tag{4.32}$$

It follows from (4.32) that $V(1, \mathcal{A}) \subset B_{R_1}^3(0) = \{\rho \in H_0^3(S) : \|\rho\|_3 \le R_1\}$. By the invariance of the attractor, one has also $\mathcal{A} = V(1, \mathcal{A}) \subset B_{R_1}^3(0)$, i.e. (4.30) holds.

Lemma 4.5. *The estimate is valid*

$$\|V_t(t, \psi_0)\|_2 \le R_3, \quad \forall\, \psi_0 \in \mathcal{A}. \tag{4.33}$$

Proof. By taking the scalar product of (4.23) with $\Delta^2\psi$ in $L^2(S)$, we obtain

$$\frac{d}{dt} \|\psi\|_3^2 + 2\sigma\|\psi\|_3^2 + \nu\|\psi\|_4^2 \le \frac{2}{\nu} \|f\|^2 + \frac{27}{\nu^3} \|\psi\|_2^4 \|\psi\|_3^2.$$

As above, integrating the inequality over time between 0 and 1 and passing to the solving operator $V(t, \psi_0)$, we get

$$\nu \int_0^1 \|V(t, \psi_0)\|_4^2 \, dt \leq \|\psi_0\|_3^2 + \frac{2}{\nu} \|f\|^2$$

$$+ \frac{27}{\nu^3} \max_{0 \leq t \leq 1} \|V(t, \psi_0)\|_2^4 \int_0^1 \|V(t, \psi_0)\|_3^2 \, dt.$$

If $\psi_0 \in \mathcal{A}$, then according to (4.27), (4.30) and (4.31), we have

$$\int_0^1 \|V(t, \psi_0)\|_4^2 \, dt \leq \frac{1}{\nu} \left(R_1^2 + \frac{2}{\nu} \|f\|^2 + \frac{27}{\nu^3} R_0^4 R_2^2 \right) \equiv R_4^2, \ \forall \, \psi_0 \in \mathcal{A}.$$

$$(4.34)$$

It follows immediately from (4.22) that

$$\int_0^1 \|\Delta \psi_t\|^2 \, dt = \int_0^1 \|f - J(\psi, \Delta \psi + 2\mu) - \sigma \Delta \psi + \nu \Delta^2 \psi\|^2 \, dt.$$

Taking into account (4.27), (4.30) and (4.34), we obtain the estimate

$$\int_0^1 \|V_t(t, \psi_0)\|_2^2 \, dt \leq R_5^2, \quad \forall \, \psi_0 \in \mathcal{A}.$$

$$(4.35)$$

We now differentiate (4.23) over time:

$$\Delta \psi_{tt} + J(\psi_t, \Delta \psi + 2\mu) + J(\psi, \Delta \psi_t) + \sigma \Delta \psi_t - \nu \Delta^2 \psi_t = 0. \quad (4.36)$$

By taking the scalar product of (4.36) by $t \Delta \psi_t$ in $L_0^2(S)$, we find

$$\frac{1}{2} \frac{d}{dt} (t \|\psi_t\|_2^2) + \sigma t \|\psi_t\|_2^2 + \nu t \|\psi_t\|_3^2 = \frac{1}{2} \|\psi_t\|_2^2 - t(J(\psi_t, \Delta \psi), \Delta \psi_t).$$

As before, we get the estimate

$$| J(\psi_t, \Delta \psi), \Delta \psi_t) | \leq \|\psi_t\|_{L^4(S)} \|\nabla \Delta \psi\|_{L^4(S)} \|\psi_t\|_3$$

$$\leq \|\psi_t\|_1 \|\psi\|_4 \|\psi_t\|_3 \leq 2^{-1/2} \|\psi_t\|_2 \|\psi\|_4 \|\psi_t\|_3$$

$$\nu \|\psi_t\|_3^2 + \frac{1}{8\nu} \|\psi_t\|_2^2 \|\psi\|_4^2.$$

Consequently,

$$\frac{d}{dt} (t \|\psi_t\|_2^2) + 2\sigma t \|\psi_t\|_2^2 \leq \|\psi_t\|_2^2 + \frac{t}{4\nu} \|\psi\|_4^2 \|\psi_t\|_2^2.$$

By virtue of uniform Gronwall Lemma we get

$$\|\psi_t\|_2^2 \,|_{t=1} \leq \int_0^1 \|\psi_t\|_2^2 \, dt \, e^{\frac{1}{4\nu} \int_o^1 \|\psi\|_4^2 \, dt}$$

or in terms of the solving operator $V(t, \psi_0)$:

$$\|V_t(1, \psi_0)\|_2^2 \leq \int_0^1 \|V_t(t, \psi_0)\|_2^2 \, dt \, \exp(\frac{1}{4\nu} \int_0^1 \|V(t, \psi_0)\|_4^2 \, dt).$$

Using (4.34) and (4.35), we find that

$$\|V_t(1, \psi_0)\|_2^2 \leq R_5^2 \, e^{R_4^2/(4\nu)} \equiv R_3^2, \quad \forall \, \psi_0 \in \mathcal{A}.$$

By the invariance of the attractor it follows that

$$\max_{\psi_0 \in \mathcal{A}} \|V_t(t, \psi_0)\|_2 = \max_{\psi_0 \in \mathcal{A}} \|V_t(1, \psi_0)\|_2 \leq R_3, \quad \forall t,$$

hence (4.33) holds.

Lemma 4.6. *The attractor $\mathcal{A}$ is bounded in $H_0^4(S)$, that is*

$$\|\psi_0\|_4 \leq R_6, \quad \forall \, \psi_0 \in \mathcal{A}. \tag{4.37}$$

Proof. It follows from (4.23) that

$$\nu \|\Delta^2 \psi\| = \|J(\psi, \Delta\psi + 2\mu) + \sigma\Delta\psi + \Delta\psi_t - f\|$$
$$\leq \|\psi_t\|_2 + \sigma\|\psi\|_2 + 2\|\psi\|_1 + \|J(\psi, \Delta\psi)\| + \|f\|.$$

As before, we estimate the norm

$$\|J(\psi, \Delta\psi)\| \leq \|\nabla\psi\|_{L^4(S)} \|\nabla\Delta\psi\|_{L^4(S)} \leq 2^{1/4}\|\psi\|_2 \|\psi\|_3^{1/2} \|\psi\|_4^{1/2}$$
$$\leq \frac{\nu}{2} \|\psi\|_4 + \frac{1}{\sqrt{2}\nu} \|\psi\|_2^2 \|\psi\|_3.$$

That is

$$\|\psi\|_4 \leq \frac{2}{\nu} \left(\|\psi_t\|_2 + (\sigma + \sqrt{2}) \|\psi\|_2 + \frac{2^{-1/2}}{\nu}\|\psi\|_2^2\|\psi\|_3 + \|f\| \right)$$

or passing to operator $V(t, \psi_0)$,

$$\|V(t, \psi_0)\|_4 \leq \frac{2}{\nu}[\|V_t(t, \psi_0)\|_2 + (\sigma + \sqrt{2}) \|V(t, \psi_0)\|_2$$
$$+ \frac{2^{-1/2}}{\nu} \|V(t, \psi_0)\|_2^2 \|V(t, \psi_0)\|_3 + \|f\|].$$

If $\psi_0 \in \mathcal{A}$, then according to (4.27), (4.31) and (4.33) we have

$$\|V(t,\psi_0)\|_4 \leq \frac{2}{\nu}(R_3 + (\sigma + \sqrt{2})R_0 + \frac{2^{-1/2}}{\nu}R_0^2 R_1 + \|f\|) \equiv R_6, \ \forall \psi_0 \in \mathcal{A}.$$

Thus,

$$V(t,\mathcal{A}) \subset B_{R_6}^4(0) = \{\rho \in H_0^4(S) : \|\rho\|_4 \leq R_6\}.$$

Taking into account $\mathcal{A} = V(t,\mathcal{A})$, we arrive at (4.37).

Lemma 4.7. *The estimates are valid*

$$\|V_t(t,\psi_0)\|_3 \leq R_7, \quad \forall \psi_0 \in \mathcal{A}, \tag{4.38}$$

$$\|V_{tt}(t,\psi_0)\|_1 \leq R_8, \quad \forall \psi_0 \in \mathcal{A}. \tag{4.39}$$

Proof. We take the scalar product of (4.36) with $\Delta\psi_t$ in $L^2(S)$. With respect to the calculations of Lemma 1.2, we find

$$\frac{d}{dt}\|\psi_t\|_2^2 + 2\sigma\|\psi_t\|_2^2 + \nu\|\psi_t\|_3^2 \leq \frac{1}{2\nu}\|\psi\|_4^2\|\psi_t\|_2^2.$$

Integrating over time between 0 and 1 and writing the result in terms of operator $V(t,\psi_0)$, we obtain

$$\nu \int_0^1 \|V_t(t,\psi_0)\|_3^2 dt \leq \|V_t(0,\psi_0)\|_2^2$$

$$+\frac{1}{2\nu} \max_{0 \leq t \leq 1} \|V_t(t,\psi_0)\|_2^2 \int_0^1 \|V(t,\psi_0)\|_4^2 dt.$$

Using (4.33) and (4.34), we get the inequality

$$\int_0^1 \|V_t(t,\psi_0)\|_3^2 \, dt \leq \frac{R_3^2}{\nu}\left(1 + \frac{R_4^2}{2\nu}\right) \equiv R_9^2, \quad \forall \psi_0 \in \mathcal{A}. \tag{4.40}$$

Let us take the scalar product of (4.36) with $t\Delta^2\psi_t$ in $L^2(S)$:

$$\frac{1}{2}\frac{d}{dt}(t\|\psi_t\|_3^2) + \sigma t\|\psi_t\|_3^2 + \nu t\|\psi_t\|_4^2 = \frac{1}{2}\|\psi_t\|_3^2$$

$$+t(J(\psi_t, \Delta\psi) + J(\psi, \Delta\psi_t), \Delta^2\psi_t).$$

We estimate the values in the right side of the above equality

$$|(J(\psi_t, \Delta\psi), \Delta^2\psi_t)| \leq \|\nabla\psi_t\|_{L^4(S)}\|\nabla\Delta\psi\|_{L^4(S)}\|\psi_t\|_4$$

$$\leq \|\psi_t\|_2\|\psi\|_4\|\psi_t\|_4 \leq \frac{\nu}{2}\|\psi_t\|_4^2 + \frac{1}{2\nu}\|\psi_t\|_2^2\|\psi\|_4^2,$$

$$|(J(\psi, \Delta\psi_t), \Delta^2\psi_t)| \leq \max_S |\nabla\psi| \|\psi_t\|_3\|\psi_t\|_4$$

$$\leq \frac{1}{\sqrt{4\pi}}\|\psi\|_3\|\psi_t\|_3\|\psi_t\|_4 \leq \frac{\nu}{2}\|\psi_t\|_4^2 + \frac{1}{8\pi\nu}\|\psi_t\|_3^2\|\psi\|_3^2.$$

Thus,

$$\frac{d}{dt}\left(t\|\psi_t\|_3^2\right) + 2\sigma t\|\psi_t\|_3^2 \le \|\psi_t\|_3^2 + \frac{t}{\nu}\|\psi_t\|_2^2\|\psi\|_4^2 + \frac{t}{4\pi\nu}\|\psi_t\|_3^2\|\psi\|_3^2.$$

Integration between 0 and 1 gives

$$\|V_t(1,\psi_0)\|_3^2 \le \int_0^1 \|V_t(t,\psi_0)\|_3^2\,dt + \frac{1}{\nu}\max_{0\le t\le1}\|V_t(t,\psi_0)\|_2^2\int_0^1\|V(t,\psi_0)\|_4^2 dt$$

$$+\frac{1}{4\pi\nu}\max_{0\le t\le1}\|V(t,\psi_0)\|_3^2\int_0^1\|V_t(t,\psi_0)\|_3^2 dt.$$

If $\psi_0 \in \mathcal{A}$, then according to (4.31), (4.33), (4.34) and (4.40) we obtain

$$\|V_t(1,\psi_0)\|_3^2 \le R_9^2 + \frac{1}{\nu}R_3^2 R_4^2 + \frac{1}{4\pi\nu}R_1^2 R_9^2 \equiv R_7^2, \quad \forall\,\psi_0 \in \mathcal{A}.$$

By the invariance of the attractor one has

$$\max_{\psi_0\in\mathcal{A}}\|V_t(t,\psi_0)\|_3 = \max_{\psi_0\in\mathcal{A}}\|V_t(1,\psi_0)\|_3 \le R_7,$$

that is (4.38) holds. By taking the scalar product of (4.36) with ψ_{tt} in $L_0^2(S)$, we obtain

$$\|\psi_{tt}\|_1^2 = (J(\psi_t,\Delta\psi + 2\mu) + J(\psi,\Delta\psi_t) + \sigma\Delta\psi_t$$
$$-\nu\Delta^2\psi_t,\ \psi_{tt}) \le| (J(\psi_{tt},\psi_t),\Delta\psi)\,|$$
$$+2\,|\,(\psi_{\lambda t},\psi_{tt})\,| + |\,(J(\psi_{tt},\psi),\Delta\psi_t)\,|$$
$$+\sigma\,|\,(\Delta\psi_t,\psi_{tt})\,| +\nu\,|\,(\Delta^2\psi_t,\psi_{tt})\,|\le [\|\psi_t\|_1\max_S\,|\,\Delta\psi\,|$$
$$+(1+\frac{\sigma}{\sqrt{2}})\|\psi_t\|_2 + \max_S\,|\,\nabla\psi\,|\,\|\psi_t\|_2 + \nu\|\psi_t\|_3]\,\|\psi_{tt}\|_1.$$

Using the inequalities [46]

$$\max_S\,|\,u\,|\le \frac{1}{\sqrt{4\pi}}\|u\|_{H_0^2(S)}, \quad \max_S\,|\,\nabla u\,|\le \frac{1}{\sqrt{4\pi}}\|u\|_{H_0^3(S)}$$

and passing to operator $V(t,\psi_0)$, one can find

$$\|V_{tt}(t,\psi_0)\|_1 \le \frac{1}{\sqrt{8\pi}}\|V_t(t,\psi_0)\|_2\|V(t,\psi_0)\|_4 + (1+\frac{\sigma}{\sqrt{2}})\|V_t(t,\psi_0)\|_2$$

$$+\frac{1}{\sqrt{4\pi}}\|V(t,\psi_0)\|_3\|V_t(t,\psi_0)\|_2 + \nu\|V_t(t,\psi_0)\|_3.$$

If $\psi_0 \in \mathcal{A}$, then taking into account (4.31), (4.33), (4.37) and (4.38), we get

$$\|V_{tt}(t, \psi_0)\|_1 \leq \left(1 + \frac{\sigma}{\sqrt{2}} + \frac{1}{\sqrt{4\pi}} R_1 + \frac{1}{\sqrt{8\pi}} R_6 \right) R_3$$

$$+\nu R_7 \equiv R_8, \quad \forall \, \psi_0 \in \mathcal{A},$$

this means that (4.39) holds.

Thus, we have proved the smoothness of function, that form the attractor $\mathcal{A}$. These results will be used in the sequel for the estimate the rate of the convergence of the difference scheme on the attractor. We now proceed to the consideration of the semidiscrete numerical method for obtaining of the approximate solution of the problem (4.23)-(4.24).

Time difference scheme and its attractor. Let $\tau > 0$ be a time step, $t_k = k\tau$, $k = 0, 1 \dots$. For the simplicity we shall assume sometimes that

$$\tau = 1/N, \quad N \in \mathbf{N}. \tag{4.41}$$

We consider the difference scheme

$$\frac{\Delta \psi^k - \Delta \psi^{k-1}}{\tau} + J\left(\psi^{k-1}, \Delta \psi^k \right) +$$

$$2\psi_\lambda^k + \sigma \Delta \psi^k - \nu \Delta^2 \psi^k = f, \quad k = 1, 2, .., \tag{4.42}$$

$$\int_S \psi^k \, ds = 0, \quad \psi^0 = \psi_0 \, (\lambda, \mu). \tag{4.43}$$

It should be noted that (4.42) represents the linear equation with respect to the unknown function ψ^k and that the unique solvability of the problem (4.42)-(4.43) for all $\tau > 0$ and for every $\psi_0 \in H_0^2(S)$ follows from the theory of linear operators. Let $V_\tau(t_k, \cdot)$, $t_k = k\tau$ denote the solving operators of this problem.

Theorem 4.6. *The estimates hold*

$$\|V_\tau(t_k, \psi^0)\|_2^2 \leq \frac{1}{(1 + \tau\delta)^k} \|\psi^0\|_2^2 + \left(1 - \frac{1}{(1 + \tau\delta)^k} \right) R_0^2, \tag{4.44}$$

$$\tau \sum_{k=1}^N \|V_\tau(t_k, \psi^0)\|_3^2 \leq \frac{1}{\nu} \|\psi^0\|_2^2 + \frac{N\tau}{\nu\delta} \|f\|^2 = \frac{1}{\nu} \|\psi^0\|_2^2 + \frac{1}{\nu\delta} \|f\|^2 \tag{4.45}$$

and for all $t_k \geq 1$ the inequality is valid

$$\|V_\tau(t_k, \psi^0)\|_3 \leq r_\tau(\max\{\|\psi^0\|_2, R_0\}) \equiv r_\tau(a), \tag{4.46}$$

where δ and R_0 are constants from (4.27), while $r_\tau(\cdot)$ is the continuous nondecreasing function, which depends on $\nu, \sigma, \|f\|$ and τ.

Proof. By taking the scalar product of (4.42) with $\Delta\psi^k$ in $L_0^2(S)$, we have

$$\frac{1}{2\tau}(\|\psi^k\|_2^2 - \|\psi^{k-1}\|_2^2) + \frac{1}{2\tau}\|\psi^k - \psi^{k-1}\|_2^2 + \sigma\|\psi^k\|_2^2$$
$$+\nu\|\psi^k\|_3^2 = (f, \Delta\psi^k) \le \frac{\delta}{2}\|\psi^k\|_2^2 + \frac{1}{2\delta}\|f\|. \qquad (4.47)$$

Since $\|\psi^k\|_3^2 \ge \lambda_1\|\psi^k\|_2^2$, from (4.47) the inequality follows

$$\frac{1}{\tau}(\|\psi^k\|_2^2 - \|\psi^{k-1}\|_2^2) + \delta\|\psi^k\|_2^2 \le \frac{1}{\delta}\|f\|^2,$$

that is

$$\|\psi^k\|_2^2 \le \frac{1}{1+\tau\delta}\|\psi^{k-1}\|_2^2 + \frac{1}{\delta^2}(1 - \frac{1}{1+\tau\delta})\|f\|^2,$$

this gives (4.44). We find from (4.47) that

$$\|\psi^k\|_2^2 - \|\psi^{k-1}\|_2^2 + \tau\nu\|\psi^k\|_3^2 \le \frac{\tau}{\delta}\|f\|^2.$$

Summing over k from 1 to N and taking into account (4.41), we obtain (4.45).

Let $\tau \le 1/2$(or $N \ge 2$). By taking the scalar product of (4.42) with $t_{k-1}\Delta^2\psi^k$ in $L_0^2(S)$, we have

$$\frac{1}{2\tau}(t_{k-1}\|\psi^k\|_3^2 - t_{k-2}\|\psi^{k-1}\|_3^2) + \sigma t_{k-1}\|\psi^k\|_3^2 + \nu t_{k-1}\|\psi^k\|_4^2$$
$$\le t_{k-1}(J(\psi^{k-1}, \Delta\psi^k) - f, \Delta^2\psi^k) + \frac{1}{2}\|\psi^{k-1}\|_3^2, \quad k = 2, 3, \ldots.$$

Estimating values in the right side of the above inequality as in proof of Lemma 4.4, we find

$$t_{k-1}\|\psi^k\|_3^2 - t_{k-2}\|\psi^{k-1}\|_3^2 \le \tau\|\psi^{k-1}\|_3^2$$
$$+\frac{27\tau}{8\nu^3}\|\psi^{k-1}\|_2^4\|\psi^k\|_3^2 t_{k-1} + \frac{\tau}{\nu}\|f\|^2 t_{k-1}.$$

It follows from (4.44) that

$$\|\psi^k\|_2 = \|V_\tau(t_k, \psi^0)\|_2 \le \max\{\|\psi^0\|_2, R_0\} \equiv a, \quad \forall t_k, \qquad (4.48)$$

that is

$$t_{k-1}\|\psi^k\|_3^2 - t_{k-2}\|\psi^{k-1}\|_3^2 \le \tau\|\psi^{k-1}\|_3^2 + \frac{27\tau}{8\nu^3}a^4\|\psi^k\|_3^2 t_{k-1} + \frac{\tau}{\nu}\|f\|^2 t_{k-1}.$$

Let us sum over k from 2 to N in this inequality:

$$\tau(N-1)\|\psi^N\|_3^2 \le \left(1 + \frac{27}{8\nu^3}a^4\right)\tau\sum_{k=1}^N \|\psi^k\|_3^2 + \frac{(N-1)\tau}{2\nu}\|f\|^2.$$

Taking into account that $\tau N = 1$, we get

$$(1-\tau)\|\psi^N\|_3^2 \le \left(1 + \frac{27}{8\nu^3}a^4\right)\tau\sum_{k=1}^N \|\psi^k\|_3^2 + \frac{(1-\tau)}{2\nu}\|f\|^2.$$

Applying (4.45) for the estimating of sum in the right side of the inequality, we find that

$$\|V_\tau(1,\psi^0)\|_3^2 \le \frac{1}{(1-\tau)\nu}\left(1 + \frac{27}{8\nu^3}a^4\right)\left(a^2 + \frac{1}{\delta}\|f\|^2\right)$$
$$+ \frac{1}{2\nu}\|f\|^2 \equiv r_\tau^2(a), \ \tau \le \frac{1}{2}. \tag{4.49}$$

If $\tau = 1$, then (4.45) gives

$$\|V_1(1,\psi^0)\|_3^2 \le \nu^{-1}(a^2 + \delta^{-1}\|f\|^2) \equiv r_1^2(a). \tag{4.50}$$

We have for any $t_k = k\tau > 1$

$$V_\tau(t_k,\psi^0) = V_\tau(1, V_\tau(t_k - 1, \psi^0)).$$

According to (4.48), one has

$$\|V_\tau(t_k,\psi^0)\|_3^2 = \|V_\tau(1, V_\tau(t_k - 1, \psi^0))\|_3^2$$
$$\le r_\tau^2(\max\{\|V_\tau(t_k - 1, \psi^0)\|_2, R_0\}) \le r_\tau^2(a),$$

i.e. the estimate (4.46) hold for all $t_k \ge 1$.

Lemma. 4.8. *Operators $V_\tau(t_k, \cdot)$ are continuous in $H_0^2(S)$ for any* $t_k = k\tau > 0$.

Proof. Let $\psi^1 = V_\tau(\tau, \psi^0)$, $\phi^1 = V_\tau(\tau, \phi^0)$ and let $g = \phi^1 - \psi^1$. Then for g the equation fulfills

$$\frac{\Delta g - \Delta g^0}{\tau} + J(g^0, \Delta\phi^1) + J(\psi^0, \Delta g) + 2g_\lambda + \sigma\Delta g - \nu\Delta^2 g = 0, \tag{4.51}$$

where $g^0 = \phi^0 - \psi^0$.

By taking the scalar product of (4.51) with Δg in $L_0^2(S)$, we obtain

$$\frac{1}{2\tau}(\|g\|_2^2 - \|g^0\|_2^2) + \frac{1}{2\tau}\|g - g^0\|_2^2 + \sigma\|g\|_2^2 + \nu\|g\|_3^2 = (J(\Delta g, g^0), \Delta\phi^1)$$

$$\|g^0\|_2\|g\|_3\|\Delta\phi^1\|_{L^4(S)} \le \nu\|g\|_3^2 + \frac{1}{2\sqrt{2\nu}}\|g^0\|_2^2\|\phi^1\|_2\|\phi^1\|_3,$$

that is,

$$\|g\|_2^2 \leq \|g^0\|_2^2 \left(1 + \frac{\tau}{\sqrt{2}\nu}\|\phi^1\|_2\|\phi^1\|_3\right) \to 0 \text{ for } \|g^0\|_2 = \|\phi^0 - \psi^0\|_2 \to 0,$$

i.e. the operator $V_\tau(\tau, \cdot)$ is continuous on $H_0^2(S)$.

The operator $V_\tau(t_k, \cdot)$ represents the superposition of k operators $V_\tau(\tau, \cdot)$, therefore $V_\tau(t_k, \cdot)$ is continuous on $H_0^2(S)$ for any $k \in \mathbf{N}$. We get the estimation

$$\|V_\tau(N\tau, \psi^0) - V_\tau(N\tau, \phi^0)\|_2^2 \leq \|\psi^0 - \phi^0\|_2^2 \left(1 + \frac{\tau}{\nu\sqrt{2}}ar_\tau(a)\right)^N.$$

Theorem 4.7. *The difference scheme* (4.42)-(4.43) *possesses a global attractor* $\mathcal{A}_\tau$ *in the space* $H_0^2(S)$ *and the estimates hold*

$$\max_{\rho \in \mathcal{A}_\tau} \|\rho\|_2 \leq R_0, \tag{4.52}$$

$$\max_{\rho \in \mathcal{A}_\tau} \|\rho\|_3 \leq R^1(\tau) \equiv r_\tau(R_0), \tag{4.53}$$

where R_0 *is the constant from* (4.27), *while the functions* $r_\tau(\cdot)$ *are defined in* (4.49) − (4.50).

Proof. By Lemma 4.8. the operators $V_\tau(t_k, \cdot)$ are continuous on $H_0^2(S)$ for all $t_k > 0$. Let $X = \{\phi \in H_0^2(S) : \|\phi\|_2 \leq R_0\}$ be the ball of radius R_0 in $H_0^2(S)$ centered at zero and let $Q_\varepsilon(X)$ be ε-neighborhood X in $H_0^2(S)$, i.e. the ball of radius $R_0 + \varepsilon$. We shall show that for any $\varepsilon > 0$ set $Q_\varepsilon(X)$ is absorbing set for the semigroup of operators $V_\tau(t_k, \cdot)$. Indeed, (4.44) gives

$$V_\tau(t_k, \psi^0) \in Q_\varepsilon(X) \quad \forall t_k, \text{ if } \psi^0 \in Q_\varepsilon(X);$$

$V_\tau(t_k, \psi^0) \in Q_\varepsilon(X)$ for k

$$\geq \ln\left[\frac{\|\psi^0\|_2^2 - R_0^2}{\varepsilon(2R_0 + \varepsilon)}\right] / \ln(1 + \tau\delta),$$

if $\psi^0 \notin Q_\varepsilon(X)$.

Let ϕ_n be a bounded sequence in $H_0^2(S)$ such that $\|\phi_n\|_2 \leq b, \forall n$, and $t_{k_n} = k_n\tau \to \infty$ as $n \to \infty$.

It follows from (4.46) that $\|V_\tau(t_{k_n}, \phi_n)\|_3 \leq r_\tau(\max\{b, R_0\})$ for all sufficiently large n. This means that the sequence $V_\tau(t_{k_n}, \phi_n)$ is precompact in $H_0^2(S)$. Thus, we have shown that for the semigroup of operators $V_\tau(k\tau, \cdot)$, that acts on the space $H_0^2(S)$, all conditions of Theorem 1.1 are satisfied, i.e. the problem (4.42)-(4.43) possesses the nonempty compact attractor $\mathcal{A}_\tau$, which is the maximal invariant bounded set in $H_0^2(S)$.

It follows from the above reasoning that $\mathcal{A}_\tau \subset Q_\varepsilon(X)$, $\forall\, \varepsilon > 0$, i.e. $\mathcal{A}_\tau \subset X$, that is (4.52) holds. Let $X_1(\tau) = \{\phi \in X : \|\phi\|_3 \le r_\tau(R_0)\}$. According to (4.46) we have $V_\tau(1, X) \subset X_1(\tau)$.

Hence $V_\tau(1, \mathcal{A}_\tau) \subset X_1(\tau)$, that is $\mathcal{A}_\tau = V_\tau(1, \mathcal{A}_\tau) \subset X_1(\tau)$, i.e. (4.53) holds. In the sequel we shall need the propositions about the properties of the difference scheme (4.42)-(4.43).

Theorem 4.8. *For every $\psi^0, \phi^0 \in X_1(1/2)$ and $\tau \le \tau_1^*$ the estimates hold*

$$\max_{1 \le k \le N} \|V_\tau(t_k, \psi^0)\|_3 \le C_1, \tag{4.54}$$

$$\tau \sum_{k=1}^N \|V_\tau(t_k, \psi^0)\|_4^2 \le C_2, \tag{4.55}$$

$$\max_{1 \le k \le N} \left\| \frac{V_\tau(t_k, \psi^0) - V_\tau(t_{k-1}, \psi^0)}{\tau} \right\|_1 \le C_3, \tag{4.56}$$

$$\|V_\tau(1, \phi^0) - V_\tau(1, \psi^0)\|_2 \le L_0\|\phi^0 - \psi^0\|_2, \tag{4.57}$$

where C_1, C_2, C_3 and L_0 are the constants that depend on σ, ν and $\|f\|$,

$$\tau_1^* = \frac{1}{2}\left(\frac{27}{8\nu^3} R_0^4 - 2\sigma \right)^{-1}.$$

Proof. By taking the scalar product of (4.42) with $\Delta^2\psi^k$ in $L_0^2(S)$, we obtain

$$\frac{1}{2\tau}\|\psi^k - \psi^{k-1}\|_3^2 + \frac{1}{2\tau}(\|\psi^k\|_3^2 - \|\psi^{k-1}\|_3^2) + \sigma\|\psi^k\|_3^2 + \nu\|\psi^k\|_4^2$$
$$= (J(\psi^{k-1}, \Delta\psi^k) - f, \Delta^2\psi^k).$$

Estimating the expression in the right side of the equality as in Lemma 1.1, we see that (4.54) and (4.55) hold. The inequality (4.56) follows immediately from the equation (4.74) and it is established by taking the scalar product of (4.42) with $\frac{\psi^k - \psi^{k-1}}{\tau}$ in $L_0^2(S)$. In proving Lemma 4.8. we arrived at the next relation

$$\|V_\tau(\tau, \phi^0) - V_\tau(\tau, \psi^0)\|_2^2 \le \left(1 + \frac{\tau}{\sqrt{2\nu}}\|\phi^1\|_2\|\phi^1\|_3 \right) \|\phi^0 - \psi^0\|_2^2.$$

Taking into account (4.54), we find that

$$\|V_\tau(1, \phi^0) - V_\tau(1, \psi^0)\|_2^2 \le \left(1 + \frac{\tau}{\sqrt{2\nu}} R_0 C_1 \right)^N \|\phi^0 - \psi^0\|_2^2$$

$$= \left(1 + \frac{\tau}{\sqrt{2\nu}} R_0 C_1 \right)^{1/\tau} \|\phi^0 - \psi^0\|_2^2 \le e^{\frac{R_0 C_1}{\sqrt{2\nu}}} \|\phi^0 - \psi^0\|_2^2.$$

Thus, (4.57) is true with the constant $L_0 = \exp\left(\frac{R_0 C_1}{2\sqrt{2}\nu}\right)$.

Theorem 4.9. *For all sufficiently small $\tau \leq \tau_0$ the inequality holds*

$$\|V_\tau(\tau, \phi^0) - V_\tau(\tau, \psi^0)\|_1 \leq (1 + \tau\alpha)\|\phi^0 - \psi^0\|_1, \ \forall\, \phi^0, \psi^0 \in X, \quad (4.58)$$

where

$$\alpha = \max\left\{0, -2\sigma + \frac{R_0^2}{4\nu^2}\left(R_0^2 + \frac{\|f\|^2}{2\delta}\right) + \frac{27}{8\nu^3}R_0^4\right\},$$

$$\tau_0 = \min\{\frac{1}{2}, \frac{1}{2\alpha}\}, \ \text{if } \alpha > 0; \ \tau_0 = 1/2 \ \text{as } \alpha = 0.$$

Proof. Let $g(\lambda, \mu)$ be the function introduced in proving Lemma 2.1. By taking the scalar product of (4.51) with g in $L_0^2(S)$, we obtain

$$\frac{1}{2\tau}(\|g\|_1^2 - \|g^0\|_1^2) + \frac{1}{2\tau}\|g - g^0\|_1^2 + \sigma\|g\|_1^2$$
$$+ \nu\|g\|_2^2 = (J(g^0, \Delta\phi^1) + J(\psi^0, \Delta g), g).$$

Let us estimate the expression

$$|\,(J(g^0, \Delta\phi^1), g)\,| = |\,(J\,(g - g^0, g), \Delta\phi^1)\,|$$
$$\leq \|g - g^0\|_1 \|\nabla g\|_{L^4(S)} \|\Delta\phi^1\|_{L^4(S)} \leq \frac{1}{2\tau}\|g - g^0\|_1^2$$
$$+ \tau\|g\|_1\|g\|_2\|\phi^1\|_2\|\phi^1\|_3 \leq \frac{1}{2\tau}\|g - g^0\|_1^2 + \frac{\nu}{2}\|g\|_2^2 + \frac{\tau^2}{2\nu}\|g\|_1^2\|\phi^1\|_2^2\|\phi^1\|_3^2.$$

Further,

$$|\,(J(\psi^0, \Delta g), g)\,| \leq \|\nabla\psi^0\|_{L^4(S)}\|\nabla g\|_{L^4(S)}\|g\|_2 \leq 2^{1/4}\|\psi^0\|_2\|g\|_1^{1/2}\|g\|^{3/2}$$
$$\leq \frac{\nu}{2}\|g\|_2^2 + \frac{27}{16\nu^3}R_0^4\|g\|_1^2.$$

Thus,

$$\frac{1}{2\tau}(\|g\|_1^2 - \|g^0\|_1^2) + \sigma\|g\|_1^2 \leq \frac{\tau^2}{2\nu}\|\phi^1\|_2^2\|\phi^1\|_3^2\|g\|_1^2 + \frac{27}{16\nu^3}R_0^4\|g\|_1^2.$$

It follows from Theorem 2.1 and the definition of the set X that

$$\|\phi^1\|_2 \leq R_0, \ \tau\|\phi^1\|_3^2 \leq \frac{1}{\nu}\left(R_0^2 + \frac{\tau}{\delta}\|f\|^2\right),$$

that is

$$\|g\|_1^2[1 - \tau\gamma(\tau)] \leq \|g^0\|_1^2,$$

where

$$\gamma(\tau) = -2\sigma + \frac{\tau^2}{\nu^2}\, R_0^2\, (R_0^2 + \frac{\tau}{\delta}\, \|f\|^2) + \frac{27}{8\nu^3}\, R_0^4 \leq \alpha \text{ for } \tau \leq \frac{1}{2},$$

$$\frac{1}{1 - \tau\gamma(\tau)} \leq 1 + \frac{\alpha\tau}{1 - \alpha\tau} \leq 1 + 2\alpha\tau \text{ for } \tau \leq \tau_0.$$

Thus,

$$\|g\|_1 \leq (1 - \tau\gamma(\tau))^{-1/2}\|g^0\|_1 \leq (1 + \alpha\tau)\|g^0\|_1$$

and by the definition of the function g (4.58) holds.

Convergence of difference scheme. We now consider the convergence of the solutions obtained by the difference scheme (4.42)-(4.43) to the solution of the initial problem (4.23)-(4.24), when τ tends to zero. The previous reasoning leads to the following

Theorem 4.10. *For any $\psi_0 \in X$ the estimates hold*

$$\max_{0 \leq t \leq 1} \|V(t, \psi_0)\|_2 \leq R_0, \quad \int_0^1 \|V(t, \psi_0)\|_3^2\, dt \leq R_2^2,$$

$$\int_0^1 \|V_t(t, \psi_0)\|_1^2\, dt \leq \bar{R}_1^2, \quad \max_{0 \leq t \leq 1} \|V_t(t, \psi_0)\| \leq \bar{R}_0,$$

and if $\psi_0 \in X_1(1/2)$, then

$$\max_{0 \leq t \leq 1} \|V(t, \psi^0)\|_3 \leq \bar{C}_1, \quad \int_0^1 \|V(t, \psi_0)\|_4^2\, dt \leq \bar{C}_2,$$

$$\max_{0 \leq t \leq 1} \|V_t(t, \psi_0)\|_1 \leq \bar{C}_3, \quad \int_0^1 \|V_t(t, \psi_0)\|_2^2\, dt \leq \bar{C}_4.$$

where R_0 and R_2 are defined in (4.27) and (4.31), $\bar{R}_0, \bar{R}_1, \bar{C}_i$ are constants that depend on ν, σ and $\|f\|$.

Let

$$\phi^k(\lambda, \mu) = \frac{1}{\tau} \int_{t_{k-1}}^{t_k} V(t, \psi_0)\, dt, \quad k = \overline{1, N}, \quad \phi^0 = \psi_0. \tag{4.59}$$

Lemma 4.9. *For all $\psi_0 \in X$ the inequality holds*

$$\tau \sum_{k=2}^N \|\frac{\phi^k - \phi^{k-1}}{\tau}\|_1^2 \leq 2\bar{R}_1^2. \tag{4.60}$$

Hereafter, we denote by c the different constants that depend on ν, σ and $\|f\|$.

Theorem 4.11. *For*

$$\tau \leq \tau^* = \frac{1}{2}\left(-2(\sigma + \nu) + \frac{27 R_0^4}{\nu^3}\right)^{-1}$$

the estimates are valid

$$\|V(1,\psi_0) - V_\tau(1,\psi_0)\|_1 \leq c\tau^{1/2}, \qquad \forall\, \psi_0 \in X, \qquad (4.61)$$

$$\|V(1,\psi_0) - V_\tau(1,\psi_0)\|_2 \leq c\tau^{1/2}, \qquad \forall\, \psi_0 \in X_1(1/2). \qquad (4.62)$$

Proof. Let $\psi(t) = V(t,\psi_0)$, $\psi^k = V_\tau(t_k,\psi_0)$, ϕ^k be the functions defined in (4.59) and let $g^k = \phi^k - \psi^k$. We integrate (4.23) over time from t_{k-1} to t_k and devide the result by τ. Subtracting from it the equation (4.42), we find that

$$\frac{\Delta g^k - \Delta g^{k-1}}{\tau} + J(g^{k-1}, \Delta\psi^k) + J(\phi^{k-1}, \Delta g^k) + 2g_\lambda^k + \sigma\Delta g^k$$

$$-\nu\Delta^2 g^k = \frac{a_k - a_{k-1}}{\tau} + b_k + d_k, \quad g^0 = 0, \qquad (4.63)$$

where

$$a_k = -\frac{1}{\tau}\int\limits_{t_{k-1}}^{t_k} \Delta\psi_t(t - t_{k-1})dt, \, k = \overline{1,N}, \quad a_0 = 0,$$

$$b_k = -\frac{1}{\tau}\int\limits_{t_{k-1}}^{t_k} J(\int\limits_{t_{k-1}}^{t} \psi_t(t')dt', \, \Delta\psi(t))dt,$$

$$d_k = \frac{1}{\tau^2}J(\int\limits_{t_{k-2}}^{t_{k-1}} \psi_t(t - t_{k-2})dt, \, \int\limits_{t_{k-1}}^{t_k} \Delta\psi\, dt), \, k = \overline{2,N}, \quad d_1 = 0.$$

By taking the scalar product of (4.63) with g^k in $L_0^2(S)$, we obtain

$$\frac{1}{2\tau}\left(\|g^k\|_1^2 - \|g^{k-1}\|_1^2\right) + \frac{1}{2\tau}\|g^k - g^{k-1}\|_1^2 + \sigma\|g^k\|_1^2$$

$$+\nu\|g^k\|_2^2 = (J(g^{k-1}, \Delta\psi^k) + J(\phi^{k-1}, \Delta g^k), g^k)$$

$$-\left(\frac{a_k - a_{k-1}}{\tau} + b_k + d_k, \, g^k\right). \qquad (4.64)$$

Estimate the values in the right side of the equality:

$$|\,(J(g^{k-1}, \Delta\psi^k), \, g^k)\,| \leq \frac{1}{2\tau}\|g^k - g^{k-1}\|_1^2 + \frac{\nu}{4}\|g^k\|_2^2 + \frac{27\tau^2}{\nu^3}R_0^4\|\psi^k\|_3^2$$

$$| (J(\phi^{k-1}, \Delta g^k), g^k) | \leq \frac{\nu}{4} \|g^k\|_2^2 + \frac{27}{2\nu^3} R_0^4 \|g^k\|_1^2$$

$$| (b_k, g^k) | \leq \sqrt{2} R_0 \left(\int_{t_{k-1}}^{t_k} \|\psi\|_3^2 dt \int_{t_{k-1}}^{t_k} \|\psi_t\|_1^2 dt \right)^{1/2},$$

$$| (d_k, g^k) | \leq \frac{2R_0}{\sqrt{3}} \left(\int_{t_{k-2}}^{t_{k-1}} \|\psi_t\|_1^2 dt \int_{t_{k-1}}^{t_k} \|\psi\|_3^2 dt \right)^{1/2}.$$

Combining these inequalities with (4.64) and summing over k, we get

$$\|g^k\|_1^2 + 2\sigma\tau \sum_{l=1}^{k} \|g^l\|_1^2 + \nu\tau \sum_{l=1}^{k} \|g^l\|_2^2 \leq \frac{27}{\nu^3} R_0^4 \tau \sum_{l=1}^{k} \|g^l\|_1^2$$

$$+ c\tau^3 \sum_{l=1}^{k} \|\psi^l\|_3^2 + c\tau \left(\int_0^{t_k} \|\psi_t\|_1^2 dt \right)^{1/2} \left(\int_0^{t_k} \|\psi\|_3^2 dt \right)^{1/2}$$

$$+ 2 \sum_{l=1}^{k-1} (a_l, g^{l+1} - g^l) - 2(a_k, g^k). \tag{4.65}$$

From (4.44)-(4.45) and (4.42) it follows that

$$\tau \sum_{k=1}^{N} \|\frac{\psi^k - \psi^{k-1}}{\tau}\|_1^2 \leq c, \text{ if } \psi_0 \in X.$$

Taking into account (4.92) and the definition g^k, we conclude that

$$\tau \sum_{k=2}^{N} \|\frac{g^k - g^{k-1}}{\tau}\|_1^2 \leq c.$$

Then

$$| \sum_{l=1}^{k-1} (a_l, g^{l+1} - g^l) | \leq \frac{\tau}{\sqrt{3}} \left(\int_0^{t_{k-1}} \|\psi_t\|_1^2 dt \right)^{1/2} \left(\tau \sum_{l=2}^{k} \|\frac{g^l - g^{l-1}}{\tau}\|_1^2 \right)^{1/2}$$

$$\leq c\tau \left(\int_0^{t_{k-1}} \|\psi_t\|_1^2 dt \right)^{1/2}.$$

We have for the last term in the right side of (4.65)

$$| (a_k, g^k) | \leq \frac{1}{\tau} \| \int_{t_{k-1}}^{t_k} \psi_t(t - t_{k-1}) dt \| \|g^k\|_2 \leq \frac{2R_0}{\sqrt{3}} \tau \max_{0 < t \leq 1} \|\psi_t(t)\|.$$

Taking into account (4.45) and the estimates of Theorem 3.1, we find that for all $\psi_0 \in X$ and $t_k \leq 1$

$$\|g^k\|_1^2 + 2\sigma\tau \sum_{l=1}^{k} \|g^l\|_1^2 + \nu\tau \sum_{l=1}^{k} \|g^l\|_2^2 \leq \frac{27}{\nu^3} R_0^4 \tau \sum_{l=1}^{k} \|g^l\|_1^2 + c\tau. \quad (4.66)$$

It is easy to see that for $\tau \leq \tau^*$ from (4.66) the inequalities follow

$$\max_{1 \leq k \leq N} \|g^k\|_1^2 \leq c\tau, \quad \tau \sum_{k=1}^{N} \|g^k\|_2^2 \leq c\tau, \quad \forall \psi_0 \in X. \quad (4.67)$$

For the function ϕ^N from (4.59) the identity is true

$$\phi^N = \psi(1) - \frac{1}{\tau} \int_{1-\tau}^{1} \psi_t(t - 1 + \tau) dt,$$

hence,

$$\|V(1, \psi_o) - \phi^N\|_1 = \frac{1}{\tau}\| \int_{1-\tau}^{1} V_t(t, \psi_0)(t - 1 + \tau) dt\|_1 \leq \frac{\bar{R}_1}{\sqrt{3}} \tau^{1/2},$$

this estimate with (4.67) gives (4.61). The inequality (4.62) is established by the product of (4.95) with Δg^k.

Theorem 4.12. *For*

$$\tau \leq \tau_1^* = \frac{1}{2} \left(\frac{27}{8\nu^3} R_0^4 - 2\sigma \right)^{-1}$$

the estimate is valid

$$\|V(\tau, \psi_0) - V_\tau(\tau, \psi_0)\|_1 \leq c_0 \tau^{3/2}, \quad \forall \psi_0 \in \mathcal{A}, \quad (4.68)$$

where c_0 is a constant that depends on ν, σ and $\|f\|$.

Proof. Let ψ_0 belong to the attractor $\mathcal{A}$ of the problem (4.23)-(4.24), $\psi(t) = V(t, \psi_0)$, $\psi_1 = \psi(\tau) = V(\tau, \psi_0)$, $\psi^1 = V_\tau(\tau, \psi_0)$ and $g = \psi_1 - \psi^1$. Integrate (4.23) over between 0 and τ and devide the result by τ and subtracting from it the equation (4.42), we get

$$\frac{\Delta g}{\tau} + J(\psi_0, \Delta g) + 2g_\lambda + \sigma\Delta g - \nu\Delta^2 g = \sum_{i=1}^{5} a_i. \quad (4.69)$$

Here

$$a_1 = -\frac{1}{\tau} \int\limits_0^\tau J\left(\int\limits_0^t \psi_t(t')\,dt', \Delta\psi(t) \right) dt,$$

$$a_2 = \frac{1}{\tau} J\left(\psi_0, \int\limits_0^\tau \Delta\psi_t\, t\, dt \right), \quad a_3 = \frac{2}{\tau} \int\limits_0^\tau \psi_{\lambda t} t\, dt,$$

$$a_4 = \frac{\sigma}{\tau} \int\limits_0^\tau \Delta\psi_t\, t\, dt, \quad a_5 = -\frac{\nu}{\tau} \int\limits_0^\tau \Delta^2\psi_t\, t\, dt.$$

From (4.33) and (4.56), it follows that

$$\left\| \frac{g}{\tau} \right\|_1 \le \left\| \frac{V(\tau,\psi_0) - \psi_0}{\tau} \right\|_1 + \left\| \frac{V_\tau(\tau,\psi_0) - \psi_0}{\tau} \right\|_1 \le C_3 + \frac{R_3}{\sqrt{2}}. \quad (4.70)$$

By taking the scalar product of (4.69) with g in $L_0^2(S)$, we obtain

$$\frac{1}{\tau}\|g\|_1^2 + \sigma\|g\|_1^2 + \nu\|g\|_2^2 = \left(J(\psi_0, \Delta g) - \sum_{i=1}^5 a_i, g \right).$$

Using (4.70) and the above results, we estimate the values on the right side of

$$|\, (J(\psi_0, \Delta g), g)\,| \le 2^{1/4}\|\psi_0\|_2 \|g\|_1^{1/2} \|g\|_2^{3/2} \le \nu\|g\|_2^2 + \frac{27\tau^2}{128\nu^3} R_0^4 \left\| \frac{g}{\tau} \right\|_1^2$$

$$\le \nu\|g\|_2^2 + c\tau^2,\ |\,(a_1 g)\,| \le \int\limits_0^\tau \left\| \int\limits_0^t \psi_t(t')dt' \right\|_2 \|\psi(t)\|_3 dt\, \frac{\|g\|_1}{\tau} \le c\tau^2,$$

$$|\,(a_2, g)\,| \le \|\psi_0\|_2 \left\| \int\limits_0^\tau \psi_t\, t\, dt \right\|_3 \frac{\|g\|_1}{\tau} \le c\tau^2,$$

$$|\,(a_3, g)\,| \le 2\frac{\|g\|}{\tau} \left\| \int\limits_0^\tau \psi_{\lambda t} t\, dt \right\| \le c\tau^2,$$

$$|\,(a_4, g)\,| \le \sigma\frac{\|g\|}{\tau} \left\| \int\limits_0^\tau \Delta\psi_t t\, dt \right\| \le c\tau^2,\ |\,(a_5, g)\,|$$

$$\le \nu\left\| \frac{g}{\tau} \right\|_1 \left\| \int\limits_0^\tau \psi_t\, t\, dt \right\|_3 \le c\tau^2.$$

Thus, $\|g\|_1^2 \le c\tau^3$ and by the definition of the function g (4.68) holds.

Global stability of scheme (4.42)-(4.43). It was shown above that the discrete time scheme (4.42)-(4.43) possesses the attractor $\mathcal{A}_\tau$ and that the solutions obtained by this scheme approach for the small time step τ to the solutions of the barotropic vorticity equation (4.23).

Now we show that the attractors $\mathcal{A}_\tau$ also approach to the attractor $\mathcal{A}$ of the initial problem. Namely, we shall establish the theorem on the global stability of the scheme (4.42)-(4.43). The global stability means that for all sufficiently small τ the attractors $\mathcal{A}_\tau$ enter any given before neighborhood $\mathcal{A}$. Let us denote by

$$\mathrm{dist}_{H_0^2(S)}(f,\mathcal{A}) = \inf_{\psi_0 \in \mathcal{A}} \|f - \psi_0\|_2, \quad \forall f \in H_0^2(S)$$

the distance of function f to the attractor $\mathcal{A}$;

$$Q_\varepsilon(\mathcal{A}) = \{f \in H_0^2(S) : \mathrm{dist}_{H_0^2(S)}(f,\mathcal{A}) \le \varepsilon\}$$

ε-neighborhood $\mathcal{A}$ in the space $H_0^2(S)$;

$$\mathrm{dist}_{H_0^2(S)}(B,\mathcal{A}) = \sup_{f \in B} \mathrm{dist}_{H_0^2(S)}(f,\mathcal{A})$$

is the distance of the bounded closed set $B \subset H_0^2(S)$ to $\mathcal{A}$.

Theorem 4.13. *Let $\tau = 1/N$, $N \in \mathbf{N}$, then for any $\varepsilon > 0$ there can be found N_ε such that*

$$\mathcal{A}_\tau \subset Q_\varepsilon(\mathcal{A}), \quad \forall N \ge N_\varepsilon.$$

Proof. Assume the converse. Let for some $\varepsilon > 0$ there exist a sequence $N_n \to +\infty$ such that

$$\mathrm{dist}_{H_0^2(S)}(\mathcal{A}_{\tau_n}, \mathcal{A}) > \varepsilon,$$

where $\tau_n = 1/N_n \to 0$ as $n \to \infty$. Denote by $V_n(t,\cdot) = V_{\tau_n}(t,\cdot)$, $t \in \mathbf{Z}_+$ and consider the sequence of SDS $\{V_n, \mathbf{Z}_+, H_0^2(S)\}$, in which to the value of the parameter $n = \infty$ there corresponds SDS $\{V, \mathbf{Z}_+, H_0^2(S)\}$. In accordance with Theorem 2.2 the union of attractors $\cup_{n \in \mathbf{N}} \mathcal{A}_{\tau_n}$ is bounded in $H_0^3(S)$, i.e. it is the precompact set in $H_0^2(S)$. Let $\rho_n \in \mathcal{A}_{\tau_n}$ and $\rho_n \to \psi_0$ strongly in $H_0^2(S)$. By the use (4.57) and (4.62), we find that for all sufficiently large n

$$\|V_n(1,\rho_n) - V(1,\psi_0)\|_2 \le L_0 \|\rho_n - \psi_0\|_2 + c\tau_n^{1/2},$$

i.e. $V_n(1,\rho_n) \to V(1,\psi_0)$ strongly in $H_0^2(S)$ as $n \to \infty$.

Thus, the sequence of SDS under consideration satisfies the conditions of Theorem 1.2 and $\mathrm{dist}_{H_0^2(S)}(\mathcal{A}_{\tau_n},\mathcal{A}) \to 0$ as $n \to \infty$. The contradiction so obtained proves the theorem.

It should be noted that all above results are true also for completely implicit scheme

$$\frac{\Delta\psi^k - \Delta\psi^{k-1}}{\tau} + J(\psi^k, \Delta\psi^k + 2\mu) + \sigma\Delta\psi^k - \nu\Delta^2\psi^k = f,$$

which also is the globally stable time difference scheme for the barotropic vorticity equation on sphere.

Approximation of the attractor $\mathcal{A}$ by uniformly asymptotically stable sets of the difference scheme. We define on the set $X = \{\psi(\lambda,\mu) \in H_0^2(S) : \|\psi\|_2 \leq R_0\}$ the metric

$$\operatorname{dist}(\phi,\rho) = \|\phi - \rho\|_1, \quad \forall\,\phi,\rho \in X.$$

Then X becomes the convex compact metric space.

For the compact sets $A, B \subset X$ we denote by

$$\operatorname{dist}_H(A, B) = \max\{\operatorname{dist}(A, B), \operatorname{dist}(B, A)\}$$

the Hausdorff metric, where

$$\operatorname{dist}(\phi, A) = \inf_{u \in A} \|\phi - u\|_1, \quad \operatorname{dist}(B, A) = \sup_{\rho \in B} \operatorname{dist}(\rho, A),$$

$\mathcal{O}(A; r) = \{\phi \in X : \operatorname{dist}(\phi, A) < r\}$ r-neighborhood A in X,

is the distance of A to B.

Definition 4.1. The compact set $M \subset X$ is a uniformly asymptotically stable set for SDS $\{W, \mathcal{T}_+, X\}$, if

1) M is positively invariant, i.e. $W(t, M) \subset M, \forall t \in \mathcal{T}_+$;

2) M is uniformly stable, i.e. $\forall\,\varepsilon > 0$ there exists $\delta = \delta(\varepsilon) > 0$ such that

$$W(t, \mathcal{O}(M; \delta)) \subset \mathcal{O}(M; \varepsilon), \qquad \forall\,t \in \mathcal{T}_+;$$

3) M uniformly attracts the trajectories from some its neighborhood, i.e there exists $\delta_0 > 0$ such that

$$\forall\,\varepsilon > 0 \,\exists\, T = T(\varepsilon) : \quad W(t, \mathcal{O}(M; \delta_0)) \subset \mathcal{O}(M; \varepsilon), \,\forall\,t \in \mathcal{T}_+, t > T(\varepsilon).$$

The solving operators $V(t, \cdot)$ of the problem (4.23)-(4.24) map X into X by virtue of (4.29). It is easy to verify that the restriction to X of these operators form the uniformly Lipschitz continuous semigroup, i.e. for any $t \in \mathbf{R}_+$ there can be found a constant $L(t)$ such that

$$\|V(t, u') - V(t, u)\|_1 \leq L(t)\|u' - u\|_1, \,\forall\,u', u \in X. \tag{4.71}$$

The attractor $\mathcal{A}$ of the problem (4.23)-(4.24) is contained in X by (4.27). Consequently, $\mathcal{A}$ also is the attractor of SDS $\{V, \mathbf{R}_+, X\}$.

Lemma 4.10. *The attractor $\mathcal{A}$ is uniformly asymptotically stable set for SDS $\{V, \mathbf{R}_+, X\}$.*

Proof. Condition 1) of Definition 5.1 is fulfilled, since $\mathcal{A}$ is an invariant set. By definition $\mathcal{A}$ attracts any bounded set of X and the X itself. We show that $\mathcal{A}$ is an uniformly stable set.

Assume the converse, let for some $\varepsilon > 0$ there exist sequences $\{u_k\} \subset X$ and $\{t_k\} \subset \mathbf{R}_+$ such that dist $(u_k, \mathcal{A}) \to 0$ as $k \to \infty$ and dist$(V(t_k, u_k), \mathcal{A}) \geq \varepsilon$. Then the sequence $\{t_k\}$ is bounded. Indeed, $\mathcal{A}$ attracts X, i.e. there can be found $T(\varepsilon)$ such that dist $(V(t, X), \mathcal{A}) < \varepsilon, \forall t > T(\varepsilon)$. Therefore $t_k \leq T(\varepsilon)$. We choose the subsequence $u_{k_i} \to u_0$ as $i \to \infty$. Then $u_0 \in \mathcal{A}$, that is $V(t_k, u_0) \in \mathcal{A}, \forall t_k$. Using (4.103), we find

$$\text{dist}(V(t_{k_i}, u_{k_i}), \mathcal{A}) \leq \text{dist}(V(t_{k_i}, u_{k_i}), V(t_{k_i}, u_0))$$
$$\leq L(t_{k_i}) \|u_{k_i} - u_0\|_1 \leq \max_{0 \leq t \leq T(\varepsilon)} L(t) \|u_{k_i} - u_0\|_1 \to 0 \text{ as } i \to \infty.$$

The contradiction so obtained proves that the set $\mathcal{A}$ is uniformly stable.

The inequality (4.44) shows that the solving operators $V_\tau(t_k, \cdot)$ of the difference scheme (4.42)-(4.43) also map X into X. We ensure that the restriction to X of these operators form the uniformly continuous semigroup in X.

Lemma 4.11. *For any $\tau = 1/N$ and any $t \in \mathbf{Z}_+\tau$ the operators $V_\tau(t, \cdot)$ are the uniformly continuous operators on X, i.e. for any $\varepsilon > 0$ there can be found $\delta_\varepsilon = \delta_\varepsilon(\tau, t) > 0$ such that*

$$\text{dist}(V_\tau(t, u'), V_\tau(t, u)) \leq \varepsilon, \ \forall u', u \in X : \text{dist}(u', u) \leq \delta_\varepsilon. \quad (4.72)$$

Proof. It follows from Theorem 2.4 that the operators $V_\tau(t, \cdot)$ are uniformly Lipschitz continuous on X for $\tau \leq \tau_0$. Let $\tau_0 < \tau \leq 1$. We show that the operator $V_\tau(\tau, \cdot)$ is continuous on X. Consider the sequence $u_k \to u$ in X and denote by $\psi_k = V_\tau(\tau, u_k) \in X$.

As in the proof of Theorem 2.1, one can derive from (4.42) the estimate

$$\tau\nu\|\psi_k\|_3^2 \leq \|u_k\|_2^2 + \frac{\tau}{\delta}\|f\|^2 \leq R_0^2 + \frac{\tau}{\delta}\|f\|^2, \quad \|\psi_k\|_3 \leq \tilde{c} = \tilde{c}(\tau),$$

i.e. the sequence ψ_k is bounded in $H_0^3(S)$. We choose the subsequence $\psi_{k_n} \to v \in X$, weakly convergent in $H_0^3(S)$ and strongly convergent in $H_0^2(S)$, then for any $\rho \in H_0^1(S)$ we have

$$| (J(u, \Delta v) - J(u_{k_n}, \Delta\psi_{k_n}), \rho) | \leq | (J(u - u_{k_n}, \Delta v), \rho) |$$
$$+ | (J(u, \Delta(v - \psi_{k_n})), \rho) | \leq \|\nabla(u - u_{k_n})\|_{L^4(S)}\|v\|_3\|\rho\|_{L^4(S)}$$
$$+ \|\nabla u\|_{L^4(S)}\|\rho\|_1\|\Delta(v - \psi_{k_n})\|_{L^4(S)}$$
$$\leq 2^{1/4}\sqrt{2R_0}\|u - u_{k_n}\|_1^{1/2}\|v\|_3\|\rho\|_1$$
$$+ R_0\|\rho\|_1 2^{1/4}\sqrt{2\tilde{c}}\|v - \psi_{k_n}\|_2^{1/2} \to 0 \text{ as } n \to \infty.$$

Using this relation it is easy to verify that

$$\left(\frac{\Delta v - \Delta u}{\tau} + J(u, \Delta v) + 2v_\lambda + \sigma \Delta v - \nu \Delta^2 v - f, \rho \right) = 0,$$

$$\forall \rho \in H_0^1(S),$$

that is $v = V_\tau(\tau, u)$. Since all subsequences $\{\psi_k\}$ converge to v, one has that $\psi_k \to v$ also strongly in $H_0^2(S)$. Thus, the operator $V_\tau(\tau, \cdot)$ is continuous on X. In accordance with Lemma 4.8. the operators $V_\tau(t_k, \cdot)$ are continuous on $H_0^2(S)$, hence for any $t \in \mathbf{Z}_+\tau$, $t > \tau$ we have

$$V_\tau(t, u_k) = V_\tau(t - \tau, \psi_k) \to V_\tau(t - \tau, v) = V_\tau(t, u), \; k \to \infty$$

strongly in $H_0^2(S)$. This leads to the continuity of $V_\tau(t, \cdot)$ on X. Let us assume now that (4.72) does not hold. Then for some $t \in \mathbf{Z}_+\tau$ and $\varepsilon > 0$ there exist sequences u_k and u_k' such that $u_k \in X$, $u_k' \in X$, $\mathrm{dist}(u_k, u_k') \to 0$ as $k \to \infty$ and $\mathrm{dist}(V_\tau(t, u_k'), V_\tau(t, u_k)) > \varepsilon$. We choose the subsequence $u_{k_n} \to \rho \in X$ strongly in $H_0^1(S)$, then one has $u_{k_n}' \to \rho$. By the continuity of the operator $V_\tau(t, \cdot)$ in X we find $\mathrm{dist}(V_\tau(t, u_{k_n}'), V_\tau(t, u_{k_n})) \to 0$ as $n \to \infty$. The contradiction so obtained proves the lemma.

The attractor $\mathcal{A}_\tau$ of the difference scheme (4.42)-(4.43) is contained in X by virtue of Theorem 2.2, consequently, $\mathcal{A}_\tau$ is also the attractor of SDS $\{V_\tau, \mathbf{Z}_+\tau, X\}$.

We now prove that SDS $\{V_\tau, \mathbf{Z}_+\tau, X\}$ possesses the uniformly asymptotically stable sets M_τ that approximate the attractor $\mathcal{A}$ for sufficiently small τ. In proving we shall give the method of the construction of the above sets (see also [83,84]).

Theorem 4.14. *For all* $\tau = 1/N$, $N \in \mathbf{N}$ *SDS* $\{V_\tau, \mathbf{Z}_+\tau, X\}$ *possesses the uniformly asymptotically stable set* M_τ *such that* $\mathcal{A} \subset M_\tau$ *for* $\tau \leq \tau_1^*$,

$$\mathrm{dist}_H(M_\tau, \mathcal{A}) \to 0 \quad \text{as} \quad \tau \to 0$$

and M_τ *absorbs* X *in the finite time* $T^*(\tau)$, *i.e.*

$$V_\tau(n\tau, X) \subset M_\tau, \quad \forall n\tau > T^*(\tau), \tag{4.73}$$

where τ_1^* *is the constant from Theorem 3.3.*

Proof. Let $\widehat{X}$ be the set of all nonempty closed subsets X. We define the metric in $\widehat{X}$

$$\widehat{R}(A, B) = \mathrm{dist}_H(A, B), \quad \forall A, B \in \widehat{X},$$

then $\widehat{X}$ becomes a compact metric space ([83]).

We give on $\widehat{X}$ the semigroup of operators $W_\tau(t, \cdot)$, $t \in \mathbf{Z}_+\tau$, setting

$$W_\tau(0, A) = V_\tau(0, A), \quad W_\tau(\tau, A) = \overline{\mathcal{O}(V_\tau(\tau, A); \chi(\tau))}, \quad \forall A \in \widehat{X},$$

where the bar denotes the closure in X, while $\chi(\tau) = c_0\tau^{3/2}$ is the function from the estimate (4.68).

Furthermore we extend on all $t \in \mathbf{Z}_+\tau$ the operators W_τ by the use of the group property $W_\tau(t+\tau, A) = W_\tau(t, W_\tau(\tau, A))$. The continuity of the operators $W_\tau(t, \cdot)$ on $\widehat{X}$ for $\forall t \in \mathbf{Z}_+\tau$ follows from the uniform continuity of the operators $V_\tau(t, \cdot)$ on X and from the inequality

$$\mathrm{dist}_H \overline{(\mathcal{O}(A; r), \mathcal{O}(B; r))} \leq \mathrm{dist}_H(A, B),$$

which is valid in the convex metric space X ([7]), see Lemma 2.7). By the compactness $\widehat{X}$ it follows that SDS $\{W_\tau, \mathbf{Z}_+\tau, \widehat{X}\}$ possesses the attractor $\mathcal{M}_\tau$.

For any $\Xi \in \widehat{X}$ we denote by $E(\Xi)$ the union in X of all elements of Ξ, i.e.

$$E(\Xi) = \bigcup_{A \in \Xi} A.$$

Let $M_\tau = E(\mathcal{M}_\tau)$. By the construction of operators W_τ the inclusion holds

$$V_\tau(t, A) \subset W_\tau(t, A), \quad \forall A \in \widehat{X}, \quad \forall t \in \mathbf{Z}_+\tau, \qquad (4.74)$$

then for the attractor $\mathcal{A}_\tau$ one has $\mathcal{A}_\tau \subset W_\tau(t, \mathcal{A}_\tau)$, $\forall t \in \mathbf{Z}_+\tau$. Further, passing to the limit as $t \to \infty$, we see that $\mathcal{A}_\tau \in \mathcal{M}_\tau$. Consequently,

$$\overline{\mathcal{O}(\mathcal{A}_\tau; \chi(\tau))} = W_\tau(\tau, \mathcal{A}_\tau) \in W_\tau(\tau, \mathcal{M}_\tau) = \mathcal{M}_\tau,$$

that is

$$\overline{\mathcal{O}(\mathcal{A}_\tau; \chi(\tau))} \subset E(\mathcal{M}_\tau) = M_\tau \qquad (4.75)$$

Since $\mathcal{A}_\tau$ attracts X, there can be found the time $T^*(\tau)$ such that $V_\tau(n\tau, X) \subset \mathcal{O}(\mathcal{A}_\tau; \chi(\tau)) \subset M_\tau$, $\forall n\tau > T^*(\tau)$, i.e. (4.73) holds.

It follows from the inclusion (4.74) and the definition of the attractor $\mathcal{M}_\tau$ that

$$V_\tau(\tau, M_\tau) = V_\tau(\tau, E(\mathcal{M}_\tau))$$
$$= E(V_\tau(\tau, \mathcal{M}_\tau)) \subset E(W_\tau(\tau, \mathcal{M}_\tau)) = E(\mathcal{M}_\tau) = M_\tau,$$

i.e. $V_\tau(\tau, M_\tau) \subset M_\tau$, then $V_\tau(n\tau, M_\tau) \subset V_\tau((n-1)\tau, M_\tau) \subset \ldots \subset M_\tau$, that is the set M_τ is positively invariant. We now show that

$$M_\tau = \overline{\mathcal{O}(V_\tau(\tau, M_\tau); \chi(\tau))}. \qquad (4.76)$$

Indeed,

$$M_\tau = E(\mathcal{M}_\tau) = E(W_\tau(\tau, \mathcal{M}_\tau)) = \overline{\mathcal{O}(E(V_\tau(\tau, \mathcal{M}_\tau)); \chi(\tau))}$$
$$= \overline{\mathcal{O}(V_\tau(\tau, E(\mathcal{M}_\tau)); \chi(\tau))} = \overline{\mathcal{O}(V_\tau(\tau, M_\tau); \chi(\tau))}.$$

By the uniform continuity of the operator $V_\tau(\tau, \cdot)$ on X there can be found $r_0 = r_0(\tau) > 0$ such that for any $u \in \mathcal{O}(M_\tau; r_0)$ the inequality holds

$$\mathrm{dist}(V_\tau(\tau, u), V_\tau(\tau, M_\tau)) < \chi(\tau),$$

i.e.

$$V_\tau(\tau, \mathcal{O}(M_\tau; r_0)) \subset \overline{\mathcal{O}(V_\tau(\tau, M_\tau); \chi(\tau))}.$$

According to (4.76) and by the positive invariance of M_τ, one can get

$$V_\tau(n\tau, \mathcal{O}(M_\tau; r_0)) \subset M_\tau, \quad \forall n \in \mathbf{N}.$$

We have thus shown that M_τ is the uniformly asymptotically stable set for SDS $\{V_\tau, \mathbf{Z}_+\tau, X\}$. We prove now the following additional assertions.

Lemma 4.12. *If $\tau \leq \tau_1^*$, then for any $u \in \mathcal{A}$ and any $A \in \widehat{X}$ from the fact that $u \in A$ it follows that $V(t, u) \in W_\tau(t, A), \forall t \in \mathbf{Z}_+\tau$.*

Proof. If $\tau \leq \tau_1^*$ and $u \in \mathcal{A}$, then according to Theorem 3.3

$$\mathrm{dist}(V(\tau, u), V_\tau(\tau, A)) \leq \|V(\tau, u) - V_\tau(\tau, u)\|_1 \leq \chi(\tau).$$

This means that

$$V(\tau, u) \in \overline{\mathcal{O}(V_\tau(\tau, A); \chi(\tau))} = W_\tau(\tau, A).$$

Since $u' = V(\tau, u) \in \mathcal{A}$, repeating n times such reasoning, we n are ensured that $V(n\tau, u) \in W_\tau(n\tau, A), \forall n \in \mathbf{N}$.

Lemma. 4.13. *Let $\tau_k = 1/N_k \to 0$ as $k \to \infty$, $A_k \in \mathcal{M}_{\tau_k}$ and $\widehat{R}(A_k, A) \to 0$. Then $\widehat{R}(W_{\tau_k}(1, A_k), V(1, A)) \to 0$ as $k \to \infty$.*

Proof. By the triangle inequality we have

$$\widehat{R}(W_{\tau_k}(1, A_k), V(1, A)) \leq \widehat{R}(W_{\tau_k}(1, A_k), V_{\tau_k}(1, A_k))$$
$$+ \widehat{R}(V_{\tau_k}(1, A_k), V_{\tau_k}(1, A)) + \widehat{R}(V_{\tau_k}(1, A), V(1, A)).$$

We estimate the values on the right side of the inequality by (4.61)

$$\widehat{R}(V_{\tau_k}(1, A), V(1, A)) \leq \sup_{u \in X} \|V_{\tau_k}(1, u) - V(1, u)\|_1 \to 0.$$

Theorem 2.4 implies that for all sufficiently small τ_k

$$\widehat{R}(V_{\tau_k}(\tau_k, A_k), V_{\tau_k}(\tau_k, A)) \leq (1 + \alpha\tau_k)\widehat{R}(A_k, A).$$

That is,

$$\widehat{R}(V_{\tau_k}(1, A_k),\, V_{\tau_k}(1, A)) \le (1 + \alpha\tau_k)^{1/\tau_k} \widehat{R}(A_k, A) \to 0 \ \text{ as } \ k \to \infty.$$

By the construction of the operators $W_\tau(t, \cdot)$ we have

dist $(V_{\tau_k}(1, A_k),\, W_{\tau_k}(1, A_k)) = 0$ (see (4.74))

   dist $(W_{\tau_k}(\tau_k, A_k),\, V_{\tau_k}(\tau_k, A_k)) \le \chi(\tau_k).$

Then, for all sufficiently small τ_k, we find

dist $(W_{\tau_k}(2\tau_k, A_k),\, V_{\tau_k}(2\tau_k, A_k))$

   $\le$ dist $(W_{\tau_k}(2\tau_k, A_k),\, V_{\tau_k}(\tau_k, W_{\tau_k}(\tau_k, A_k)))$

   $+$dist$(V_{\tau_k}(\tau_k,\, W_{\tau_k}(\tau_k, A_k)),\, V_{\tau_k}(2\tau_k, A_k)) \le \chi(\tau_k) + (1 + \alpha\tau_k)\chi(\tau_k).$

Using the induction, we get

dist$(W_{\tau_k}(n\tau_k, A_k),\, V_{\tau_k}(n\tau_k, A_k))$

   $\le$ dist$(W_{\tau_k}(n\tau_k, A_k),\, V_{\tau_k}(\tau_k, W_{\tau_k}((n-1)\tau_k, A_k)))$

   $+$dist$(V_{\tau_k}(\tau_k, W_{\tau_k}((n-1)\tau_k, A_k)),\, V_{\tau_k}(n\tau_k, A_k))$

   $\le \chi(\tau_k) + (1 + \alpha\tau_k)$dist$(W_{\tau_k}((n-1)\tau_k, A_k),\, V_{\tau_k}((n-1)\tau_k, A_k))$

   $\le \chi(\tau_k)(1 + (1 + \alpha\tau_k) + \ldots + (1 + \alpha\tau_k)^{n-1})$

   $= \chi(\tau_k)\dfrac{(1 + \alpha\tau_k)^n - 1}{\alpha\tau_k} = c_0\tau_k^{1/2}\dfrac{(1 + \alpha\tau_k)^n - 1}{\alpha}.$

Consequently,

$$\text{dist}(W_{\tau_k}(1, A_k),\, V_{\tau_k}(1, A_k)) \le \frac{c_0\tau_k^{1/2}}{\alpha}[(1 + \alpha\tau_k)^{1/\tau_k} - 1] \to 0 \text{ as } k \to \infty.$$

If the estimate (4.58) holds for the constant $\alpha = 0$, then in the similar way one can get

dist $(W_{\tau_k}(1, A_k),\, V_{\tau_k}(1, A_k)) \le \chi(\tau_k)\, N_k$

   $= c_0\, \tau_k^{3/2}\, N_k = c_0\, \tau_k^{1/2} \to 0$ as $k \to \infty.$

By the compactness of $\widehat{X}$ the union of the attractors on $\widehat{X}$

$$\widehat{\bigcup_{\tau_k > 0}} \mathcal{M}_{\tau_k}$$

is a precompact in $\widehat{X}$ set.

   This and the assertions of Lemmas 5.3 and 5.4 shows that at $\tau \le \tau_1^*$ for SDS $\{V, \mathbf{Z}_+, X\}$ and $\{W_\tau, \mathbf{Z}_+, \widehat{X}\}$ all conditions of the following theorem are satisfied.

Theorem 4.15.([84]) *Let the bounded uniformly continuous SDS* $\{V, T_+, X\}$ *possesses a compact attractor* $\mathcal{A} \subset X$. *Let* D *be a metric compact, let* d_0 *be its nonisolated point and let* $\{W_d, T_+, \widehat{X}\}$ *be the family of SDS possessing compact attractors* $\mathcal{M}_d \subset \widehat{X}$, *hence :*

1) $W_d\left(\overline{E(\Xi)}\right) = \overline{\bigcup_{A \in \Xi} W_d(A)} \qquad \forall \Xi \in \widehat{X};$

2) *the set* $\bigcup_{d \in D \setminus \{d_0\}} \mathcal{M}_d \subset \widehat{X}$ *is precompact;*

3) *if* $d_k \to d_0$, $A_k \in \mathcal{M}_k$ *and* $\widehat{R}(A_k, A) \to 0, as\, k \to \infty$, *then* $\widehat{R}\left(W_{d_k}(t, A_k), \overline{V(t, A)}\right) \to 0$ *for some* $t > 0$;

4) *at some* $t > 0$ *for any* $x \in A$, $d \in D \setminus \{d_0\}$, $B \in \widehat{X}$ *from* $x \in B$ *follows that* $V(t, x) \in W_d(t, B)$.

Then $\mathcal{A} \subset \overline{E(\mathcal{M}_d)}$ *and* $\text{dist}_X\left(\overline{E(\mathcal{M}_d)}, \mathcal{A}\right) \to 0$ *for* $d \to d_0$.

Thus, $\mathcal{A} \subset M_\tau$ for $\tau \geq \tau_1^*$ and $\text{dist}_H(\mathcal{A}, M_\tau) \to 0$ for $\tau \to 0$. Theorem 4.14 is proved.

Thus, we have proved the global stability of the time difference scheme for the barotropic vorticity equation on rotating sphere and the existence, for this scheme, of uniformly asymptotically stable sets approximating the attractor of the initial problem.

Remark. The results of the section are valid for the case of time-dependent forcing. These results are described in work [52].

Chapter 5

Numerical Study of Structure of Attractor Generated by Barotropic Equations on Sphere

With the existence theorems concerning an attractor generated by a system of barotropic equations on sphere and those concerning an inertial manifold and an invariant measure at our disposal, we, unfortunately, have nothing to say at present about the structure of the attractor. We can only estimate the attractor dimension.

As shown in preceding chapter, the existence theorems are valid if one considers Galerkin approximation with any truncation number instead of the original system of barotropic equations. However, every truncated system, generally speaking, may have its own attractor.

In this chapter we shall carry out the numerical investigation of the structure of the attractor generated by the finite-dimensional system of Galerkin approximations of the barotropic vorticity equation with the time discretization.

Further, we shall derive the estimates for the attractor dimension and calculate the global and local Lyapunov exponents. There will be also studied the role of the stationary points in the generation of the statistical stationary solution and the first order moments of the system. We then shall make an attempt to estimate the mean life time, which the trajectory spends in the neighborhood of the stationary points and shall link this time with the stability properties of these points. For a more detailed results of calculations the reader is referred to the works [35,38].

5.1   Equations and Parameters of Model. Methods of Solving of Stationary and Nonstationary Problems

We write the dynamic equation of viscous incompressible barotropic fluid on sphere with respect to the orographic inhomogeneity of the underlying surface, in the usual way, in terms of the streamfunction. All variables will be parametrized in terms of Earth's radius $R = 6.371 * 10^6$ m and the angular velocity of the Earth $\Omega = 7.27 \cdot 10^{-5} c^{-1}$. This gives

$$\frac{\partial \Delta \psi}{\partial t} + \rho J(\psi, \delta \psi) + J(\psi, l + h) + \sigma \Delta \psi + \nu \Delta^2 \psi = f, \qquad (5.1)$$

where ψ is the streamfunction, Δ is the Laplacian operator on the unit sphere

$$\Delta \psi = \frac{1}{\cos(\varphi)} \frac{\partial}{\partial \varphi} \cos(\varphi) \frac{\partial \psi}{\partial \varphi} + \frac{1}{\cos^2(\varphi)} \frac{\partial^2 \psi}{\partial \lambda^2};$$

J is the Jacobian

$$J(\psi, \Omega) = \frac{1}{\cos(\varphi)} \left[\frac{\partial \psi}{\partial \lambda} \frac{\partial \Omega}{\partial \varphi} - \frac{\partial \psi}{\partial \varphi} \frac{\partial \Omega}{\partial \lambda} \right];$$

l is the Coriolis parameter, $l = 2\Omega \sin \varphi$, h are orographic inhomogeneities of the underlying surface. In our model the orography is presented in the simplified form as two continents which are separated by two oceans [35]. The heights of the continents lower toward the poles and equatorward

$$h = h_0 \cos 2\lambda \sin^2 2\varphi,$$

where h_0 is the amplitude of inhomogeneities which is taken equal to 0.1 of the atmosphere height.

The term $\sigma \Delta \psi$ in the first approximation describes the dissipation in the planetary atmosphere boundary layer. Here we have chosen as $\sigma = 1/20$ days. The term $\nu \Delta^2 \psi$ describes the turbulent viscous dissipation. The external forcing is presented by the right-hand side of the problem $f = \sigma \Delta \psi^*$, where the streamfunction ψ^* was chosen according to [35] as a pure zonal flow in the form

$$\psi^* = -\psi_0 \sin^3 \varphi. \qquad (5.2)$$

The coefficient ψ_0 and velocity of the forcing flow (the radius of the sphere is to be equal to the unit) are connected by the relation

$$u = -\frac{\partial \psi}{\partial \varphi} = \psi_0 3 \sin^2 \varphi \cos \varphi. \qquad (5.3)$$

The velocity u approaches its maximum at latitude $\varphi = \arctan \sqrt{2}$ $\cong 55^0$ and its magnitude at this latitude is $u_{\max} \cong 1.15\psi_0$ in dimensionless units. Hereafter, estimating the value of the forcing we shall use the value of the maximal velocity. The equation (5.1) is considered in the northern hemisphere, thereby as the boundary conditions on the equator, there are taken the nonslip conditions $\partial\psi/\partial\lambda = 0$ for $\varphi = 0$. With respect to the arbitrary choice of the constant for the definition of ψ, this condition may be written in the form

$$\psi|_{\varphi=0} = 0. \tag{5.4}$$

At the pole we have the condition of boundedness for the function ψ and its derivatives. At the coordinate λ the periodicity condition will be applied. For the space discretization of (5.1) we have used the Galerkin method with the spherical harmonics which are taken as the base functions. Such a discretization assumes that the application of the boundary condition (5.4) reduces to the use of spherical harmonics which are antisymmetric relative to the equator, i.e. there are taken only those $Y_{m,n}$ that have odd sum of indices $m + n$:

$$\psi(\lambda,\varphi,t) \cong \sum_{n=0}^{N} \sum_{m=-n}^{n} \psi_{m,n}(t)Y_{m,n}(\lambda,\varphi) = \sum_{\beta}^{N} \psi_\beta(t)Y_\beta(\lambda,\varphi), \tag{5.5}$$

$$\beta = (m,n), \quad m + n \text{ is odd.}$$

As a maximal truncation number in the problem there is chosen $N = 7$. Thus, in the calculation one uses 28 Fourier coefficients.

Since we simulate only the large-scale processes, we omit the term $\nu\Delta^2\psi$ in calculations, taking into account that the general dissipation for such processes is occurred in the boundary layer and this term does not play a significant role.

For the integration of the equation over time, we have used the Crank–Nicolson scheme with the time step $\tau = 0.25$ days. We have chose this scheme for its absolute stability.

We have applied the Newton method for the inversion of the operator formed at every step of the integration over time. The value of the streamfunction from the preceding time step was used as the initial approximation for the Newton iteration process. Under such conditions the process, practically, converges after 3–4 iterations (the norm of residual term decreases in 8–10 orders). The stationary solution of (5.1) satisfy the relation (if $\mu = 0$):

$$\rho J(\bar\psi, \Delta\bar\psi) + J(\bar\psi, l + h) + \sigma\Delta\bar\psi = f. \tag{5.6}$$

To find the stationary solutions $\bar\psi$ of (5.6), we have used the procedure proposed in [105] and which is as follows.

We rewrite (5.6), separating the linear and nonlinear portions of the operator

$$\rho J(\bar{\psi}, \Delta\bar{\psi}) + A\bar{\psi} = f,$$
$$A\bar{\psi} = J(\bar{\psi}, l+h) + \sigma\Delta\bar{\psi}, \bar{\psi} = \bar{\psi}(\rho, \lambda, \varphi). \tag{5.7}$$

For $\rho = 0$, the stationary solution of (5.7) may be easily obtained by solving the system of linear equations $A\bar{\psi} = f$. To find the stationary solution for $\rho \neq 0$, we shall use the following procedure. In the finite-dimensional notation the operator A is

$$\overset{N}{A_{\gamma,\nu}} = -\sum_{\beta}^{N}\langle J(Y_{\beta}, Y_{\nu}), Y_{\gamma}\rangle h_{\beta} + \sum_{\beta}^{N}\langle Y_{\beta}Y_{\nu}, Y_{\gamma}\rangle\sigma_{\beta}\chi_{\nu} + l2m_{\gamma}\delta_{\gamma,\nu},$$

where $\chi_{\gamma} = -n_{\gamma}(n_{\gamma}+1)$ are the eigenvalues of the Laplacian operator, h_{β} are Fourier coefficients in the expansion of the function h, while $\sigma_{\beta} \equiv \sigma_{m,n}$ are the coefficients in the expansion of σ in series of spherical harmonics. Let

$$G(\rho, \psi) = \rho J(\psi, \Delta\psi) + A\psi - f. \tag{5.8}$$

We now formulate the successive two-step iteration process of the predictor-corrector type, the first step of which represents the tangential motion to the curve $G\psi = 0$ in the coordinates (ρ, ψ)

$$\left.\frac{\partial G}{\partial\psi}\right|_{\rho_n\psi_n}\dot{\psi} + \left.\frac{\partial G}{\partial\rho}\right|_{\rho_n\psi_n}\dot{\rho} = 0,$$

$$\dot{\rho} + \|\dot{\psi}\|^2 = 1; \quad \psi_0 = \psi(0, \lambda, \varphi); \quad \rho_0 = 0;$$

$$\dot{\psi} = \frac{d\psi}{ds}; \quad \dot{\rho} = \frac{d\rho}{ds}; \quad ds^2 = \|d\psi\|^2 + d\rho^2;$$

$$\rho^* = \rho_n + G\dot{\rho}; \quad \psi^* = \psi_n + \sigma\dot{\psi}, \tag{5.9}$$

where σ is the step in the curvilinear coordinates s.

The second step is to obtain (ψ_{n+1}, ρ_{n+1}) by solving the equation $G(\rho_{n+1}, \psi_{n+1}) = 0$ through the use of the Newton method with the initial approximation (ρ^*, ψ^*):

$$\left.\frac{\partial G}{\partial\psi}\right|_{\rho^k,\psi^k}\left(\psi^{k+1} - \psi^k\right) + \left.\frac{\partial G}{\partial\rho}\right|_{\rho^k,\psi^k}\left(\rho^{k+1} - \rho^k\right) = G,$$

$$\psi^0 = \psi^*, \quad \rho^0 = \rho^*, \quad (\rho^{k+1} - \rho^k)^2 + \|\psi^{k+1} - \psi^k\|^2 = 1. \tag{5.10}$$

As a criterion of the completion of the iteration process of (5.10) the following condition serves $\|G(\rho_k, \psi_k)\| \leq \varepsilon$, where ε is a given value (here $\varepsilon = 10^{-12}$ was used). The obtained solution $(\rho^k \psi^k)$ is considered as the next iteration of the process (5.9): $(\rho_{n+1}, \psi_{n+1}) = (\rho^k \psi^k)$.

It is worthwhile to consider the recurrent turning points, where the curve $G(\rho, \psi) = 0$ is not smooth. In such points the application of the process (5.9) is not possible. Therefore, to find the values (ρ^*, ψ^*), the random perturbation of the state (ρ_n, ψ_n) was used. Thereby the process (5.10) with some degree of probability reduces to a new solution lying near the turning recurrent point. With such a choice of the initial approximation three variants of the process (5.10) are possible: the process diverges, the process converges to (ρ_n, ψ_n) and the process converges to a new solution. With realization of one of the first two variants the initial approximation is considered to be unsuccessful and the result does not taken into account.

The same procedure was applied for finding the isolated branches of the stationary solutions $\bar{\psi} = \bar{\psi}(\rho)$. Namely, when the main branch (going from $\rho = 0$ to $\rho = \infty$) was found, there were chosen on it the points $(\rho, \bar{\psi})$ with the step $\Delta\rho = 0.1$. The value of $\bar{\psi}$ at these points was randomly disturbed and value of ψ^* so obtained was then used as the initial approximation for the process (5.10) in accordance with the procedure described above.

To get the chaotic motion of the trajectory, we choose the parameter u which answers for the forcing intensity to be equal to 180 m/s. The dependence of the energy of the stationary solution on the parameter of nonlinearity ρ is shown in Fig. 1.

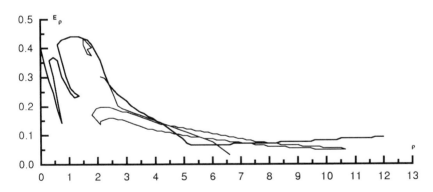

Fig.1. Dependence of the energy of stationary solution E_ρ on the parameter ρ.

It is easy to see that for small ρ the problem resembles the linear one and it has one unique stable stationary solution. When ρ is increasing, the solution becomes nonunique and unstable. Moreover, for ρ exceeding 1.5, there appear the isolated closed branches of the stationary point.

For very large ρ (in our experiment $\rho > 12$) the stationary solution again becomes the unique one (that fact is emphasized in [106]).

In our experiments ρ was taken on the most interesting interval, on which for a given ρ there exists the largest number of points, say, $\rho = 3$. For such ρ the system has 5 stationary solutions. Thus, for the numerical experiments there were chosen the following parameters of the problem

$$\rho = 3, \quad u = 180, \quad \sigma = 1/20 \text{ days}^{-1}, \quad \mu = 0. \qquad (5.11)$$

Given parameters of the model, the integration over time for 4000 model days with the time step of 0.25 days was carried out. The results of the initialization of the model were used as the initial value, i.e. the results of 200-days integration from one of the stationary solutions. The output of results for the successive data processing was performed at every 2 the model steps, i.e. within 0.5 days interval

5.2 Statistical Stationary Solution and Stationary Points

As noted in 3.4., we choose the expression as an measure on attractor

$$\mu(u) = \frac{1}{T} \int\limits_0^T \mu(t,\omega)dt, \quad T \to \infty,$$

$$\mu(t,\omega) = \delta_{Ss_t\varphi_o(\omega)} = \left\{ \begin{array}{ll} 1 & \text{for } S_t\varphi_0 \in \omega, \\ 0 & \text{for } S_t\varphi_0 \notin \omega. \end{array} \right.$$

Since we shall deal with only the crude form of the statistical stationary solution, we first must choose a neighborhood of the stationary points, i.e. we eventually must devide the phase space into the small volumes. If we assume that our system is ergodic (we need such a assumption to state the uniqueness of the statistical stationary solution), then there exists a range of the changes of the radius of a neighborhood, within which the share of time that the trajectory spends in the neighborhood k-th stationary point is connected with its radius by the relation

$$T_k = a_k r^d, \quad r_1 < r < r_2,$$

where d is the dimension of the attractor of the system.

Choosing the values of the neighborhood radius from the interval $[r_1, r_2]$, we get the relations

$$T_k^{(1)} = a_k r_1^d, \quad T_k^{(2)} = a_k r_2^d = a_k(r_1 r_2/r_2)^d = T_k^{(1)}(r_2/r_1)^d = C T_k^{(1)}.$$

It follows from them that within this interval the increasing of the radius of the neighborhood leads to the increasing of the share of time, but the ratio of these shares remains constant.

The numerical calculations show that this fact takes place and following these calculations we choose to the rough presentation the neighborhoods of radius $r = 0.15$.

To obtain the quantitative estimate of the life time of the trajectories in the neighborhood of the stationary points we shall project the phase portrait onto two-dimensional space of the natural orthogonal components and introduce the distribution function by the means of the following relation

$$dT = F_{km}(\xi_k, \xi_m) \, d\xi_k d\xi_m,$$

where dT means the summary time, in which the projection of the solution belongs to the set $(\xi_i, \xi_k + d\xi_k) \times (\xi_m, \xi_m + d\xi_m)$.

In [35] shown that the main time the solution spends in the neighborhood of the stationary points. This being the case, we can attempt to project, say, the first order moment of the solution onto a subspace spanned by the stationary points.

The above reasonings show that the expansion coefficients must be proportional to the mean life time, that the trajectory spends in the small neighborhoods of the corresponding stationary points. This mean time can be calculated as the product of the mean life time of one trajectory in a neighborhood with the average number of enters into the neighborhood.

Let u_i be the stationary solution of (5.7) and let M be the first order moment of the solution. We shall seek M in the form of the expansion

$$M \stackrel{\sim}{\div} \sum_1^k c_i u_i.$$

The expansion coefficients of the first order moment over the system of the stationary solutions may be found by the method of least squares

$$\left\| M - \sum_1^k c_i u_i \right\|^2 \to \min.$$

In our problem the angle between the vectors $\sum c_i u_i$ and M is a value of the order 10^0, while the relative error is $\| M - \sum c_i u_i \| / \| M \| \sim 0.3$.

It should be noted that the obtained solution of the above problem may show a bad agreement with the above given concept of measure, i.e. some additional requirements imposed on coefficients may not fulfill. Because of this given problem can be reformulated as follows [64].

We assume, as before, that the system spends most part of time in small neighborhoods of the stationary points, i.e. the support of measure is concentrated in the neighborhoods of the points (u_i). Consequently, the approximate relation is valid

$$\hat{\mu} = \sum_1^k p_j \hat{\mu}_j. \tag{5.12}$$

Here $\hat{\mu}_j$ are the probability measures, thereby

$$\hat{\mu}_j(Q_j) = 1, \quad \hat{\mu}_j(R^n - Q) = 0,$$

where Q_j is the essentially small neighborhood of the stationary point u_j. It follows from the relation (5.12) that

$$\int (\varphi - u_i) \hat{\mu}(d\varphi) = M - u_i$$

and

$$\int (\varphi - u_i) \hat{\mu}(d\varphi) \approx \sum_1^k (u_j - u_i) p_j,$$

i.e. the approximate equality holds

$$M - u_i \approx \sum_1^k (u_j - u_i) p_j, \quad i = 1, \ldots, k,$$

which leads to the extremal problem

$$\sum_{i=1}^k \left\| M - u_i - \sum_1^k p_j(u_j - u_i) \right\|^2 \to \min.$$

The problem has an unique solution if the exact equality holds

$$\sum p_j = 1.$$

In this case the problem reduces to solving the system of equations

$$\left(M - \sum_1^k p_j u_i, \sum_1^k u_i - k u_s \right) = 0, \quad \sum_1^k p_j = 1, \ s = 1, \ldots k.$$

In the table there are shown the values p_i and $h_i = (M_j, u_i)/\|u_i\|^2$ and the mean time which the trajectory spends in the neighborhood of every stationary point. It is seen that the correlation coefficient between $(p_i), (h_i)$ and (T_i) is essentially high, but it is still not equal to the unit. This means that the system spends in the neighborhood of the stationary points large but not the most share of time. It is possible that in the system there are occurred the unstable limit cycles, that along with the stationary points contribute to the generation of the first order moment.

Number of stationary point	p_i	h_i	T_i
1	0.094	0.413	0.11
2	0.340	0.318	0.04
3	0.026	0.456	0.10
4	0.547	0.514	0.24
5	-0.008	0.369	0.05
Correlation coefficient	0.7	0.9	

Following the preceding reasoning, we shall make an attempt to calculate the expansion coefficients over characteristics of stable and unstable manifold of the stationary solutions of the original problem.

The first our hypothesis is that the mean life time of the trajectory in the neighborhood of the stationary point is defined by characteristics of its unstable manifold. It is clear that the life time of every (individual) enter into the neighborhood of the stationary point will depend on the initial value of projection of the solution onto the unstable manifold and on the characteristics of its stability. However, averaging the mean life time over all enters, we must, in principle, have dependency of the life time only on characteristics of the stability. The choice of the characteristics of degree of the instability of the stationary point place here the important role.

Since the trajectory spends in the neighborhood the finite time and the operator of the problem linearized with respect to the stationary point is essentially nonsymmetric, the growth of the amplitude of the initial perturbation is not exponential as it were asymptotically at the large times. Therefore, it seems natural to use in this case the local Lyapunov exponents or characteristics that are approximate to them. Since our estimate must be a priori estimate, we can choose from all local exponents only instantaneous Lyapunov exponents [1,37] or the values equivalent to them in some sense.

We consider again the relative vorticity equation

$$\frac{\partial \Delta \psi}{\partial t} + \rho J(\psi, \Delta \psi + h + l) + \sigma \Delta \psi - \nu \Delta^2 \psi = f. \qquad (5.13)$$

Let

$$\psi = \bar{\psi} + \psi', \quad \rho J(\bar{\psi}, \Delta \bar{\psi} + h + l) + \sigma \Delta \bar{\psi} - \nu \Delta^2 \bar{\psi} = f.$$

Then we rewrite the equation

$$\frac{\partial \Delta \psi'}{\partial t} + \underbrace{\rho J(\psi', \Delta \bar{\psi} + h + l) + J(\bar{\psi}, \Delta \psi') + \sigma \Delta \psi' - \nu \Delta^2 \psi'}_{A\psi'}$$

$$+\rho J(\psi', \Delta \psi) = 0. \qquad (5.14)$$

The disturbance energy equation is

$$\frac{\partial E'}{\partial t} = (A\psi', \psi'), \qquad (5.15)$$

where the scalar product is defined in the real Hilbert space

$$(\eta, \varphi) = \int\limits_{\lambda} \int\limits_{\varphi} \eta \, \varphi \cos \varphi \, d\lambda \, d\varphi.$$

Since $A = S + K$, where $S = \dfrac{A + A^*}{2}$, $K = \dfrac{A - A^*}{2}$, the equation (5.15) may be rewritten in the form

$$\frac{\partial E}{\partial t} = (S \psi', \psi'). \tag{5.16}$$

Hereafter, we shall apply the next procedure. Although the analyzed system is finite-dimensional, for the simplicity, we shall use first for performing the transformations its infinite-dimensional notation, passing to the finite-dimensional notation if we need it.

We consider the model problem, when the stationary solution does not depend on the zonal coordinate, i.e. we put $\bar{\psi} = \bar{\psi}(\varphi)$. To shorten the writing, we choose the Cartesian coordinates. It is seen from (5.16) that the stationary solution will be stable, if all the eigenvalues of the symmetric portion of the operator A linearized with respect to the stationary solution are negative. We shall take the sum of the positive eigenvalues of the operator S as the characteristics of the instability of the stationary point. We shall show that for our model problem this sum will be connected with the enstrophy of the stationary solution. Let us simplify the problem, setting first $\sigma = \mu = 0$; and let us take as the boundary conditions the periodicity conditions for x and y. This gives

$$A\psi' = \bar{u} \frac{\partial \Delta \psi'}{\partial x}$$

and

$$2 \cdot S \equiv A + A^* = \bar{u} \frac{\partial}{\partial x} \Delta - \Delta \left(\bar{u} \frac{\partial}{\partial x} \right).$$

The problem of the eigenvalues of the operator S has the form

$$\bar{u} \frac{\partial}{\partial x} \Delta \tilde{\varphi} - \Delta \left(\bar{u} \frac{\partial \tilde{\varphi}}{\partial x} \right) = \lambda \tilde{\varphi}.$$

Since $\bar{u} = \bar{u}(y)$, we shall look for the solution $\tilde{\varphi}$ in the form $\tilde{\varphi} = \tilde{\varphi}(y)e^{ikx}$ so that

$$\Delta \tilde{\varphi} = \left(\frac{\partial^2}{\partial y^2} - k^2 \right) \tilde{\varphi}, \quad \frac{\partial \tilde{\varphi}}{\partial x} = ik\tilde{\varphi},$$

and after making the simple transformations, one has

$$\bar{u} \frac{\partial^2 \tilde{\varphi}}{\partial y^2} - \frac{\partial^2 (\bar{u} \tilde{\varphi})}{\partial y^2} = \frac{\lambda}{ik} \tilde{\varphi},$$

or

$$\frac{\partial \bar{\omega}\tilde{\varphi}}{\partial y} + \bar{\omega}\frac{\partial \tilde{\varphi}}{\partial y} = \frac{\lambda}{ik}\tilde{\varphi}, \quad \bar{\omega} = -\frac{\partial \bar{u}}{\partial y}. \tag{5.17}$$

Since the initial operator by the definition was a symmetric one, but on the left side of (5.17) there is a skew-symmetric operator, hence

$$S = ikK. \tag{5.18}$$

It follows from (5.18) that the eigenvalues of the operator are symmetric with respect to 0, since K is a real skew-symmetric operator. Because of this strong symmetry property of the spectrum of the operator S we shall instead of the spectrum of the operator S investigate the spectrum of the operator K^2, the eigenvalues of which are equal to the squares of the eigenvalues of the operator S.

To exclude singularities, we proceed now to the finite-dimensional approximation of the operator K. If we use the symmetric approximations of derivatives, then the skew-symmetric property will remain in force. Then the matrix of the finite-dimensional operator has the form

$$\frac{\bar{\omega}_{i+1}\tilde{\varphi}_{i+1} - \bar{\omega}_{i-1}\tilde{\varphi}_{i-1}}{2\Delta y} + \bar{\omega}_i\frac{\tilde{\varphi}_{i+1} - \tilde{\varphi}_{i-1}}{2\Delta y} = \frac{\lambda}{ik}\tilde{\varphi}_i.$$

We use the known algebraic relation

$$\sum_i \lambda_i(A^*A) = \sum_{ij}(a_{ij})^2.$$

Since $a_{ij}(K) = \dfrac{\bar{\omega}_{ij}}{\Delta y}$, the relation follows

$$\sum \lambda_i^2(-S) = \frac{2}{\Delta y}\sum_{ij}\bar{\omega}^2.$$

Thus, we have shown that the sum of squares of the positive eigenvalues of the symmetric part of the operator A for our model problem is proportional to the enstrophy of the stationary solution.

The correlation coefficient between so introduced characteristics and the mean life time of the trajectory in the neighborhood of the stationary points for our problem is the value of order 0.95.

The calculation of the characteristics defining the number of entrances of the trajectory into the neighborhood of the stationary points is the more complicated procedure. Since the unstable manifold of one stationary point intersects the stable manifolds of other stationary points, it seems reasonable to construct such a characteristic, taking into account the properties of the stable manifolds of points i.e. their dimension, the mean coefficient of attraction in arbitrary directions. So constructed characteristics for a given problem [35] in fact show the high correlation with the mean number of entrances into the neighborhood of the stationary points, this correlation is approximately 0.95.

5.3   Lyapunov Exponents and Attractor Dimension

Once again we write down the relative vorticity equation

$$\frac{\partial \Delta \psi}{\partial t} + \rho J(\psi, \Delta \psi + l + h) + \sigma \Delta \psi - \nu \Delta^2 \psi = f. \qquad (5.19)$$

Let $\bar{\psi}(t)$ be a solution of the system. Linearizing of (5.19) with respect to this solution yields

$$\frac{\partial \Delta \psi'}{\partial t} + \underbrace{\rho J(\psi', \Delta \bar{\psi} + l + h) + \rho J(\bar{\psi}, \Delta \psi') + \sigma \Delta \psi' - \nu \Delta^2 \psi'}_{A \Delta \psi'} = 0.$$
$$(5.20)$$

(We assume that in system (5.20) the Galerkin approximation via spherical harmonics was carried out (in the way similar to the procedure of solving the stationary problem)).

We use the Crank–Nicolson scheme to solve (5.20) over time, putting $\Delta \psi' = \xi'$

$$\frac{\xi^{|j+1}\xi^i}{\tau} + A(t_j)\frac{\xi^{j+1} + \xi^j}{2} = 0, \quad A(t_j) \equiv A_J,$$

or

$$\xi^{j+1} = \left(E + \frac{\tau}{2}A_j\right)^{-1}\left(E - \frac{\tau}{2}A_j\right)\xi^j \equiv B_j \xi^j.$$

Carrying out the subsequent exclusion of the intermediate steps, we get

$$\xi^{j+1} = \prod_{j=1}^{j} B_j \xi^0 \equiv M_j \xi^0.$$

In accordance with the multiplicative ergodic theorem [109], the global Lyapunov exponents may be calculated as the logarithms of the eigenvalues of the limit operator

$$M = \lim_{j \to \infty} (M_j^* \cdot M_j)^{1/2j}, \quad \lambda_i = \ln \nu_i(M).$$

The sum of the positive global Lyapunov exponents in accordance with [110] define the Kolmogorov entropy, i.e. the value showing the mean divergence of the trajectories on the attractor. The Lyapunov exponents allow to derive the estimates of the attractor dimension by the Kaplan–Yorke formula [77].

Let J be such that $\sum_1^J \lambda_i > 0$ and let $\sum_1^{J+1} \lambda_i < 0$. Then one can derive the upper bound of the attractor dimension by the formula

$$d = J + \sum_1^J \lambda_i/|\lambda_{J+1}|. \qquad (5.21)$$

In (5.21) the arrangement of the Lyapunov exponent is carried out as follows

$$\lambda_1 \geq \lambda_2 \geq \lambda_3 \geq \cdots \geq \lambda_N.$$

If we arrange the Lyapunov exponents so that the positive exponents will be arranged decreasingly, while those negative arranged increasingly, then we get the upper bound of the attractor dimension.

Along with global Lyapunov exponents, one can introduce the concept of the local Lyapunov exponents which according to [1,2] may be defined as the logarithms of the eigennumbers of the operator $(M_j^* M_j)^{1/2j}$ for a given finite j. It is clear that in the general case the value of the local Lyapunov exponents will depend on the piece of the trajectory, on which they (exponents) are calculated. If we average the local Lyapunov exponents over measure (in our case over the lengths of the segments of the trajectories), then we obtain the averaged local Lyapunov exponents which are the function of the length of the piece only. We make an important, from our point of view, notion. By virtue of the existence of the weak limits in the definition of invariant measure and global Lyapunov exponents with the ergodicity conditions there always exists the time T_* such that the local Lyapunov exponents will approach the global ones to accuracy as we need. This means that the relations between the local and global Lyapunov exponents may be replaced by the relationships between only the local Lyapunov exponents.

If we choose now instead of the infinite sampling the finite one corresponding to T_* (local Lyapunov exponents corresponding to this sampling we shall identificate with the global ones), then the relations between the averaged local and global Lyapunov exponents may be established by means of theorems of linear algebra.

Let A, B, C be the square matrices of order N, where $C = AB$. Let ρ_i, μ_i, τ_i be their singular numbers enumerated in the nonincreasing order. Then for any $k : 1 \leq k \leq N$ the relations are valid [131]:

$$\sum_{i=k}^N \ln \tau_i \leq \sum_{i=k}^N \ln \rho + \sum_{i=k}^N \ln \mu_i, \qquad (5.22)$$

thereby for $k = 1$ there can be the identity obtained. We divide the trajectory segment of the length T_* into two segments of length $T/2$.

(We shall assume that the segment of the length T_* contains $M = 2L$ time steps.) Let

$$A = \prod_{j=1}^{L} B_j$$

for the first segment, and

$$B = \prod_{j=L+1}^{2L} B_j$$

for the second segment. Since the operator defining the global Lyapunov exponents is equal to $B \cdot A$, we get by virtue of the inequality (5.22)

$$\frac{1}{2L} \sum_{i=k}^{N} \ln \tau_i \leq \left(\frac{1}{L} \sum_{i=k}^{N} \ln \rho_i + \frac{1}{L} \sum_{i=k}^{N} \ln \mu_i \right) / 2.$$

By induction, continuing to divide each of segments as above, we can assert that:

1. When the lengths of the segments of the trajectory, on which averaged local Lyapunov exponents are calculated, increase the maximal averaged Lyapunov exponent does not increase.

2. When the lengths of segments of trajectory increase the sum of averaged positive Lyapunov exponents does not increase.

This assertions also mean that the averaged local Lyapunov exponents are the upper bound for the global Lyapunov exponents.

We note that from the numerical calculations standpoint the problem of calculation of the local and global Lyapunov exponents is a nontrivial one, since to solve it the large amount of matrices should be multiplied together. Therefore it is convenient to use in this case special algorithms based on QR-expansion [1,2].

The attractor dimension of the considered problem according to the formula (5.21), derived in [35] via the above procedure, is found to be equal to the value of order 6.

5.4 Analysis of Analytical Estimates of Attractor Dimension of Barotropic Atmospheric Equations

As we have noted above, we understand under the climate of system all main characteristics of attractor, the attractor dimension is among them. In this sense, when speaking about the sensibility of climate to the external parameters changes, we mean the dependence of these characteristics on changes of the external parameters.

In Chapter 3 there were given the upper bounds of the attractor dimension for the barotropic vorticity equation on the sphere. The principal feature of these upper bounds is the fact that all constants in them are calculated. In this connection there arise the questions concerning the application of these upper bounds for the calculation of attractor dimension with the really observed in the atmosphere values of parameters, the study of continuous dependence of the attractor dimension on the parameters of the problem etc. Some of such questions may be answered.

Let us have the barotropic vorticity equation with the zero dissipation in the boundary layer and the zero height of inhomogeneites of underlying surface. Then, for the Hausdorff attractor dimension of this equation the estimate (3.29) is valid:

$$\dim_H A \leq \left(\frac{12}{\sqrt{\pi}}\right)^{2/3} G^{2/3} \leq \left(\frac{1}{2} + \ln \frac{3\sqrt{2}}{\sqrt{\pi}} + \ln G\right)^{1/3}.$$

Here G is the generalized Grasshoff number, $G = \frac{\|f\|_{-1}}{|\lambda_1| \nu^2}$, λ_1 is the minimal over module eigenvalue of the Laplacian operator on a sphere.

Let us consider the case of the small Grasshoff numbers. This case is connected with the small norm forcing or the sufficiently large horizontal diffusion coefficient. Let G_0 is the solution of the equation

$$\frac{1}{2} + \ln \frac{3\sqrt{2}}{\sqrt{\pi}} + \ln G_0 = 0. \tag{5.23}$$

Evidently, such a solution exists always. Therefore the following assertion is true: *if $G \leq G_0$, then the attractor dimension of the barotropic vorticity equation on sphere is equal to 0.*

This assertion means that only the stationary point may represent the attractor of system. If φ is the stationary solution of the barotropic equation for the small Grasshoff numbers, then the estimate holds [75]:

$$\|\varphi\| \leq G \leq G_0 = \frac{1}{3}\sqrt{\frac{\pi}{2e}} \tag{5.24}$$

The more complicated is the investigation of the asymptotics of the large Grasshoff numbers. The matter is that we do not know a priori the accuracy of the estimate for a given specific f. It is clear that for the different f possessing one the same norm the structure of attractor may be of any form. If one supposes that there exists a nonempty set $\{f\}$ for which the estimate (3.29), for large G, is practically accurate, then the analysis of asymptotics for this set, given the large G, is not complicated procedure. It is easily seen that in this case as $G \to \infty$

the following asymptotics takes place:

$$\delta\left(\dim_H A\right) \sim \frac{\delta G}{G^{1/3}}, \tag{5.25}$$

where δ is the symbol of the increment. Hence a given characteristic of attractor for the large G will be stable one in the sense that there exists a continuous dependence of the change of attractor dimension on the change of Grasshoff number with the proportionality coefficient tending to 0 as $G \to \infty$. The physical sense of such a stability will be elucidated in the next chapters, when we shall consider the connection between the attractor dimension and the number of statistically independent degrees of freedom.

The second problem which will be considered in the present section is the analysis of the dependence of attractor dimension of barotropic atmospheric model on the degree of the Laplacian operator in the expression for the horizontal diffusion. This problem is actual and real since at present the idea of using the Laplacian operator having the degree more than first one for the description of the horizontal diffusion is realized in many models of general atmospheric circulation.

Let us have the following equation for the barotropic vorticity on sphere

$$\frac{\partial \omega}{\partial t} + J(\psi, \omega + l) = -\nu_s(-\Delta)^s \omega + f,$$

$$\omega|_{t=0} = \omega_0, \quad \omega = \Delta \psi. \tag{5.26}$$

The estimate of the Hausdorff dimension of the attractor for this problem has the form [75]:

$$\dim_h(A) \leq \left(\frac{2^s(s+1)}{\sqrt{\pi}}\right)^{\frac{2}{2s+1}} \left(\frac{2s+2}{2s+1}\right)^{1/2s+1}.$$

$$G^{\frac{2}{2s+1}}\left(\ln G + \frac{1}{2s+1}\ln(2^s(s+1)) - \frac{1}{2}\ln\pi\right)^{1/2s+1}. \tag{5.27}$$

The first problem to be solved is the problem of choice of the coefficient ν_s as the function of s. It may have the next solution. We shall choose the coefficient ν_s so that the characteristic time of dissipation for the spherical harmonic with number n_0 is constant and does not depend on s. Let φ_{n_0} be a given spherical harmonic. Thus, we have

$$-\nu\Delta\varphi_{n_0} = \nu n_0(n_0+1)\varphi_{n_0},$$

$$-\nu_s(-\Delta)^s\varphi_{n_0} = -(n_0(n_0+1))^s\nu_s\varphi_{n_0}.$$

The characteristic time of dissipation for a given spherical harmonic may be expressed as follows

$$T = \frac{1}{\nu(n_0+1)n_0} = \frac{1}{\nu_s[n_0(n_0+1)]^s}.$$

Hence, we have for the large n_0 and $s > 1$:

$$\nu_s = \frac{\nu(n_0 + 1)n_0}{[n_0(n_0 + 1)]^s} \approx \nu n_{n_0}^{2-2s}. \qquad (5.28)$$

Since the Grasshoff number in (5.27) has the form [75]

$$G = \frac{\|f\|_1}{\nu_s^2 \lambda_1^{2s-1}},$$

and in the dimensionless case $\lambda_1 = 2$, one has

$$G = \frac{\|f\|_1}{\nu^2 n_0^{2-2s} 2^{2s-1}} = \frac{\|f\|_1}{\nu^2 \lambda_1} \left(\frac{n_0}{2}\right)^{2s-2} \equiv G_1 \left(\frac{n_0}{2}\right)^{2s-2}. \qquad (5.29)$$

We consider the asymptotics of the estimate (5.27) for sufficiently large G_1 and $s \to \infty$. Since

$$\lim_{s \to \infty} \left(\frac{2^s(s+1)}{\sqrt{\pi}}\right)^{2/2s+1} = 2, \quad \lim_{s \to \infty} \left(\frac{2s+2}{2s+1}\right)^{1/2s+1} = 1$$

$$\lim_{s \to \infty} (\ln G)^{1/2s+1} = 2, \quad \lim_{s \to \infty} \left[G_1(\frac{n_0}{2})^{2s-2}\right]^{2/2s+1} = \left(\frac{n_0}{2}\right)^2,$$

the equality holds

$$\dim_H(A) \le \frac{(n_0)^2}{4} 4 = n_0^2 \qquad (5.30)$$
$$s \to \infty.$$

The result (5.30) is not surprising one, since the dissipation described by the Laplacian operator of the degree s, where s is very large, acts as the rectangular spatial filter cutting off all harmonics with $n > n_0$ so that the system must behave itself as the finite-dimensional Hamilton system.

3. Let us proceed to the problem of application of the estimates obtained to the calculation of the dimension of attractors of the barotropic equations with the real value of forcing. If we assume that the turbulent diffusion coefficient for the large-scale quasibarotropic processes in the atmosphere is the value $\nu \sim 10^5 \mathrm{m}^2/\mathrm{sec}$. and calculate the norm of the "real" baroclinic forcing, then the estimate of attractor dimension for such a model will have the value $\sim 10^4$.

To calculate the real value of attractor dimension of the system (5.1) for fixed ν, σ, h, ρ and l, we must consider the sequence of the set of equations with continuously increasing number of Galerkin basis functions. Each of these systems will possess its own attractor. It was

shown in [4] that the limit attractor of this sequence of systems will
be embedded into the attractor of the original system

$$\lim_{m \to \infty} A_m = A_\infty \subset A, \tag{5.31}$$

It follows from (5.31) that

$$\dim A_\infty \leq \dim A.$$

The numerical calculations made by A.S.Gritsun (the personal
communication) with respect to the system (5.1), where the right-
hand side was calculated as the remainder term of the equation (5.1)
for ψ which was equal to the winter mean climatic value of the
streamfunction at 500 mb level,

$$\sigma \sim \frac{1}{14\text{days}}, \quad \nu \approx 2 \cdot 10^5 \text{m}^2/\text{sec}, \quad \rho = 1, \quad h_m = 3\text{km}$$

for the maximal number of spherical harmonics $\sim 10^3$ the attractor
had the value of dimension less than 40. This results shows that in
this case the analytical estimate is rough one, since it has the value
$\sim 10^4$.

Chapter 6

Two-Layer Baroclinic Model

6.1 Two-Layer Baroclinic Model

Let us consider the baroclinic atmosphere equations in p-system of coordinates:

$$\frac{du}{dt} - lv = -\frac{\partial \varphi}{\partial x} + \frac{\partial}{\partial p} \nu \frac{\partial u}{\partial p} + \mu \Delta u,$$

$$\frac{dv}{dt} + lu = -\frac{\partial \varphi}{\partial y} + \frac{\partial}{\partial p} \nu \frac{\partial v}{\partial p} + \mu \Delta v,$$

$$\frac{\partial \Phi}{\partial p} = -\frac{RT}{p} \qquad \frac{\partial u}{\partial x} + \frac{\partial v}{\partial y} + \frac{\partial \tau}{\partial y} = 0,$$

$$\frac{dT}{dt} = \frac{RT}{pc_p}\tau + \frac{\partial}{\partial p}\nu_1\frac{\partial T}{\partial p} + \mu_1 \Delta T + \varepsilon/c_p,$$

$$\frac{d}{dt} = \frac{\partial}{\partial t} + u\frac{\partial}{\partial x} + v\frac{\partial}{\partial y} + \tau\frac{\partial}{\partial p},$$

or if we set $T = T' + \bar{T}(p)$:

$$\frac{dT'}{dt} = \frac{R\bar{T}}{pg}(\gamma_a - \bar{\gamma})\tau + \frac{\partial}{\partial p}\nu_1\frac{\partial T'}{\partial p} + \mu \Delta T' + \varepsilon/c_p.$$

Here T is the temperature, u, v, τ are components of the velocity vector in p-system of coordinates, ϕ the geopotential, γ_a the dry-adiabatic temperature gradient, $\bar{\gamma}$ the gradient of standard distribution of temperature in z-system of coordinates, c_p the specific heat of air at constant pressure, ε represents diabatic heating, R the gas constant.

We rewrite the first two equations for the vorticity

$$\omega = \frac{\partial v}{\partial x} - \frac{\partial u}{\partial y},$$

neglecting the terms of vertical transport

$$\frac{d}{dt}(\omega + l) + (\omega + l)\left(\frac{\partial u}{\partial x} + \frac{\partial v}{\partial y}\right) = \frac{\partial}{\partial p}\nu\frac{\partial \omega}{\partial p} + \mu\Delta\omega,$$

or

$$\frac{d}{dt}(\omega + l) = (\omega + l)\frac{\partial \tau}{\partial p} + \frac{\partial}{\partial p}\nu\frac{\partial \omega}{\partial p} + \mu\Delta\omega.$$

We make the next simplifications. We shall assume that $\omega \ll l$ and $l = l_0 + l_1(\varphi)$, so that $l_1(\varphi) \ll l_0$. Introduce the geostrophic stream-function

$$\psi = \phi'/l_0$$

and the geostrophic velocities

$$u = -\frac{\partial \psi}{\partial y}, \quad v = \frac{\partial \psi}{\partial x}.$$

The equation for hydrostatics can be rewritten as follows

$$\frac{\partial \psi}{\partial p} = -\frac{RT'}{pl_0}.$$

Finally, we write the equation for ω and T' in the form

$$\frac{\partial}{\partial t}\Delta\psi + J(\psi, \Delta\psi + l) = l_0\frac{\partial \tau}{\partial p} + \frac{\partial}{\partial p}\nu\frac{\partial}{\partial p}\Delta\psi + \mu\Delta^2\psi$$

$$\frac{\partial}{\partial t}T' + J(\psi, T') = \gamma_0\tau + \frac{\partial}{\partial p}\nu_1\frac{\partial T'}{\partial p} + \mu_1\Delta T' + \frac{\varepsilon}{c_p}$$

$$\gamma_0 = \frac{R\bar{T}}{pg}(\gamma_a - \bar{\gamma}). \tag{6.1}$$

We shall use the following vertical structure of the grid

———————————— $p = 0,$ Plot 1.
———————————— $p = 250$ mb $(1/2)$,
———————————— $p = 500$ mb (1),
———————————— $p = 750$ mb $(3/2)$,
———————————— $p = 1000$ mb.

As the boundary dynamical conditions there will be taken $\tau = 0$ for $p = 0$ and $p = 1000$ mb

$$\left.\nu\frac{\partial\omega}{\partial p}\right|_{p=0} = 0, \quad \left.\nu_1\frac{\partial T'}{\partial p}\right|_{p=0} = 0,$$

$$\left.\nu\frac{\partial\omega}{\partial p}\right|_{p=1000} = -\sigma_0\omega_{1000} \approx -\sigma_i\omega_{750}, \quad \sigma = c|\bar{u}|g\rho > 0,$$

$$\left.\nu_1\frac{\partial T'}{\partial p}\right|_{p=1000} = -\hat{\sigma}(T - T_0).$$

We write down the equations for $\Delta\psi$ at the levels $1/2$ and $3/2$

$$\frac{\partial}{\partial t}\Delta\psi_{1/2} + J(\psi_{1/2}, \Delta\psi_{1/2} + l) = l_0\frac{\tau_1}{\Delta p} + \mu\Delta^2\psi_{1/2},$$

$$\frac{\partial}{\partial t}\Delta\psi_{3/2} + J(\psi_{3/2}, \Delta\psi_{3/2} + l) = -l_0\frac{\tau_1}{\Delta p} - \frac{\sigma_0}{\Delta p}\Delta\psi_{3/2} + \mu\Delta^2\psi_{3/2}. \quad (6.2)$$

The temperature equation for the layer (1) has the form:

$$\frac{\partial T'_1}{\partial t} + J(\psi_1, T'_1) = \gamma_0\tau_1 + \mu_1\Delta T'_1 + \frac{\nu_1}{2\Delta p}\left.\frac{\partial T'}{\partial p}\right|_{1000} + \frac{\varepsilon_1}{c_p},$$

$$\left.\nu_1\frac{\partial T'}{\partial p}\right|_{1000} = -\hat{\sigma}_1(T_1 - T_0) = -\hat{\sigma}_1(T'_1 + \bar{T}_1 - T_0) = -\hat{\sigma}_1 T'_1 + \hat{\sigma}_1(T_0 - \bar{T}_1).$$

Finally, we have

$$\frac{\partial T'_1}{\partial t} + J(\psi_1, t'_1) = \gamma_0\tau_1 + \mu_1\Delta T'_1 - \sigma_1 T'_1 + \sigma_1(T_0 - \bar{T}_1) + \frac{\varepsilon_1}{c_1},$$

where

$$\sigma_1 = \frac{\hat{\sigma}_1}{2\Delta p}.$$

Introduce now the baroclinic and barotropic components of relative vorticity

$$u_1 = \frac{\psi_{1/2} + \psi_{3/2}}{2}, \quad u_2 = \frac{\psi_{3/2} - \psi_{1/2}}{2}.$$

Setting $\sigma/2 = \sigma_0/2\Delta p$, we obtain

$$\frac{\partial\Delta u_1}{\partial t} + J(u_1, \Delta u_1 + l) + J(u_2, \Delta u_2) = \mu\Delta^2 u_1 - \frac{\sigma}{2}\Delta(u_1 + u_2),$$

$$\frac{\partial\Delta u_2}{\partial t} + J(u_1, \Delta u_2) + J(u_2, \Delta u_1 + l) = \mu\Delta^2 u - \frac{\sigma}{2}\Delta(u_1 + u_2) - \frac{l_0\tau_1}{\Delta p}.$$

Substitute now the temperature expressed by ψ

$$T_1' = -\frac{\partial \psi_1}{\partial p}\frac{p_1 l_0}{R} = -\frac{\psi_{3/2} - \psi_{1/2}}{\Delta p}\frac{p_1 l_0}{R} = -u_2 \frac{p_1 l_0}{R\delta p}$$

into the temperature equation. Then we get

$$-\frac{p_1 l_0}{R\Delta p}\left(\frac{\partial u_2}{\partial t} + J(u_1, u_2)\right) = \gamma_0 \tau_1 + \mu_1 \left(-\frac{p_1 l_0}{R\Delta p}\right)\Delta u_2$$

$$-\sigma_1 \left(-\frac{p_1 l_0}{R\Delta p}\right) u_2 + f_1, \quad f_1 \equiv \sigma_1(T_0 - \bar{T}_1) + \frac{\varepsilon}{c_p},$$

or

$$\frac{\partial u_2}{\partial t} + J(u_1, u_2) = -\frac{R\Delta p \gamma_0}{p_1 l_0}\tau_1 + \mu_1 \Delta u_2 - \sigma_1 u_2 - f,$$

$$f = \frac{R\Delta p}{p_1 l_0}f_1.$$

We denote

$$\frac{R\Delta p \gamma_0}{p_1 l_0} = a, \quad \text{and} \quad \frac{1}{a}\frac{l_0}{\Delta p} = \alpha^2.$$

Then, we eliminate τ_1 from the equation for u_2 and obtain the system of equations

$$\frac{\partial \Delta u_1}{\partial t} + J(u_1, \Delta u_1 + l) + J(u_2, \Delta u_2) = \mu\Delta^2 u_1 - \frac{\sigma}{2}\Delta(u_1 - u_2),$$

$$\frac{\partial \Delta u_2}{\partial t} + J(u_1, \Delta u_2) + J(u_2, \Delta u_1 + l) = -\frac{\sigma}{2}\Delta(u_1 + u_2) + \mu\Delta^2 u_2$$

$$+\alpha^2\left[\left(\frac{\partial u_2}{\partial t} + J(u_1, u_2) - \mu_1 \Delta u_2 + \sigma_1 u_2 + f\right)\right]. \tag{6.3}$$

In the sequel we shall consider this model in spherical system of coordinates, denoting $\alpha^2 \mu_1$ by μ_1 and $\alpha^2 \sigma_1$ by σ_1 and $\alpha^2 f$ by f. All parameters $\sigma, \mu, \sigma_1, \mu_1, \alpha^2$ of (6.3) are positive and have the next dimensions $[\sigma] = T^{-1}$, $[\mu] = [L^2 T^{-1}]$, $[\sigma_1] = T^{-1}L^2$, $[\mu_1] = T^{-1}$, $[\alpha^2] = L^{-2}$. It is convenient to write (6.3) in the vector-matrix form by introducing the matrices-operators L_1, L_2 and vectors l_1, l_2 and F:

$$L_1 = \begin{pmatrix} \Delta & 0 \\ 0 & \Delta - \alpha^2 \end{pmatrix},$$

$$L_2 = \begin{pmatrix} -\mu\Delta^2 + \dfrac{\sigma}{2}\Delta & \dfrac{\sigma}{2}\Delta \\ \dfrac{\sigma}{2}\Delta & -\mu\Delta^2 + \left(\dfrac{\sigma}{2} + \mu_1\right)\Delta - \sigma_1 \end{pmatrix},$$

$$l_1(u) = \begin{pmatrix} J(u_1, l) \\ J(u_2, l) \end{pmatrix},$$

$$l_2(u) = \begin{pmatrix} J(u_1, \Delta u_1) + J(u_2, \Delta u_2) \\ J(u_1, \Delta u_1) + J(u_2, \Delta u_1) - \alpha^2 J(u_1, u_2) \end{pmatrix},$$

$$u = \begin{pmatrix} u_1 \\ u_2 \end{pmatrix}, \quad F = \begin{pmatrix} 0 \\ f \end{pmatrix}. \tag{6.4}$$

Taking into account (6.4), we write (6.3) as

$$\frac{\partial}{\partial t} L_1 u + L_2 u + l_1(u) + l_2(u) = F, \quad u|_{t=0} = u_0. \tag{6.5}$$

It is useful to solve the system (6.5) with respect to $\partial u/\partial t$. This can be done in two ways: one either multiplies (6.5) by L_1^{-1}, or introduces a new variable $\psi = L_1 u$, $u = L_1^{-1}\psi$. In the first case one has the equation

$$\frac{\partial u}{\partial t} + L_1^{-1} L_2 u + L_1^{-1} l_1(u) + L_1^{-1} l_2(u) = L_1^{-1} F, \quad u|_{t=0} = u_0. \tag{6.6}$$

In the second case (6.5) takes the form

$$\frac{\partial}{\partial t} \psi + L_2 L_1^{-1} \psi + l_1(L_1^{-1}\psi) + l_2(L_1^{-1}\psi) = F, \quad \psi_{t=0} = L_1 u_0 = \psi_0. \tag{6.7}$$

We introduce bilinear forms $a_1(u, v)$, $a_2(u, v)$, $b_1(u, v)$ and trilinear one $b_2(u, v, \omega)$:

$$a_1(u, v) = \int_{S^2} (\Delta u_1 \cdot \Delta v_1 + \Delta u_2 \cdot \Delta v_2 + \alpha^2 u_2 v_2) ds,$$

$$a_2(u, v) = \int_{S^2} \left\{ \mu(\Delta u_1 \cdot \Delta v_1 + \Delta u_2 \cdot \Delta v_2) + \mu_1 \Delta u_2 \cdot \Delta v_2 \right.$$

$$\left. + \frac{\sigma}{2} \Delta(u_1 + u_2) \cdot \Delta(v_1 + v_2) + \sigma_1 u_2 v_2 \right\} ds,$$

$$b_1(u, v) = \int_{S^2} (J(u_1, l)v_1 + J(u_2, l)v_2) ds,$$

$$b_2(u, v, \omega) = \int_{s^2} J \left\{ \omega_1, v_1 \right) \Delta u_1 + J(\omega_2, v_2) \Delta u_1 + J(\omega_1, v_2) \Delta u_2$$

$$+ J(\omega_2, v_1) \Delta u_2 + \alpha^2 J(\omega_2, v_2) u_1 \right\} ds.$$

Here the following lemmas are valid.

Lemma 6.1. *Bilinear form* $a_1 : V_1 \times V_1 \to R$ *is continuous and coercive, i.e.*

$$|a_1(u,v)| \leq c_1 \|u\|_{V_1} \|v\|_{V_1}, \quad \|u\|_{V_1}^2 \leq a_1(u,u) \leq c_1 \|u\|_{V_1}^2, \ c_1 = 1 + \frac{\alpha^2}{\lambda_1}.$$

Lemma 6.2. *Bilinear form* $a_2 : V_2 \times V_2 \to R$ *is continuous and coercive, i.e.*

$$|a_2(u,v)| \leq c_2 \|u\|_{V_2} \cdot \|v\|_{V_2},$$

$$\mu \|u\|_{V_2}^2 \leq a_2(u,u) \leq c_2 \|u\|_{V_2}^2, \quad c_2 = \mu_1 + \frac{\mu_1 + 2\sigma}{\lambda_1} + \frac{\sigma_1}{\lambda_1^2}.$$

Lemma 6.3. *Bilinear form* $b_1 : v_1 \times v_1 \to R$ *is continuous and skew-symmetric. Here*

$$|a_1(u,v)| \leq \begin{cases} \max\limits_{S^2} |l| \cdot \|u\|_{v_1} \cdot \|v\|_{v_1}, & u,v \in V_1, \\ \max\limits_{S^2} |\Delta l| \cdot \|u\|_{v_1} \cdot \|v\|_{v_0}, & u \in V_1, \ v \in V_0, \\ \max\limits_{S^2} |l| \lambda_1^{-1/2} \|u\|_{v_1} \cdot \|v\|_{v_2}, & u \in V_1, \ v \in V_2; \end{cases}$$

$$b_1(u,v) = -b_1(v,u), \quad b_1(u,u) = 0.$$

Lemma 6.4. *Trilinear form* $b(u,v,\omega) : V_2 \times V_2 \times V_2 \to R$ *is continuous and admits the following estimates:*

$$b_2(u,v,\omega) \leq \begin{cases} \lambda_1^{-1/2} c \|u\|_{V_2} \cdot \|v\|_{V_2} \cdot \|\omega\|_{V_2}, \\ c \|v\|_{V_3}^{1/2} \cdot \|u\|_{V_3}^{1/2} \cdot \|v\|_{V_3}^{1/2} \cdot \|v\|_{V_1}^{1/2} \cdot \|\omega\|_{V_1}, \\ c \|u\|_{V_4}^{1/2} \cdot \|u\|_{V_3}^{1/2} \cdot \|v\|_{V_2}^{1/2} \cdot \|v\|_{V_1}^{1/2} \cdot \|\omega\|_{v_0}; \end{cases}$$

$$b_2(u,v,\omega) = -b(u,\omega,v,), \quad c = \sqrt{2}\left(4 + \frac{\alpha^2}{\lambda_1}\right).$$

Lemmas 6.1 and 6.2 allow us to introduce the operators A_1 and A_2:

$$\langle A_1 u, v \rangle = a_1(u,v), \quad A_1 : V_1 \to V_1',$$
$$\langle A_2 u, v \rangle = a_2(u,v), \quad A_2 : v_2 \to v_2'.$$

The operator A_1 is the isomorphism between V_1 and V_1', thereby

$$\|A_1 u\|_{V_1'} \leq c_1 \|u\|_{V_1}, \quad \|A_1\|_{z(V_1, V_1')} \leq c_1,$$

$$\|A_1^{-1}\|_z(V_1', V_1) \leq 1, \quad c_1 = 1 + \frac{\alpha^2}{\lambda_1}.$$

This operator can be extended as unbounded positive selfadjoint operator in V_0, if one sets

$$D(A_1) = \{u : u \in V_1, \quad A_1 u \in V_0\}.$$

Here A_1 will have the compact inverse operator $A_1^{-1} : V_0 \to V_0$. Similarly, the operator A_2 is the isomorphism between V_2 and V_2', hence

$$\|A_2 u\|_{V_2'} \le c_2 \|u\|_{V_2}, \quad \|A_2\|_{L(V_2,V_2')} \le c_2, \quad \|A^{-1}\|_{L(V_2',V_2)} \le \frac{1}{\mu},$$

$$c_2 = \mu_1 + \frac{\mu_1 + 2\sigma}{\lambda_1} + \frac{\sigma_1}{\lambda_1}.$$

The operator A_2 can be extended as unbounded positive operator in V_0, if one sets

$$D(A_2) = \{u : u \in V_2, \quad A_2 u \in V_0\}.$$

The operator A_2^{-1} will have the compact inverse one $A_2^{-1} : V_0 \to V_0$. Eigenfunctions ω_{jk} of operators A_k, $k = 1, 2$ form basises in V_0:

$$A_k \omega_{jk} = \mu_{jk} \omega_{jk}, \quad k = 1, 2,$$

$$a_k(\omega_{jk}, \omega_{ik}) = \langle A_k \omega_{jk}, \omega_{ik} \rangle = \mu_{jk} \delta_{jk},$$

$$0 < \mu_{1k} \le \mu_{2k} \le \cdots, \quad k = 1, 2.$$

In the usual way one can determine the degrees A_k^α, $\alpha \in R$, for $k = 1, 2$, of the operators A_k. In a given case

$$A_k^\alpha u = \sum_{j=1}^\infty \mu_{jk}^\alpha (u, \omega_{jk}) \omega_{jk}, \quad k = 1, 2.$$

The operators A_k^α, for $k = 1, 2$ are unbounded positive selfadjoint operators in V_0 with the dense in V_0 domain of definition

$$D(A_k^\alpha) = \{u : u \in V_0, \quad \sum_{j=1}^\infty \mu_{jk}^{2\alpha}(u, \omega_{jk})^2 < \infty\},$$

where A_k^α is the isomorphism $D(A^\alpha)$ in V_0.

The spaces $D(A_k^\alpha)$, $\alpha \in R$, for $k = 1, 2$, may be combined with the scalar product

$$(u, v)_{D(A_k^\alpha)} = (A_k^\alpha u, A_k^\alpha v) = \sum_{j=1}^\infty \mu_{jk}^{2\alpha}(u, \omega_{jk})(v, \omega_{jk}),$$

$$\|u\|_{D(A_k^\alpha)} = \left(\sum_{j=1}^\infty \mu_j^{2\alpha}(u, \omega_{jk})^2 \right)^{1/2}.$$

We notice that the embeddings $D(A_k^\alpha) \subset (A_k^{\alpha-\epsilon})$ are compact for $\forall \alpha \in R$ and $\forall \epsilon > 0$.

Lemma 6.5. *Eigenfunctions and eigenvalues of the operator A_1 have the form*

$$\omega_{1mn} = \begin{pmatrix} Y_{mn} \\ 0 \end{pmatrix}, \quad \hat{\omega}_{1mn} = \begin{pmatrix} 0 \\ Y_{mn} \end{pmatrix},$$

$$\mu_{j1} = \begin{pmatrix} \lambda_j \\ \lambda_j + \alpha^2 \end{pmatrix} \equiv \begin{pmatrix} \mu_{j1} \\ \hat{\mu}_{j2} \end{pmatrix}.$$

If $f \in D(A_1^\alpha)$ and

$$f = \begin{pmatrix} f_1 \\ f_2 \end{pmatrix} = \begin{pmatrix} \sum\limits_{n=1}^{\infty} \sum\limits_{|m|\leq n} f_{1mn} Y_{mn} \\ \sum\limits_{n=1}^{\infty} \sum\limits_{|m|\leq n} f_{2mn} Y_{mn} \end{pmatrix},$$

then

$$A_1^\alpha f = \begin{pmatrix} \sum\limits_{n=1}^{\infty} \sum\limits_{|m|\leq n} \lambda_n^\alpha f_{1mn} Y_{mn} \\ \sum\limits_{n=1}^{\infty} \sum\limits_{|m|\leq n} (\lambda_n + \alpha^2)^\alpha f_{2mn} Y_{mn} \end{pmatrix}.$$

Lemma 6.6. *If $\beta \geq 0$, then*

$$\|u\|_{V_\beta} \leq \|A_1^{\beta/2} u\|_{V_0} \leq \left(1 + \frac{\alpha^2}{\lambda_1}\right)^{\beta/2} \|u\|_{V_\beta}.$$

If $\beta < 0$, then

$$\left(1 + \frac{\alpha^2}{\lambda_1}\right)^{\beta/2} \|u\|_{V_\beta} \leq \|A_1^{\beta/2} u\|_{V_0} \leq \|u\|_{V_\beta}.$$

Lemma 6.7. *Eigenfunctions and eigenvalues of the operator A_2 are*

$$\omega_{2mn} = \begin{pmatrix} Y_{mn} \\ 0 \end{pmatrix}, \quad \hat{\omega}_{2mn} = \begin{pmatrix} 0 \\ Y_{mn} \end{pmatrix},$$

$$\mu_{j2} = \begin{pmatrix} \mu_{j2} \\ \hat{\mu}_{j2} \end{pmatrix} = \begin{pmatrix} \mu\lambda_j^2 + a_j + b_j - \sqrt{a_j^2 + b_j^2} \\ \mu\lambda_j^2 + a_j + b_j + \sqrt{a_j^2 + b_j^2} \end{pmatrix},$$

$$a_j = \frac{\sigma\lambda_j}{2}, \quad b_j = \frac{\mu_1\lambda_j + \sigma_1}{2},$$

$$\mu\lambda_j^2 \leq \mu_{j2} \leq \mu\lambda_j^2 + (\sigma + \mu_1)\lambda_j + \sigma_1.$$

We notice that if

$$f = \begin{pmatrix} f_1 \\ f_2 \end{pmatrix} = \begin{pmatrix} \sum\limits_{n=1}^{\infty} \sum\limits_{|m| \leq n} f_{1mn} Y_{mn} \\ \sum\limits_{n=1}^{\infty} \sum\limits_{|m| \leq n} f_{2mn} Y_{mn} \end{pmatrix}, \quad f \in D(A_2^\alpha),$$

then

$$A_2^\alpha f = \begin{pmatrix} \sum\limits_{n=1}^{\infty} \sum\limits_{|m| \leq n} \mu_{n2}^\alpha f_{1mn} Y_{mn} \\ \sum\limits_{n=1}^{\infty} \sum\limits_{|m| \leq n} \hat{\mu}_{nj}^\alpha f_{2mn} Y_{mn} \end{pmatrix}.$$

Lemma 6.8. *If $\beta \geq 0$, then*

$$\mu^\beta \|u\|_{V_{4\beta}} \leq \|A_2^\beta u\| \leq \left(\mu + \frac{\sigma + \mu_1}{\lambda_1} + \frac{\sigma_1}{\lambda_1^2} \right)^\beta \|u\|_{V_{4\beta}}.$$

If $\beta < 0$, then

$$\left(\mu + \frac{\sigma + \mu_1}{\lambda_1} + \frac{\sigma_1}{\lambda_1^2} \right)^\beta \|u\|_{V_{4\beta}} \leq \|A_2^\beta u\| \leq \mu^\beta \|u\|_{V_{4\beta}}.$$

Lemmas 6.3 and 6.4 allow us to introduce the operators $B_1(u), B_2(u,v)$ according to the equalities

$$\langle B_1(u), v \rangle = b_1(u,v), \quad u \in V_1, \quad v \in V_0, \quad B_1 : V_1 \to V_0',$$

$$\langle B_2(u,v), \omega \rangle = b_2(u,v,\omega), \quad B_2 : V_2 \times V_2 \to V_2', \quad u, v, \omega \in V_2.$$

It should also be noted that $B_1(u) : V_1 \to V_1'$ and

$$\|B_1(u)\|_{V_0'} \leq \max_{s^2} |\nabla l| \|u\|_{V_1},$$

$$\|B_1(u)\|_{V_2'} \leq \max_{s^2} |l| \lambda_1^{-1/2} \|u\|_{V_1}.$$

Evidently,

$$\|B_2(u,v)\|_{V_2'} \leq c \lambda_1^{-1/2} \|u\|_{V_2} \|v\|_{V_2},$$

$$\|B_2(u,v)\|_{V_1'} \leq c \lambda_1^{-1/2} \|u\|_{V_3} \|v\|_{V_2}, \quad c\sqrt{2} \left(4 + \frac{\alpha^2}{\lambda_1} \right).$$

Now one can introduce the concept of the generalized solutions of the problem (6.4).

Problem 1. Definition. Let $u_0 \in V_1$, $F \in L^2(0,T;V_2)$ and $T > 0$. The generalized solution of (6.4) is the function u such that

$$u \in L^2(0,T;V_2) \cap L^\infty(0,T;V_1),$$

$$a_1\left(\frac{\partial u}{\partial t}, v\right) + a_2(u,v) + b_1(u,v) + b_2(u,u,v) = \langle F, v\rangle,$$

$$\forall\, v \in V_2, \quad u|_{t=0} = u_0. \tag{6.8}$$

In this case

$$A_2 u(t) \in L^2(0,T;V_2'), \quad B_1 u(t) \in L^2(0,T;V_2'),$$

$$B_2(u(t), u(t)) \in L^2(0,T;V_2').$$

We write the equality (6.8) as

$$\left\langle A_1 \frac{\partial u}{\partial t}, v \right\rangle = \langle F - A_2 u - B_1(u) - B_2(u,u), v\rangle, \quad \forall\, v \in V_2$$

or

$$A_1 \frac{\partial u}{\partial t} = F - A_2 u - B_1(u) - B_2(u,u), \quad \forall\, u \in V_2,$$

hence,

$$A_1 \frac{\partial u}{\partial t} \in L^2(0,T;M_2').$$

Consequently,

$$A_1 u \in C(0,T;V_2'), \quad u \in C([0,T], V_0).$$

Thus, the initial condition $u|_{t=0} = 0$ has the sense.

Now we can give another formulation of the generalized solution.

Problem 2. Let $u_0 \in V_1$, $F \in L^2(0,T;V_1')$. We need to find the function u such that

$$u \in L^2(0,T;V_2) \cap L^\infty(0,T;V_1), \quad u' \in L^2(0,T;V_1'),$$

$$A_1 \frac{\partial u}{\partial t} + A_2 u + B_1(u) + B_2(u,u) = F, u|_{t=0} = u_0.$$

Theorem 6.1. *There exists at least one solution of Problem 1 such that*

$$u \in L^2(0,T;V_2) \cap L^\infty(0,T;V_2) \cap C([0,T], V_0), \ u' \in L^2(0,T;V_1'),$$

$$\|u\|_{V_1}^2 + \mu_0 \int_0^t \|u\|_{V_2}^2 dt \le c_1 \|u_0\|_{V_1}^2 + \frac{1}{\mu} \int_0^t \|F\|_{V_2'}^2 dt.$$

Theorem 6.2. *If $u_0 \in v_2$, then there exists a unique generalized solution, such that*

$$u \in L^2(0,T;V_3) \cap c([0,T],V_2), \ u' \in L^2(0,T;V_2).$$

If one has $u_0 \in v_3$, then

$$u \in ([0,T],V_3) \cap L^2(0,T;V_4).$$

The mapping $u \to u(t)$ for every $t \in [0,T]$ is continuous.

If $f \in V_0$ and it do not depend on t, then by virtue of this theorem the considered system generates the semigroup of continuous operators $S(t) : V_2 \to V_2$, $\forall t \geq 0$. This semigroup will possess the absorbing set $B_R(0) \subset V_2$ and the global attractor.

Theorem 6.3. *The semigroup $S(t)$ possesses: 1) a bounded absorbing set $B_R(0)$ in $V v_2$, which attracts all bounded sets of the space v_2, and 2) the global attractor $A \subset B_R(0)$.*

6.2 Estimate of Attractor Dimension

Let us estimate the attractor dimension of the semigroup $S(t)$ generated by the problem (6.5) [68]. Taking the scalar product of (6.5) by u in $L_0^2(S)^2$, we find

$$\frac{1}{2}\frac{\partial}{\partial t}\|(-L_1)^{1/2}u\|^2 + \|(-L_2)^{1/2}u\|^2$$
$$+(l_1(u),u) + (l_2(u),u) = (F,u). \tag{6.9}$$

Notice that $(l_1(u),u) + (l_2(u),u) = 0$. Therefore (6.9) takes the form

$$\frac{\partial}{\partial t}\|(-L_1)^{1/2}u\|^2 + 2\|(-L_2)^{1/2}u\|^2$$
$$= 2(F,u) \leq \|(-L_2)^{1/2}u\|^2 + \|(-L_2)^{1/2}F\|^2.$$

Taking into account that $\|u\|^2 \leq \lambda_1^{-1}\|\nabla u\|^2$, $\|\nabla u\|^2 \leq \lambda_1^{-1}\|\Delta u\|^2$, we get

$$\|(-L)^{1/2}u\|^2 = \|\nabla u_1\|^2 + \|\nabla u_2\|^2 + \alpha^2\|u_2\|^2$$
$$\leq \lambda_1^{-1}(\|\Delta u_1\|^2 + \|\Delta u_2\|^2) + \frac{\alpha^2}{\lambda_1\mu_1 + \sigma_1}(\mu_1\|\nabla u_2\|^2 + \sigma_1\|u_2\|^2)$$
$$\leq c_1[\mu(\|\Delta u_1\|^2 + \|\Delta u_2\|^2) + \mu_1\|\nabla u_2\|^2 + \sigma_1\|u_2\|^2 + \frac{\sigma}{2}\|\nabla u_1 + \nabla u_2\|^2]$$
$$= c_1\|(-L_2)^{1/2}u\|^2.$$

Here $c_1 = \max\left(1/\lambda_1\mu, \; \alpha^2/\lambda_1\mu_1 + \sigma_1^2\right)$. With respect to this inequality we obtain

$$\frac{\partial}{\partial t}\,\|(-L_1)^{1/2}\,u\|^2 + c_1^{-1}\,\|(-L_1)^{1/2}\,u\|^2 \le c_1\,\|(-L_1)^{-1/2}\,F\|^2,$$

hence it follows that

$$\|(-L_1)^{1/2}u\|^2 \le \|(-L_1)^{1/2}u\|^2 e^{-t/c_1}$$
$$+ c_1^2\|(-L_1)^{-1/2}F\|^2(1 - e^{t/c_1}). \tag{6.10}$$

Consequently, the attractor lies inside the set

$$M_1 = \{u \in V_1 : \|(-L_1)^{1/2}\,u\| \le c_1\,\|(-L)^{-1/2}\,F\|\}.$$

Multiplying now the equation (6.3) by $X\,u$, where

$$X = \begin{pmatrix} \Delta - x & 0 \\ 0 & \Delta - (x + \alpha^2) \end{pmatrix}, \quad x \ge 0,$$

we get

$$\left(\frac{\partial}{\partial t}L_1 u, X u\right) + (L_2 u, X u) + (l_1(u) + l_2(u), X u) = (F, x u). \tag{6.11}$$

We transformate the first term in the left-hand side of the equality

$$\left(\frac{\partial}{\partial t}\,L_1\,u,\, X\,u\right) = \frac{\partial}{\partial t}\,(L_1\,u,\, x\,u) - \left(L_1\,u,\, \frac{\partial}{\partial t}\,X\,u\right)$$
$$= \frac{\partial}{\partial t}\,(L_1\,u,\, X\,u) - \left(L_1\,u,\, X\,L_1^{-1}\,\frac{\partial}{\partial t}\,L_1\,u\right)$$
$$= \frac{\partial}{\partial t}\,(L_1\,u,\, X\,u) - \left(L_1^{-1}\,X\,L_1\,u,\, \frac{\partial}{\partial t}\,L_1\,u\right)$$
$$= \frac{\partial}{\partial t}\,(L_1\,u,\, X\,u) - \left(\frac{\partial}{\partial t}\,L_1\,u,\, X\,u\right),$$

that is,

$$\frac{\partial}{\partial t}\,(L_1\,u,\, X\,u) = \frac{1}{2}\,\frac{\partial}{\partial t}\,(L_1\,u,\, X\,u).$$

Further, it can be shown that $(l_1(u) + l_2(u), X u) = 0$.

We shall prove that for the corresponding choice of the parameter x the forms $(L_1\,u,\, X\,u)$ and $\|X\,u\|^2$ determine the norms that are equivalent to the norm in V_2 and the form $(L_2\,u,\, X\,u)$ determines the norm which is equivalent to the norm in V_3. Indeed,

$$\|X\,u\|^2 = \|\Delta\,u_1\|^2 + 2\,x\,\|\nabla\,u_1\|^2 + \|\Delta\,u_2\|^2$$
$$+ 2\,(x + \alpha^2)\|\nabla\,u_2\|^2 + x\,\|u_1\|^2 + (x + \alpha^2)^2\,\|u_2\|^2.$$

Therefore,

$$\|u\|_{V_2}^2 \le \|X u\|^2 \le K(x) \|u\|_{V_2}^2, \quad K(x) = (1 + \frac{X^2 + \alpha^2}{\lambda_1})^2. \quad (6.12)$$

In the similar way we find that

$$\|u\|_{V_1}^2 \le (L_1 u, X u) \le K_2(x) \|u\|_{V_2}^2,$$
$$K_2(x) = (1 + \frac{x + \alpha^2}{\lambda_1})(1 + \frac{\alpha^2}{\lambda_1}); \quad (6.13)$$

$$\mu \|u\|_{V_3}^2 \le (L_2 u, X u) \le K_3(x) \|u\|_{V_3}^2, \quad (6.14)$$

where $K_3(x)$ is some known function of x.

The inequality (6.14) fulfills if there exists the number γ satisfying the inequalities

$$\lambda_1 + x \ge \alpha^2/2\gamma,$$
$$\sigma \alpha^{2\gamma}/4 \le \lambda_1(\mu_1 + \mu(x + \alpha^2)) + \sigma_1 + \mu_1(x + \alpha^2).$$

Such a γ exists if

$$\sigma \alpha^4/8 \le (\lambda_1 + x)[(\mu_1 + \mu(x + \alpha^2))\lambda_1 + \sigma_1 + \mu_1(x + \alpha^2)]. \quad (6.15)$$

The inequality (6.15) may be met if x is sufficiently large. Taking into account the above-said, we get the following inequality

$$\frac{1}{2}\frac{\partial}{\partial t}(L_1 u, X u) + (L_2 u, X u) \le \frac{K_1(x)\|F\|^2}{2\lambda_1 \mu}$$
$$+\frac{\lambda_1 \mu \|X u\|^2}{2K_1 x)} \le \frac{K_1(x)\|F\|^2}{2\lambda_1 \mu} + \frac{(L_2 u, X u)}{2}, \quad (6.16)$$

this leads to

$$(L_1 u, X u) \le (L_1 u_0, X u_0) e^{-ct} + B(1 - e^{-ct}),$$
$$c = \lambda_1 \mu t/K_2(x), B = K_1 K_2 \|F\|^2/\lambda_1^2 \mu^2.$$

Therefore, the attractor lies inside the set

$$M_2 = \{u \in V_2 : \|u\|_{V_2} \le B^{1/2}\}.$$

It follows also from (6.16) that

$$\frac{\partial}{\partial t}(L_1 u, X u) + (L_2 u, X u) \le \frac{K_1(x)\|F\|^2}{\lambda_1 \mu}.$$

Integrating of this inequality yields

$$\int\limits_0^t (L_2\, u\,(\tau),\, X\, u\,(\tau)\, d\,\tau \le (L_1\, u\,(0),\, X\, u\,(0) + \frac{K_1\,(x)\,\|F\|^2}{\lambda_1\,\mu}\, t.$$

Using (6.15) we find that on the attractor

$$\lim\limits_{t\to\infty}\, \sup\, \frac{1}{t} \int\limits_0^t \|u\,(\tau)\|_{V_3}^2\, d\,\tau \le \frac{K_1\,(x)\,\|F\|^2}{\lambda_1\,\mu}.$$

We write the variation equation corresponding to (6.5). We have

$$\frac{\partial}{\partial t} L_1 w = -L_2 w - l_1(w) - \delta l_2(u,w) \equiv \bar{L}(t,u)w \equiv L(L_1 w), \quad (6.17)$$

where

$$\delta l_2(u,w) =$$
$$\left(\begin{array}{c} J(u_1,\Delta w_1) + J(u_2,\Delta w_2) + J(w_1,\Delta u_1) + J(w_2,\Delta u_2) \\ J(u_1,\Delta w_2) + J(u_2,\Delta w_1) + J(w_1,\Delta u_2)+ \\ +J(w_2,\Delta u_1) - \alpha^2[J(u_1,w_2) + J(w_1,u_2)] \end{array}\right).$$

Setting $L_1\, w\, =\, \varphi$, we write (6.17) as

$$\frac{\partial}{\partial t}\, \varphi\, =\, L\, \varphi.$$

Let A be the operator

$$A\, =\, \left(\begin{array}{cc} 1 - \alpha^2\Delta^{-1} & 0 \\ 0 & 1, \end{array}\right)$$

and $\{\varphi_j\}$, $j = \overline{1,n}$ is the system of orthonormal functions $(\varphi_i,\varphi_j)_A$ $= (A\varphi_i,\varphi_j) = \delta_{ij}$. We calculate the spur $Sp_n\, L$

$$Sp_n\, L\, =\, \sum\limits_{j=1}^n (L\,\varphi_j,\, \varphi_j)_A\, =\, Sp_{n1}\, L\, +\, Sp_{n2}\, L\, +\, Sp_{n3}\, L, \quad (6.18)$$

where

$$Sp_{n1}\, L\, =\, -\sum\limits_{j=1}^n (L_2\, L_1^{-1}\, \varphi_j,\, \varphi_j)_A,$$

$$Sp_{n2}\, L\, =\, -\sum\limits_{j=1}^n (\delta\, l_2\,(u,\, L_1^{-1}\, \varphi_j),\varphi_j)_A,$$

$$Sp_{n3}\, L\, =\, -\sum\limits_{j=1}^n (l_1\,(L_1^{-1}\, \varphi),\, \varphi_j)_A.$$

One can immediately see that $Sp_{n3} L = 0$. Making the sufficiently complicated calculations we can show that [68],

$$S p_{n1} L \leq \mu \lambda_1 T_n + M n,$$

where

$$T_n = -1/\lambda_1 \sum_{j=1}^{n} (\tilde{\Delta} \varphi_j, \varphi_j)_A,$$

and

$$\tilde{\Delta} = \begin{pmatrix} \Delta & 0 \\ 0 & \Delta \end{pmatrix}, \quad M = \max(0, \mu \alpha^2 - \mu_1).$$

We notice that $T_n \geq n^2/8$.

The main difficulty here is to estimate $Sp_{n2} L$. Omitting the sufficiently cumbersome calculations, we give the final result [68]:

$$Sp_{n2} L \leq 3(2\pi)^{-1/2} n^{1/2} (\ln T_n + 1)^{1/2} \left((2 + \frac{\alpha^2}{\lambda_1}) \|u\|_{V_3} + \alpha^2 \sqrt{6} \|u\|_{V_1} \right).$$

Let $u(t)$ is the solution lying on the attractor. Hence, if

$$\lim_{t \to \infty} \sup \frac{1}{t} \int_0^t S p_n L(\tau, u(\tau)) dt \leq C < 0, \qquad (6.19)$$

then the attractor dimension is less than n. For this it is sufficient to require the fulfillment of the inequality

$$T_n > \alpha_1 T_n^{1/2} + \alpha_2 T_n^{1/4} (1 + \ln T_n)^{1/2}, \qquad (6.20)$$

where

$$\alpha_2 = 3 \cdot 2^{1/4} \pi^{-1/2} / \mu \lambda_1 \times$$

$$\times \alpha_2 \times \left[(2 + \alpha^2/\lambda_1) + \sqrt{K_1/\lambda_1 \mu} \|F\| + \alpha^2 \sqrt{6} c_1 \|(-L_1)^{-1/2} F\| \right],$$

$$\alpha_1 = M \sqrt{8} / \mu \lambda_1, \quad c_1 = \max(1/\lambda_1 \mu, \alpha^2/\mu_1 \lambda_1 + \sigma).$$

We introduce the parameter ξ in the following way: $\xi = 0$, if $\alpha_1 = 0$, i.e. $M = 0$ and $\mu_1 = \mu \alpha^2$; $\xi \in (0,1)$, if $M > 0$. One can show that the inequality (6.20) will fulfill if

$$T_n \geq \max(\alpha_1/\xi, (\alpha_2/1 - \xi)^{4/3} (1 + 4 \ln \alpha_2/1 - \xi)^{2/3}),$$

where $\alpha_1 \neq 0$ (if $\alpha_1 = 0$, then $\xi = 0$).

Thus, the attractor dimension A has the upper estimate:

$$\dim A \leq 2\sqrt{2}\,\max\left\{\frac{\alpha_1}{\xi},\,(\frac{\alpha_2}{1-\xi})^{2/3}(1\,+\,4\ln\frac{\alpha_2}{1-\xi})^{1/3}\right\},\quad(6.21)$$

where $\xi \in (0,1)$. In the case, when $\alpha = 0$ the attractor dimension is estimated in this way:

$$\dim A \leq 3\cdot 2^{11/4}/\pi^{1/2}\,G^{2/3}\,(1\,+\,4\ln G)^{1/2},$$

where $G = \|F\|/\mu^2\,\lambda^{3/2}$.

6.3   Numerical Investigation of Attractor Characteristics of Two-Layer Baroclinic Model

In this section the results of numerical calculations of some characteristics of attractors generated by the two-layer baroclinic quasi-geostrophic model of atmosphere circulation on the sphere will be presented.

Since the two-layer baroclinic model in some sense is also a good approximation for primitive equations of hydrothermodynamics, first of all because of the availability of the mechanism of the baroclinic instability, we emphasize once again that it is important to obtain the numerical values of the global invariants of the system under consideration, e.g., the attractor dimension and the global Lyapunov exponents. Evidently, the logic of discussion here may be as follows.

Let the attractor dimension and the number of statistically independent degrees of freedom are connected between them in the some manner. The analysis of this connection for the attractor of two-layer baroclinic model will be given in Section 2 of Chapter 7. Then, for sufficiently large attractor dimensions and the chaotic behavior of the trajectory on it because of the availability of the positive Lyapunov exponents the probability distributions of the states on the attractor must be close to the Gauss distribution according to the central limit theorem and the assumption that the states can be expanded over the basis system of independent states. Here, of course, it is important to remember the conditions of the central limit theorem.

For the small values of dimension it is possible that there exist multi-modal distributions. The characteristic example of this assumption is the Lorenz problem, where on proper averaging of the trajectory the probability distribution of the states has the clearly defined two-modality [190]. Since the precise lower bound of the attractor dimension is the number of positive global Lyapunov exponents, with respect to the above discussions it is very important to know this value.

The Lyapunov exponents themselves give the estimate of the average predictability of the trajectory on the attractor. In the last years much attention is given to the calculation of local (nonaveraged) Lyapunov exponents at the given pieces of trajectory. This gives the possibility to investigate the predictability of trajectory at the different portions of attractor.

The one of the main problems of this section is the calculation of the above discussed characteristics of the attractor of two-layer baroclinic model for the different values of forcing. The task is to evaluate the growth rate of the attractor dimension and the growth rate of the lower bound of the attractor dimension, i.e. the number of positive Lyapunov exponents. The second important problem, which may be solved with the aid of these calculations, is connected with the possibility to use the number of statistically independent degrees of freedom to evaluate the attractor dimension. The solution of this problem appears to be crucial, since it enables to evaluate the attractor dimension of the "ideal" model of the climate. The problem is described more fully in the next chapter.

For the numerical calculations we used the two-layer baroclinic model which somewhat differs from the model investigated in the previous chapter. The main difference in this case consists in the fact that the Coriolis parameter varies. Therefore, the calculations can be carried out on the sphere without especial selection of the equatorial region.

The governing differential equations of the problem of interest are the following equations:

1. Vorticity equation

$$\frac{\partial \Delta\psi}{\partial t} + J(\psi, \Delta\psi + l) = \text{div}\left(l\nabla\frac{\partial\chi}{\partial p} + \mu\Delta^2\psi + \frac{\partial}{\partial p}\rho^2 g^2\nu\frac{\partial\Delta\psi}{\partial p}\right), \quad (6.22)$$

2. Thermodynamic equation

$$\frac{\partial T}{\partial t} + J(\psi, T) = \sigma\Delta\chi + \mu\Delta T + \frac{\partial}{\partial p}\rho^2 g^2\nu\frac{\partial T}{\partial p}, \quad (6.23)$$

3. Thermal wind balance equation

$$\text{div}\left(l\nabla\frac{\partial\psi}{\partial p}\right) = -\frac{R}{p}\Delta T, \quad (6.24)$$

Boundary conditions:

for $p = 0$, $\chi = 0$, $\rho^2 g^2\nu\frac{\partial\Delta\psi}{\partial p} = 0$, $\rho^2 g^2\nu\frac{\partial T}{\partial p} = 0$. \quad (6.25)

For $p = 1000$ mb

$$\operatorname{div}(l\nabla\chi) + J(\psi, kh) = 0, \quad \rho^2 g^2 \nu \frac{\partial \Delta\psi}{\partial p} + \alpha\Delta\psi = 0,$$

$$\rho^2 g^2 \nu \frac{\partial T}{\partial p} + \alpha T = \alpha T'_s.$$

In the system (6.22)-(6.25) the following notations are used: ψ denotes the streamfunction, χ the velocity potential ($\chi = \Delta^{-1}\tau$, $\tau = dp/dt$ is the analog of the vertical velocity in p-system of coordinates, T the temperature, ρ the density ($p = \rho RT$) and l is the Coriolis parameter, p the pressure, T_s the temperature of underlying surface, ν the coefficient of vertical mixing, μ the coefficient of horizontal diffusion, h the orographic inhomogeneites of the Earth surface, α the coefficient of Rayleigh friction and σ is the parameter of statistical stability.

In such a setting of the problem it is assumed that the main perturbation of the atmosphere circulation is due to the flux of heat from the ununiformly heated underlying surface. The temperature of underlying surface is equal to the mean climatic January temperature. It is easily understood that the amplitudes of stationary waves in the model atmosphere on such a setting of the problem will be essentially smaller than the really occurred ones, since the model does not include the additional sources of heating from the sources of the water vapor condensation which are located along the east coasts of the continents and along the storm treks. It is also clear that using for the vertical approximation only two layers, i.e. describing only the structure of the tropospheric circulation, we can not compensate deficient magnitude of the stationary waves amplitude by introducing of the suitable magnitude of the orographic inhomogeneity of surface. However, the availability of the mechanism of the baroclinic instability of zonal-symmetric flow seems to be the sufficient condition for the model can generate the chaotic behavior of the trajectory in the phase space.

For the vertical approximation of the model we have used the grid shown in the Plot 1. If we pass to the dimensionless form of equations of the model (setting as the time unit $\bar{t} = 1/\Omega$, where Ω is the angular velocity of the Earth, R radius of the Earth as the distance unit, $p = 500$ mb as the pressure unit) and put

$$Ax \equiv \nabla(\sqrt{2}\cos\varphi\,\nabla x), \quad Bx \equiv A\Delta^{-2}Ax,$$

then the model equations in the dimensionless form are:

$$\frac{\partial \Delta\bar{\psi}}{\partial t} + J(\bar{\psi}, \Delta\bar{\psi} + l + th) + J(\psi^*, \Delta\psi^* + 2th) = \mu\Delta^2\bar{\psi} - \alpha\Delta(\psi^* + \bar{\psi}/2),$$

$$\frac{\partial(\Delta\psi^* + \beta^2 B\psi^*)}{\partial t} + J(\bar{\psi}, \Delta\psi^* + th) + J(\psi^*, \Delta\bar{\psi} + 2th + l)$$
$$+\beta^2 A\Delta^{-1}J(\bar{\psi}, \Delta^{-1}A\psi^*) = \mu\Delta\psi^* - \alpha\Delta(\psi^* + \bar{\psi}/2)$$
$$-2\alpha_1\Delta\psi^* - \mu\beta^2 A\Delta^{-1}A\psi^* + \alpha_s A\Delta^{-1}\bar{T}_s + \alpha\beta^2\psi^*, \qquad (6.26)$$

where $\bar{\psi} = (\psi_{3/2} + \psi_{1/2})/2$, $\psi^* = (\psi_{3/2} - \psi_{1/2})/2$.

There were chosen the following numerical values of dimensionless parameters: $\beta = 160$, $\alpha_s = 0.005$, $\mu = 5.10^{-4}$, $t = 0.14$ (the dimension values α and α_1 were chosen such that the characteristic time of dissipation was equal to 5 and 50 days respectively).

For the numerical solution of (6.26) there was used a Galerkin method including as the basis functions the spherical harmonics, i.e. the eigenfunctions of the Laplacian operator on the sphere which correspond to the eigenvalues not exceeding 56. Thus, the system (6.26) was reduced to the system of 126 ordinary differential equations for the spectral coefficients:

$$\frac{\partial\eta}{\partial t} + F(\eta) = f, \quad \eta_{t=0} = \eta'_0. \qquad (6.27)$$

This system was solved over time by the Matsuno scheme [99] with the time step ≈ 1 hour. The calculation were carried out for the time interval ≈ 5000 days. In the surface temperature there was introduced a bifurcation parameter γ to obtain the different magnitudes of forcing: $\bar{T}_s = \gamma T_s$, where $\gamma = 1$ corresponds to the really occurred surface temperature. To characterize the quality of the model, we give the distribution of zonally-averaged zonal velocity (Fig.2).

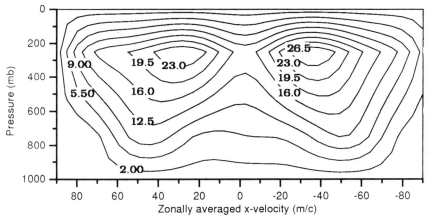

Fig.2 Isolines of zonally–averaged zonal velocity.

It is easily seen that the main disadvantage of the distribution (relative to the really observed in the atmosphere) is the lowered meridional gradient of zonal velocity. This naturally leads to the lowering of the barotropic energy transformations. The vertical gradient of velocity associated with the meridional gradient of temperature (which is responsible for the baroclinic energy transformations) is sufficiently close to those which are really occurred in the atmosphere [27].

Method of calculation of the Lyapunov exponents and attractor dimension is based on the use of the Oseledets theorem [109] and the Kaplan–Yorke formula [77], as in the case of the barotropic problem, and has the form:

1. The system of equations is integrated over L-steps. The matrix products forms

$$(D^L f(x_0))^T (D^L f(x_0)) = \prod_{i=0}^{L-1} (D^1 f(y(i)))^T \prod_{i=L-1}^{0} (D^1 f(y(i))) , \quad (6.28)$$

where f is the resolving operator of the system; $D^L f = (f^L)'(x_0)$ is the Freshet derivative of the mapping $f^L = f \circ f \ldots \circ f$ (L-times) at the point x_0; $(\cdot)^T$ is the passage to the transpose.

2. There are calculated $\lambda_j = 1/2L \, lnc_j$, where c_j is the eigenvalues of the product obtained. On certain conditions, for $L \to \infty$, the exponents λ_j coincide with the Lyapunov exponents for a given system according to the well-known Oseledets theorem [109].

3. The convergence of λ_j is studied as L increases. If the Lyapunov exponents are known then the estimate of the attractor dimension is calculated by the Kaplan–Yorke formula [77]. The lower bound is equal to the number of the positive Lyapunov exponents. The Kolmogorov entropy, according to the Pesin theorem [110], is equal to the sum of the positive Lyapunov exponents. A given method was used to calculate the Lyapunov exponents of the above presented model of the atmosphere. At first the problem of the convergence of this method, when the length of the time series for different γ increases.

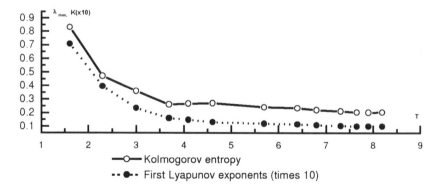

Fig.3   Dependence of first Lyapunov exponent $\lambda_{\max}$ and Kolmogorov entropy K on length of trajectory T (day's, logarithmic scale).

Fig.3. shows the curves of the convergence for the highest Lyapunov exponent and Kolmogorov entropy at $\gamma = 1$ (the curves 1 and 2 respectively). Considering the behavior of the curves, one can make the conclusion that the method converges for $T \gg 1000$ days. The similar situation takes place for other values of γ.

From the graph of the convergence of the highest Lyapunov exponent, one can obtain the estimate of the mean length of the predictability interval on the attractor, i.e. there can be estimated minimal time in which the error may grow in e-times. For the real value of forcing ($\gamma = 1$) this time is equal to 7.8 days. Further, the dependence of the attractor parameters of the right-hand side coefficient was considered. The graphs of the dependence of the dimension estimates are presented in Fig.4 (the curves 1 and 2).

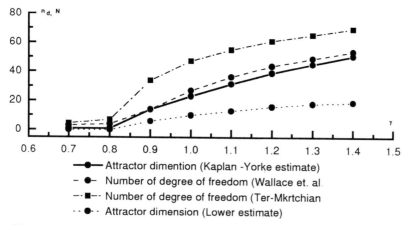

Fig.4 Dependence of attractor dimension estimates (n_d) and number of degrees of freedom (N) on the parameter of forcing (γ) for the two-layer baroclinic model.

The estimate obtained by the Kaplan–Yorke formula represents the curve 1, while the lower bound (the number of the positive Lyapunov exponents) is presented by the curve 2. For $\gamma = 0.7, 0.8$, the attractor of the system represents a limit cycle and the corresponding dimension is equal to 1. For the large values of the coefficient γ the attractor is chaotic one, since there exist the positive Lyapunov exponents. In the case $\gamma = 1$ (the situation resembles that of the real forcing) the dimension is equal to 23.1. It should be noted that the attractor parameters are quite sensitive to the right-hand side for the relatively small values of dimension. Indeed, if the coefficient γ increases from 0.9 to 1.1 (the corresponding equator-pole gradient of temperature increases in $10^0 C$) then the attractor dimension increases in 2.5 times.

Chapter 7

Investigation of Structure of Climate Attractors by Observed Data Series

7.1   Correlation Dimension of Attractor

In this chapter the theory and the results of numerical experiments on reproduction of some characteristics of attractors (there are primarily a correlation dimension and Kolmogorov entropy) from the observed data series are presented. This problem is of great importance from various point of view. First, reproducing the characteristics of attractor of the real atmosphere and assuming that the observed series of parameters of climatic system are generated by some ideal system, we have the possibility to identify from this point of view the adequacy of models in use. Second, it becomes possible to compute the predictability characteristics of quantities, for the description of which we have not even the approximate idea on the corresponding dynamical system (e.g., time and space averaged meteoparameters). The principal task here is to find the necessary length of the observed data series, as in computing the invariant measure and Lyapunov exponents for the problem with known operator. Some presently known estimates of the solution to this problem will be given below. The basis for the further discussion are the following Takkens theorems [124].

Theorem 7.1.   *Let M be a compact manifold of dimension m, $\varphi :$ $M \to M$ a smooth diffeomorphism (C^2), $y : M \to R$ a smooth (of smoothness C^2) real function on M. Then the mapping $\Phi_{\varphi,y} : M \to R^{2m+1}$ defined by the equality*

$$\Phi_{\varphi,y}(x) = (y(x), y(\varphi(x)), \ldots, y(\varphi^{2m}(x))), \quad x \in M,$$

is an embedding of the manifold M in the space R^{2m+1}.

(We note that this theorem deals with a dynamical system on M for discrete time which is given by the use of diffeomorphism φ, while $\varphi^k(x)$ defines the location of a point at the moment $t = k$.)

Theorem 7.2. *Let M be a compact manifold of dimension m, F a smooth (of smoothness C^2) vector field on M, y a smooth (of smoothness C^2) real function given on M. Then the mapping*

$$\Phi_{F,y}(x) = (y(x), y(\varphi(x)), \ldots, y(\varphi_{2m}(x))), \quad x \in M,$$

where φ_t is a flow generated by F is an embedding of the manifold M into the space R^{2m+1}. (Here one considers the dynamical system on the manifold M for continuous time.)

If we assume that the vector field F and the function y are of smoothness $2m + 1$, then the next theorem can be formulated.

Theorem 7.3. *Let M be a compact manifold of dimension m, F a smooth (of smoothness C^{2m+1}) vector field on M, y a smooth (of smoothness C^{2m+1}) real function given on M. Then the mapping $\hat{\Phi}_{F,y}$: $M \to R^{2m+1}$ defined by the equality*

$$\hat{\Phi}_{F,y}(x) = \left(y(x), \frac{d}{dt}\left((y(\varphi_t(x)))_{t=0}, \ldots, \frac{d^{2m}}{dt^{2m}}(y(\varphi_t(x))_{t=0}) \right) \right),$$

is an embedding of the manifold M into R^{2m+1}. The space R^{2m+1} is called the embedding space.

Grassberger and Procaccio in a number of their works [70] have used the above mentioned ideas and proposed the technique of computation of the attractor correlation dimension by the observed data series (or by the series generated by some dynamical system). However, we note that the strange attractors are not the smooth manifolds, therefore the Takkens theorems are not directly applicable to such objects. It is known, however, that the attractors having the fractal Hausdorff dimension may be embedded into the space R^d, where d is sufficiently large.

Let us assume that the dynamical system is defined by the system of equations

$$\frac{dx}{dt} = f(x), \quad x \in R_m, \tag{7.1}$$

and let $x(t) = (x_1(t), x_2(t), \ldots, x_m(t))$ be a vector function which represents the solution of the system. Let us take one from "typical" components of this vector $(x_i(t))$ and compose the vector $\xi(t)$ of the form

$$\xi(t) = (x_i(t), x_i(t + \tau), \ldots, x_i(t + 2m\tau)), \quad \tau > 0.$$

The vector $\xi(t)$ executes the embedding of the flow generated by the system (7.1) into R^{2m+1}.

One can compute over the series $\xi(t_k)$ the correlation integral

$$C_m(\varepsilon) = \lim_{n \to \infty} \frac{1}{N^2} \sum_{i,j}^{N} \Theta(\varepsilon - |\xi(t_i) - \xi(t_j)|),$$

where Θ is the Heaviside function.

It can be shown that [61] there takes place an asymptotic formula

$$C_m(\varepsilon) \sim e^{d_k} l^{-m\tau L}, \tag{7.2}$$

where d_k is the correlation dimension of attractor, while L is the sum of the positive Lyapunov exponents. It follows from (7.2) that the correlation dimension of attractor may be calculated by the formula

$$d_k = \lim_{\varepsilon \to \infty} \left\{ \frac{\ln C_m(\varepsilon)}{\ln \varepsilon} - \frac{m\tau L}{\ln \varepsilon} \right\}, \tag{7.3}$$

and the sum of the positive Lyapunov exponents can be obtained by the formula

$$L = \frac{1}{\tau} \ln \frac{C_m(\varepsilon)}{C_{m+1}(\varepsilon)}, \quad m \to \infty. \tag{7.4}$$

We discuss now some problems which arise when we compute the dimension and the Kolmogorov entropy according to the above formulas. (In both cases the principal problem is to find the necessary length N of the observed data series). In the work [116], for instance, it was shown that if we have at our disposal the observed data series of length N, then the computed value of the correlation dimension must satisfy the relation

$$\tilde{d}_k \leq 2 \log_{10} N. \tag{7.5}$$

This estimate was obtained following the next reasoning. The domain of variations of ε, for which the slope of the straight line has to be calculated $\ln C_m(\varepsilon) = d_k \ln \varepsilon$ must belong to the interval $(\varepsilon_{\min}, \varepsilon_{\max})$, where $\varepsilon_{\min}$ and $\varepsilon_{\max}$ are minimal and maximal distances between the pairs of vectors. It is clear that $\varepsilon_{\max}$ is upper bounded, for instance, by the attractor sizes; $\varepsilon_{\min}$ must be lower bounded, for instance, by the existence of random errors of measurements (stochastic noise). We can not also infinitely extend the length of the series on account of the interpolation, since it will lead to the strong correlation and to the false stabilization of the slope at small ε [116]. Consequently, we need to calculate the slope of the straight line by the formula

$$\tilde{d}_k = \frac{\log_{10} \tilde{C}_m(\varepsilon_2) - \log_{10} \hat{C}_m(\varepsilon_1)}{\log_{10} \varepsilon_2 - \log_{10} \varepsilon_1},$$

$$\varepsilon_{\min} \leq \varepsilon_1 < \varepsilon_2 \leq \varepsilon_{\max}. \tag{7.6}$$

(Here $\tilde{C}_m(\varepsilon)$ is calculated for finite N.) Since

$$\tilde{C}_m(\varepsilon) \geq \frac{1}{N^2}, \quad \tilde{C}_m(\varepsilon) \leq \frac{1}{2}\frac{N(N-1)}{N^2} \leq 1,$$

one has

$$\log_{10}\tilde{C}_m(\varepsilon_2) - \log_{10}\hat{C}_m(\varepsilon_1) \leq \log_{10}N^2.$$

If we choose the interval of variation of ε so that the inequality $\varepsilon_2/\varepsilon_1 \geq 10$ is satisfied, then

$$\log_{10}\varepsilon_2 - \log_{10}\varepsilon_1 \geq 1,$$

and we obtain the estimate sought.

$$\tilde{d}_k \leq 2\log_{10}N. \tag{7.7}$$

Of course, one can debate the validity of the choice of the relation between ε_2 and ε_1. However, this does not change the result. The estimate (7.7) pose the strong restriction on the lower bound of the observed data series length N. For instance, if we analyze 100-years series of mean monthly data, then we have $N \sim 10^3$ and $d_k \leq 6$. In the light of these estimates all the obtained results concerning the computation of the climate and weather attractors dimension by a given technique are doubted [116].

The procedure of computation of the sum of positive Lyapunov exponents seems to be more acceptable, for the reason that the computations results for finite N can be presented as the sum of local positive Lyapunov exponents calculated at time intervals $m\tau = T$ and a given piece of the trajectory (naturally, for a judiciously chosen ε). More justified in this sense procedure is the computation of the correlation integral for the whole vector of state and not just for one of its components.

7.2 Calculation of Lyapunov exponents

The above technique of computation of the correlation dimension shows good results for the attractors of small dimension which are similar to the Lorenz attractor [95]. For hydrodynamics problems (even when only the barotropic equations for the sufficiently large numbers of truncation of the Galerkin approximations are used) the procedure requires the reproduction of very long observed data series. This fact was pointed out when we have used a given technique to calculate the correlation dimension of attractor generated by the dynamical system described in [35]. The situation looks nicer if the sum of the positive Lyapunov exponents for this problem are calculated.

Fig.5 shows the dependencies of values of the sum of positive Lyapunov exponents, as the function m, for $N = 4000$, calculated by the two different techniques: through the singular eigenvalues of the operator M (see, Section 5) and by the above described procedure. It is seen that these curves are close and they are saturated by the judicious value of m. From the form of dependencies one can conclude that the local Lyapunov exponents for moderate m describe mainly the polynomial divergence of trajectories [35].

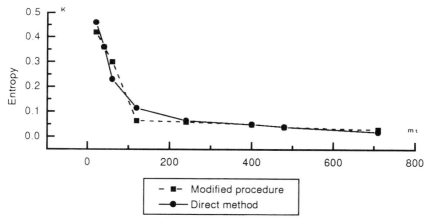

Fig.5. Curves of dependence of the sum of local positive Lyapunov exponents on $m\tau$; calculated via: 1 the direct application of the transient operators; 2 the procedure (7.4).

We note that the realization of algorithms of calculation of the sum of positive Lyapunov exponents on the base of the correlation integral has been made with the use of the procedures described in [96,97,112]. The references to the last works, made in the framework of the problem of the observed data analysis are given in [2].

7.3 Statistically Independent Degrees of Freedom and Attractor Dimension

Since the Grassberger–Procaccio technique used to evaluate the attractor dimension for the sufficiently large values of dimension requires the very long observed data series or modelling data, any another technique giving the acceptable estimate of dimension for judiciously chosen lengths of series is to be very important. It is evident that the reducing of the multi-dimensional field to the observational data at one point must lead to increasing of the length of series to keep the quantity of information.

Therefore, if we want to use the judicious values of the lengths of the observed data series, then we need to look for the methods of evaluation of the attractor dimension in the class of methods which are applicable to the multi-dimensional random vector processes.

In this section we present the technique of evaluation of the number of independent degrees of freedom for the random vector stationary process, which will be applied to evaluate the dimension of attractors of the atmosphere circulation model. These methods will be presented following the work of Ter-Mcrtchian [127].

Thus, let us have the vector N-dimensional stationary process $\xi = (\xi_1, \ldots, \xi_N)$ with the spatial autocovariance matrix V and autocorrelation matrix $R = (r_{ij})$. The problem which needs to be solved is posed in the following way: one needs to find the dimension of the random vector stationary process of dimension $n = n(R)$ which is in some sense equivalent to the initial process.

The concept of equivalency is of decisive importance here. We introduce this concept on a basis of the important characteristics of the degree of anomality of the vector field K defined in [127]

$$K = \frac{1}{N} \sum_{i=1}^{N} \frac{\xi_1^2}{\sigma_i^2} = \frac{1}{N} \sum_{i=1}^{N} \eta_i^2, \qquad (7.8)$$

where $\eta_i = \frac{\xi_i}{\sigma_i}$ is normalized component of the vector field, σ_i^2 is the dispersion of the $i - th$ component of the field. The magnitude K is some statistics, i.e. it represent the function of random values η_i. Consequently, K has as a whole definite distribution function $F_k(x)$, which is uniquely defined if there is given a distribution of the normalized values η_i.

Definition 7.1. We shall speak that two random stationary process $\{\xi_i\}_{i=1}^{N}$ and $\{\gamma_i\}_{i=1}^{N}$ are equivalent if their characteristics of anomality K_ξ and K_γ are distributed in the similar way, i.e. $F_{K_\xi}(x) = F_{K_\gamma}(x)$.

If η_i are normally distributed, then N-dimensional random vector $\vec{\eta} = \{\eta_i\}$ has the distribution density

$$f_\eta(x) = \frac{1}{(2\pi)^{N/2}|R|^{1/2}} e^{-\frac{1}{2}x'R^{-1}x}. \qquad (7.9.)$$

Here R is the autocorrelation matrix of the vector η, while $|R|$ is the determinant of R. One can show [127] that in this case the characteristic function $\varphi_{K_\eta}(t)$ for K_η has the form

$$\varphi_{K_{\eta(N)}}(t) = \Pi_{s=1}^{N} \left(1 - \frac{2}{N} \lambda_s it \right)^{-1/2}, \qquad (7.10)$$

where λ_s are the eigennumber of the matrix R.

(We remember that the characteristic function is the Fourier transformation of the probability density distribution function

$$f(x) = \frac{1}{2\pi} \int\limits_{-\infty}^{\infty} e^{-itx} \Pi_{s=1}^{N} \left(1 - \frac{2}{N}\lambda_s it\right)^{-1/2} dt. \qquad (7.11)$$

Because of that the procedure of calculation of the probability density function is very cumbersome, in the work [127] there was taken the semi-invariants (derivatives at point zero of logarithm of characteristic function), since they are easily expressed via central moments. The expression for the semi-invariants has the form

$$\chi_s(K_{\eta(N)}) = \frac{2^{s-1}(s-1)!}{N^s} \sum_{i=1}^{N} \lambda_i^s. \qquad (7.12)$$

Let us wish to calculate the semi-invariants for the vector normalized field of dimension n in which all components are statistically independent. The semi-invariants for this random field are

$$\chi_s(K_{\gamma(n)}) = 2^{s-1}(s-1)! \frac{\lambda^s}{n^{s-1}}. \qquad (7.13)$$

This formula can be easily obtained from (7.12), since for the latter case $\lambda_s = \lambda$ and $\sum \lambda = n$, while $\chi_s(k_{\gamma(n)})$ depends on two parameters n and λ. Therefore, identifying the two first semi-invariants $\chi_s(K_{\gamma(n)})$ and $\chi_s(K_{\eta(N)})$ we get

$$\lambda = \frac{\mathrm{Sp}(R)}{N}, \quad \frac{\lambda^2}{n} = \frac{\mathrm{Sp}\,R^2}{N^2}.$$

Hence,

$$n = \frac{[\mathrm{Sp}\,(R)]^2}{\mathrm{Sp}\,R^2} = \frac{N^2}{\mathrm{Sp}\,R^2} = \frac{N^2}{\sum \lambda_i^2(R)}. \qquad (7.14)$$

We note that the formula for the number of independent degree of freedom of the form (7.14) was presented also in [62]).

If to define the parameter of anomality we use the non-normalized vector fields, then it is easily seen that the expression for the number of independent degrees of freedom will take the form

$$n' = \frac{(\mathrm{Sp}\,V)^2}{\mathrm{Sp}\,V^2}, \qquad (7.15)$$

where, as noted above, V is the covariance matrix of field $\vec{\xi}$. Since

$$\mathrm{Sp}\,V^2 = \sum_{i,j}(v_{ij})^2, \quad (\mathrm{Sp}\,V)^2 = (\sigma_\xi^2)^2,$$

then

$$n' = \frac{(\sigma_\xi^2)^2}{\sum (v_{ij})^2} = \frac{1}{\sum (\frac{v_{ij}}{\sigma_\xi^2})^2}. \tag{7.16}$$

It is easily seen that if $\sigma_{xi_i} \approx \sigma_{xi_j}$ for all i, j then $n' \approx n$.

We note once again that the principal moment in the above given method for determination of the number of independent degrees of freedom is represented by the condition of equality of portions of energy which is associated with each degree of freedom.

We consider briefly another method which was introduced in [133]. The base of this method is the central limit theorem. As a random value in this work there was taken a scalar product between two vectors which are chosen from the general sampling.

Let $\vec{x}^{(i)}$ and $\vec{x}^{(j)}$ be such two vectors. Let $\{\vec{\psi}_K\}$ be some orthogonal system of basis vectors within N-dimension space. Then

$$\vec{x}^{(i)} = \sum_k a_k^{(i)} \vec{\psi}_k; \quad R_{ij} = \left(\vec{x}^{(i)}, \vec{x}^{(j)}\right) = \sum_K a_k^{(i)} a_k^{(j)}.$$

Normalize

$$R_{ij} : \frac{R_{ij}}{|\bar{S}|^2} = \sum \frac{a_k^{(i)}, a_k^{(j)}}{|\bar{S}|^2} = \sum b_k^{(i)} b_k^{(j)},$$

where

$$\bar{S} = \left(\vec{x}^{(i)}, \vec{x}^{(j)}\right)^{1/2}.$$

If in the expansion

$$R_{ij}/|\bar{S}|^2 = \sum b_k^{(i)} b_k^{(j)}$$

the number of statistically independent values tends to ∞ then according to the limit theorem (on corresponding assumptions) the distribution $R_{ij}/|\bar{S}|^2$ will tend to Gauss distribution. Similarly to the method presented in the first part of the section one can to construct the equivalent system of independent variables with unit dispersion so that the total dispersion will be equal to the dispersion $R_{ij}/|\bar{S}|^2$. Evidently, in this case the following relations hold

$$\frac{R_{ij}}{|\bar{S}|^2} = \frac{1}{K} \sum Y_k^{i,j}; \quad K = \frac{1}{\mathrm{var}(R_{ij}/\bar{S}^2)},$$

where var( ) denotes the dispersion of ( ).

Strictly speaking, if one considers the equivalency in the sense of the distribution equivalency, i.e. when considering the existence of Gauss distributions, one must a priori assume that the number of statistically independent degrees of freedom is sufficiently large.

The above considered methods for the calculation of the number of independent degrees of freedom were used to evaluate the attractor dimension, generated by the two-layer baroclinic atmosphere model which was considered in Section 3 of Chapter 6 for the different values of forcing, i.e. for the different values of attractor dimension. The results obtained give grounds to assume that for the sufficiently large attractors dimensions (about 10 and more) the number of statistically independent degrees of freedom may be used as the bound of the fractal attractor dimension (Fig. 4).

Chapter 8

Regimes of Atmosphere Circulation

8.1 Definition of Atmosphere Circulation Regimes

The term "regime of atmosphere circulation" has long been in use by the meteorologists. It will suffice to remember the classification of the circulation regimes given by Dzerdzeevsky [23], Girs and Vangengeim [67]. The concepts of the circulation regimes such as blocking and zonal flow are well known [20].

However, it should be noted that the authors of the above-mentioned classifications have not given the strict definition of the circulation regimes. This is not surprising, since these classifications, although they have a physical meaning show rather describing qualitative character than the formally quantitative one.

In the present section we shall made an attempt to describe by the formulas the concept of the regime of motion for those dissipative systems, for which it can be done. To be specific, let us assume that we deal with the regimes of atmosphere circulation. Thus, let us have a dynamical system

$$u(t) = S_t u(0), \quad u(t) \in M, \tag{8.1}$$

where M is the phase space of the system. If the semigroup S_t possesses a global attractor A, then by M is meant the set A as such. Further, we assume that on the set M (or on A) there exists an invariant measure with respect to the semigroup S_t. This invariant measure may be associated to the statistical stationary solution of the system under consideration, while the existence and uniqueness of invariant measure may be connected with the existence and uniqueness of statistical stationary solution (see, Chapter 1).

In this case, if the system is ergodic, then the corresponding invariant measure is unique. Moreover, in the ergodic case the probability of the fact that the trajectory belongs to the set $\Omega \subset A$ is equal $\mu(\Omega)$, i.e. it is equal to the measure of the set Ω, if the measure is normalized: $\mu(A) = 1$. Since we have not at the present time the algorithm of the construction of the statistical stationary solution or invariant measure for the real climatic problem, we have take as the measure of the set (if the system is ergodic) the probability of the fact that the trajectory belongs to a given set (if we assume that it is possible to make independent estimate of this probability). We notice that the problem will be strictly speaking of the same severity. It is intuitively clear that there is a good reason to introduce the concept of the regime of motion on the attractor only in the case if there are few such regimes. If it is not the case, the concept of the regime will resemble the concept of the system state as the point in the phase space of a dynamical system. To justify the introduction of the concept of the circulation regimes, one can assume that the motion on the attractor is realized so that the trajectory spends the more portion of time in the small neighborhood of some base states of the system (stationary points, limit cycle, some other solutions) [7].

We now consider this case more closely. Let $\{\Psi_i\}$ be a system of base states and let $\{\Omega_i\}$ be their small neighborhoods. Following the above-made assumptions, we shall suppose that

$$\sum_i \mu(\Omega_i) \approx 1, \quad \mu(A) = 1, \quad \Omega_i \subset A, \, i = \overline{1, N}. \tag{8.2}$$

The smallness of the neighborhood Ω_i means that if $u(t) \in \Omega_i$, then $\|u(t) - \Psi_i\| \ll \varepsilon\|\Psi_i\|$, where $\varepsilon \ll 1$, while $\|\cdot\|$ is defined according to the existence and uniqueness theorems for the solution of (8.1).

By virtue of (8.2) the motion $u(t)$ on the attractor A can be approximated as follows (the approximation means the averaging of the characteristics over the probability measure)

$$u(t) \approx \sum_i \alpha_i \, \Psi_i, \tag{8.3}$$

where

$$\alpha_i = \begin{cases} 1, & u(t) \in \Omega_i \\ 0, & u(t) \notin \Omega_i \end{cases}. \tag{8.4}$$

Definition 8.1. We shall speak that the motion $u(t)$ belongs to the i-th regime if $u(t) \in \Omega_i$.

This definition, evidently, has the sense for the systems in which the probability measure satisfies the relations (8.2). However, it is difficult to suppose that all the considered systems will satisfy this condition.

In [31] we have introduced the next (more general) definition of the regime of motion.

Definition 8.2. The regime of motion is the set of states of the dissipative system possessing some general property.

According to this definition all motions belonging to the small neighborhoods of stationary points can be associated with one quasistationary regime, those belonging to the neighborhoods of the limit cycles can be associated with quasiperiodic regime an so on. One can consider the systems possessing two regimes of motion – the laminar and turbulent ones, as it is done, for instance, in [5].

In the present chapter we shall deal with the patterns of the classification of the circulation regimes which corresponds to both abovementioned definitions.

8.2 Dynamical Theory of Two-Regime Barotropic Circulation

In this section we shall consider the simplest case when the barotropic atmosphere circulation satisfies the condition (8.2), where $N = 2$, while $\bar{\Psi}_1$ and $\bar{\Psi}_2$ are two stationary solutions of the barotropic problem.

Thus, let us have the equation of barotropic atmosphere on rotating sphere $S^2 \subset R^3$ of radius a (see, Chapter 3):

$$\frac{\partial \omega}{\partial t} + J(\Psi, \omega + l) = -\sigma \omega + \nu \Delta \omega + f,$$

$$\omega \big|_{t=0} = \omega_0, \quad \omega = \Delta \Psi. \tag{8.5}$$

In Chapter 3 for the problem (8.5) there were proved an existence theorem for the global attractor, for a statistical stationary solution and the probability invariant measure concentrated on the attractor. We shall assume that there exists f such that the probability measure is concentrated in the small neighborhoods of two stationary solutions of (8.5):

$$J(\Psi_i, \omega_i + l) = -\sigma \omega_i + \nu \Delta \omega_i + f, \quad i = 1, 2. \tag{8.6}$$

Using the results of the work [100], we put

$$\bar{\omega} = \frac{\omega_1 + \omega_2}{2}, \quad \bar{\Psi} = \frac{\Psi_1 + \Psi_2}{2}.$$

Let

$$\omega_1 = \bar{\omega} + \omega', \quad \Psi_1 = \bar{\Psi} + \Psi', \quad \omega_2 = \bar{\omega} - \omega', \quad \Psi_2 = \bar{\Psi} - \Psi'. \tag{8.7}$$

Substituting (8.7) into (8.6) we get the equation for ω', Ψ' in the form:

$$L\,(\bar{\Psi}, \bar{\omega})\,\omega' = 0, \tag{8.8}$$

where the linear operator L is

$$L\,(\bar{\Psi}, \bar{\omega})\,\omega' = J\,(\bar{\Psi}, \omega') + J\,(\Delta^{-1}\omega', \bar{\omega} + l) + \sigma\,\omega' - \nu\,\Delta\,\omega'. \tag{8.9}$$

It is easily seen that the operator L is the operator of (8.6) linearized with respect to the mean state $\bar{\omega}$. According to (8.3) we look for the solution of (8.5) in the form

$$\omega\,(t) \approx \alpha_1\,\omega_1 + \alpha_2\,\omega_2,$$

so that our task is to find the coefficients α_1 and α_2 and the stationary points ω_1 and ω_2. If we suppose that $\bar{\omega}$ is known, then the task is to find ω'. The equation (8.8) with the corresponding homogeneous boundary condition on sphere represents the eigenvalues problem, more exactly, it is the problem of finding out the kernel of not selfadjoint operator L. In [28] it was shown that the operator L generates a total system of eigenfunctions in the space L_2. It follows from (8.8) that the simmetric operator L^*L is also degenerate. Thus, the assumption that there exist two quasistationary regimes of barotropic atmosphere circulation leads to the degeneracy of the linearized operator L according to Definition 8.1. We shall speak that in this case the linearized operator of the problem (8.5) possesses a resonance mode. Evidently, the existence of the stationary resonance mode of linearized operator does not mean that there exists two-regime circulation. This conclusion was also arrived at in work [100]. The importance of the existence of stationary resonance mode for the linearized operator of the barotropic problem is supported by the following reasoning.

We average (8.5) according to Reynolds at the interval $[-T/2, T/2]$. This gives

$$\frac{\partial}{\partial t}\tilde{\omega} + \sigma\tilde{\omega} - \nu\Delta\tilde{\omega} + J(\Delta^{-1}\tilde{\omega}, \tilde{\omega} + l) = \tilde{f} - \overline{J\Delta^{-1}\omega, \omega + l)}, \tag{8.10}$$

where

$$\tilde{\omega} = \Delta\,\tilde{\Psi}, \quad \tilde{\omega} = \frac{1}{T}\int_{t-T/2}^{t+T/2} \omega\,(s)\,ds.$$

Let $\bar{\omega} = \lim_{T \to \infty} \tilde{\omega}$ and $\omega' = \tilde{\omega} - \bar{\omega}$. The equation for ω' may be written in the form:

$$\frac{\partial\omega'}{\partial t} + \sigma\omega' - \nu\Delta\omega' + J(\Delta^{-1}\omega', \bar{\omega} + l) + J(\Delta^{-1}\bar{\omega}\omega')$$

$$= F - J(\Delta^{-1}\omega, \omega'), \tag{8.11}$$

where
$$F' = \tilde{f} - \bar{f} - \overline{J\left(\Delta^{-1}, \omega + l\right)} - \overline{J\left(\Delta^{-1}\omega,\, \omega + l\right)}.$$

We shall assume that $F' = \varepsilon_1 F\left(\varepsilon_2 t\right)$, where ε_1 and ε_2 are small parameters: $0 < \varepsilon_1,\ \varepsilon_2 \ll 1,\ \varepsilon = O(\varepsilon_1)$. Then by the change of variable $\varepsilon_2 t = \tau$ the equation (8.11) reduces to the form

$$\varepsilon_2 \frac{\partial \omega'}{\partial \tau} + L\left(\bar{\omega}\omega'\right) + J\left(\Delta^{-1}\omega'\omega'\right) = \varepsilon_1 F\left(\tau\right), \tag{8.12}$$

where

$$L\left(\bar{\omega}\right)\omega' \equiv -\nu\,\Delta\omega' + \sigma\omega' + J\left(\Delta^{-1}\omega,\, \bar{\omega} + l\right) + J\left(\Delta^{-1}\bar{\omega},\, \omega'\right)$$

is the problem's operator linearized with respect to $\bar{\omega}$.

With (8.12) we consider the equation which may be obtained from it for $\varepsilon_2 = 0$. We get

$$L\left(\bar{\omega}\right)\omega_0' + J\left(\Delta^{-1}\omega_0',\, \omega_0'\right) = \varepsilon_1 F\left(\tau\right). \tag{8.13}$$

Estimate the solution of this equation. Multiplying (8.13) by ω_0' in $L_2^0\left(s\right)$, we get

$$\nu\|\nabla\omega_0'\|^2 + \sigma\|\omega_0'\|^2 \leq \max_S |\nabla\bar{\omega}| \|\nabla(\Delta^{-1}\omega_0')\| \|\nabla\omega_0'\| + \varepsilon_1\|F\| \|\omega_0'\|.$$

Taking into account that

$$\lambda_1\|\omega_0'\|^2 \leq \|\nabla\omega_0'\|^2, \quad \|\nabla\left(\Delta^{-1}\omega_0'\right)\| \leq \lambda_1^{-1}\|\nabla\omega_0'\|,$$

we obtain

$$\|\nabla\omega_0'\| \leq \frac{\varepsilon_1 \lambda_1^{1/2}\|F\|}{a}, \quad \|\omega_0'\| \leq \frac{\varepsilon_1}{a}\|F\|, \tag{8.14}$$

where

$$a = \nu\lambda_1 - \max_S |\nabla\bar{\omega}| > 0.$$

Assuming that the function F is given, we write the equation (8.12) in departures from ω_0', setting $\omega' = \omega_0' + \hat{\omega}$; we have

$$\varepsilon_2 \frac{\partial \hat{\omega}}{\partial \tau} + L\left(\bar{\omega}\right)\hat{\omega} + J\left(\Delta^{-1}\hat{\omega},\, \hat{\omega}\right) + J\left(\Delta^{-1}\hat{\omega},\, \omega_0'\right)$$
$$+ (\Delta^{-1}\omega_0',\, \hat{\omega}) = -\varepsilon_2\,\partial\omega_0'/\partial\tau. \tag{8.15}$$

The equation obtained resembles the equation (8.13). Therefore, repeating the foregoing calculations we find that

$$\|\theta\| = \|\frac{\partial\omega_0'}{\partial\tau}\| \leq \frac{\varepsilon_1\|F_\tau'\|}{b}, \quad b = a - \max|\nabla\omega_0'| > 0.$$

Now we can estimate the solution of (8.15). Multiplying it by $\hat{\omega}$ we get

$$\varepsilon_2/2 \frac{\partial}{\partial \tau} \|\hat{\omega}\|^2 + \nu \|\nabla \hat{\omega}\|^2 + \sigma \|\hat{\omega}\|^2 \leq | (J(\Delta^{-1}\hat{\omega}, \bar{\omega}), \hat{\omega}|$$

$$+ | (J(\Delta^{-1}\hat{\omega}, \omega_0'), \hat{\omega}) | -\varepsilon_2 (\frac{\partial \omega_0'}{\partial \tau}, \hat{\omega})$$

$$\leq (\max | \nabla \bar{\omega} | + \max | \nabla \omega_0' |)/\lambda_1 \|\nabla \hat{\omega}\|^2 + \varepsilon_2 \|\frac{\partial \omega_0'}{\partial \tau}\| \|\hat{\omega}\|.$$

Taking into account that $b > 0$ we obtain

$$\frac{\partial}{\partial \tau} \|\hat{\omega}\|^2 + \frac{\sigma}{\varepsilon_2} \|\hat{\omega}\|^2 \leq \frac{\varepsilon_2}{\sigma} \|\frac{\partial \omega_0'}{\partial \tau}\|^2 \leq \frac{\varepsilon_1^2 \varepsilon_2}{b^2 \sigma} \|F_\tau'\|^2.$$

Integrating of this inequality yields

$$\|\hat{\omega}(\tau)\|^2 \leq \|\hat{\omega}(0)\|^2 e^{-\sigma\tau/\varepsilon_2} + \frac{\varepsilon_1^2 \varepsilon_2^2}{b^2 \sigma^2} \|F_\tau'\|^2 (1 - e^{-\sigma\tau/\varepsilon_2}).$$

From this estimate it follows that on any interval $0 < t < T$ we have $\|\hat{\omega}(\tau)\|^2 \to 0$, $\varepsilon_2 \to 0$.

Proposition 8.1. If $b > 0$ and ε_2 is sufficiently small, than on any given interval $0 < t < T$ the solution of (8.13) will be close to that of (8.12).

We note that the nonlinear term in (8.13) is small in some integral sense, namely

$$| \int_S J(\Delta^{-1}\omega_0', \omega_0') h\, ds | \leq \max_S | h | \frac{\lambda_1 \varepsilon_1^2}{a^2} \|F\|,$$

where h is any function of the class $C(S)$. Hence it follows that the solution of (8.13) can be obtained by the expansion over the small parameter ε_1. Let $L(\tilde{\omega})$ be a nondegenerate operator. Setting $\omega_0' = \varphi_0 + \varepsilon_1 \varphi_1 + \varepsilon_1^2 \varphi_2 + \dots$, we get

$$\varphi_0 = 0, \quad L(\bar{\omega})\varphi_1 = F(\tau). \tag{8.17}$$

Thus, for the first approximation we shall have the equation

$$L\varphi_1 = F.$$

Of course, the nonlinearity in the left side of (8.13) eliminates the singularity if the operator L has a resonance mode. However, if there exists a resonance mode, then on perturbation of the right side of (8.13), in the first approximation the response of the atmosphere circulation must lie in the kernel of the operator L.

Such a situation can be occurred in the models of the general atmosphere circulation, if the operator L has a quasiresonance mode (see, e.g., [78,79]). Since ω' is defined from (8.8) with a accuracy of the multiplier (ω may be defined from the variational principle for the selfadjoint operator $L^* L$, as it is done in [100]), it is evident that we can not uniquely reproduce the circulation regime from (8.7).

We need to formulate the additional relations which will allow us to define not only ω_1 and ω_2, but the mean values $\bar{\alpha}_i$ in the relation (8.3) as well, provided that the mean state $\bar{\omega}$ is known. It follows from (8.3) that the time-averaged state in the case of the existence of two regimes is expressed as follows

$$\tilde{\omega} = \bar{\alpha}_1 \omega_1 + \bar{\alpha}_2 \omega_2, \tag{8.18}$$

where $\bar{\alpha}_1$ and $\bar{\alpha}_2$ are the probabilities of the fact that the trajectory belongs to the neighborhood of stationary solutions ω_1 and ω_2 respectively.

We emphasize once again that in the present analysis it is supposed that the mean state is known. For $\tilde{\omega}$ to coincide with $\bar{\omega}$, it is necessary that $\bar{\alpha}_1 = \bar{\alpha}_2 = 1/2$. This means that the trajectories spend the same time in the neighborhoods of every stationary point, i.e. the stationary points in some sense are equivalent. In what sense namely?

We give some euristic reasoning. Suppose that the stationary points belong to the class of saddle points. Then it is natural to assume that the trajectory approaches to stationary point along the stable manifold and departs from it along the unstable manifold. If we take the neighborhood of the stationary point, then the lifetime of the specific trajectory in this neighborhood will be defined by the minimal distance between the point and the characteristics of the unstable manifold of the stationary point (evidently, it will be the main governing characteristics [106]).

Had the problem's operator linearized with respect to the stationary point be simmetric, the rate of increasing of maximal unstable mode would have been taken as such a characteristic. However, the linearized operator, in essence, is nonsimmetric and the local Lyapunov exponents calculated for the lifetime of the trajectory (or the average local Lyapunov exponents) are unique characteristics. But since the lifetime itself is the characteristic sought, one can take as the characteristic of the stability the instantaneous Lyapunov exponents.

In the works [35,36,38] it was shown that (see also Chapter 5 of the present book), the inverse mean lifetime of the trajectory in the neighborhood of stationary points strongly correlates with instantaneous Lyapunov exponents. Theoretically it may be explained if one assumes that the probability of the trajectory's entrance the neighborhood of the stationary point with a given minimal distance from it does not depend on this distance.

Further in Chapter 5 it is shown that the instantaneous Lyapunov exponents in the neighborhood of stationary points are defined by the enstrophy of these points, so that we can write with some degree of accuracy

$$\frac{\|\omega_1\|^2}{\|\omega_2\|^2} \approx \frac{\bar{\alpha}_2}{\bar{\alpha}_1}. \tag{8.19}$$

Since we assume that the invariant measure is concentrated in the neighborhood of stationary points, i.e. $\bar{\alpha}_1 + \bar{\alpha}_2 = 1$, the relation (8.19) may be rewritten as follows

$$\frac{\|\omega_1\|^2}{\|\omega_2\|^2} = \frac{1 - \bar{\alpha}_1}{\bar{\alpha}_1}. \tag{8.20}$$

Consequently, for $\bar{\alpha}_1 = 1/2$ the relation $\|\omega_1\| = \|\omega_2\|$ holds and the equivalency of the stationary points for this case we shall understand in this sense only. It follows from (8.7) that

$$
\begin{aligned}
\|\omega_1\|^2 &= \|\bar{\omega}\|^2 + \|\omega'\|^2 + 2\,(\bar{\omega}, \omega_1), \\
\|\omega_2\|^2 &= \|\bar{\omega}\|^2 + \|\omega'\|^2 - 2\,(\bar{\omega}, \omega_1)
\end{aligned}
\tag{8.21}
$$

(since $\bar{\omega}$ and ω_1 are real). Since for $\bar{\alpha}_1 = 1/2$, $\|\omega_1\|^2 = \|\omega_2\|^2$, we get from (8.21) that $(\bar{\omega}, \omega_1) = 0$, i.e. ω_1 and $\bar{\omega}$ are orthogonal of each other. Thus, though the above technique is acceptable only for the case when $\bar{\alpha}_1 = \bar{\alpha}_2 = 1/2$, but even in this case we can not uniquely determine this regime of circulation for the above-given conditions of closure algorithm. This procedure may be realized by the use of the statistical approach, which will be presented in the next section.

8.3   Statistical Theory of Two-Regime Barotropic Circulation

For the more generality, some part of computations in this section will be given for the case of the n circulation regimes. We shall assume that the dynamics of the barotropic atmosphere is realized in the finite dimensional space of dimension N. This assumption does not pose the restrictions, since in Chapter 3 it was proved that (8.5) possesses a global attractor and an inertial manifold of finite dimension. Thus, we suppose that the trajectories of the system can be approximated by the expression

$$\omega\,(t) = \sum_{i=1}^{n} \alpha_i\,(t)\,\omega_i, \tag{8.22}$$

where $\alpha_i\,(t)$ are defined by the relations (8.4).

The system will be considered to be ergodic so that the averaging over the ensemble of states may be replaced by the averaging over time (time averaging). We calculate the mean time $\tilde{\omega}$ and noncentered autocorrelation matrix $C =< \omega(t)\omega^T(t) >$, where ω^T is the row-vector, $< \cdot >$ denotes the operation of averaging. We have

$$\tilde{\omega} = \sum_{i=1}^{n} \bar{\alpha}_i \omega_1, \quad C \leq \sum_i \sum_j \alpha_i \alpha_j \omega_i \omega_j^T \geq \sum_i \sum_j < \alpha_i, \alpha_j > \omega_i \omega_j^T. \quad (8.23)$$

According to (8.4) we shall have $< \alpha_i \alpha_j >=< \alpha_i \alpha_i >= \bar{\alpha}_i$. Consequently,

$$C = \sum_i \bar{\alpha}_i \, \omega_i \, \omega_i^T \equiv \sum_{i=1}^{n} \bar{\alpha}_i \, M_i. \quad (8.24)$$

Let $\{\omega_i\}$ be the system of linearly independent vectors. The following proposition is valid.

Proposition 8.2. The matrix C has the rank n.

If $\{\omega_i\}$ is the system of orthonormal vectors then the next proposition is valid

Proposition 8.3. The matrix C has the rank n, while $\lambda_i = \bar{\alpha}_i \|\omega_i\|^2$, where λ_i are nonzero eigenvectors of the matrix $C(\lambda_i > 0)$. The proof of Proposition 8.3 follows from the possibility of the presentation of this matrix via the matrices of rank 1. In this case, if ω_i are normalized on 1 eigenvectors then $\bar{\alpha}_i$ are eigenvalues of the matrix C. In the case if the vectors are not normalized, evidently, we shall have the relation $\lambda_i = \bar{\alpha}_i \|\omega_i\|^2$. Without going into the analysis of the solvability of the general system of algebraic nonlinear equations (8.23)–(8.24) (given $\tilde{\omega}$ and C), we return to the case of two regimes of circulation. For the purposes of comparison with results of "dynamical" analysis we reformulate equations (8.23) and (8.24) in terms of departure from the mean value and obtain in this way the centered autocorrelation matrix. The corresponding system of equations is

$$\bar{\alpha}_1 \Psi_1 + (1 - \bar{\alpha}_1)\Psi_2 = 0, \quad C_1 = \bar{\alpha}_1 \Psi_1 \Psi_1^T + (1 - \alpha_1)\Psi_2 \Psi_2^T, \quad (8.25)$$

where $\Psi_1 = \omega_1 - \tilde{\omega}$, $\Psi_2 = \omega_2 - \tilde{\omega}$. The vectors Ψ_1 and Ψ_2 are linearly dependent. Consequently, the matrix C_1 will have the rank 1. Let φ_1, φ_2 be the Fourier expansion coefficients of the vector Ψ_1 over the eigenvectors of the matrix C_1. Then in the basis of eigenvectors of this matrix the second equation of the system (8.25) has the form

$$\begin{pmatrix} \lambda & 0 \\ 0 & 0 \end{pmatrix} = \frac{\bar{\alpha}_1}{1 - \bar{\alpha}_1} \begin{pmatrix} \varphi_1^2 & \varphi_1 \varphi_2 \\ \varphi_1 \varphi_2 & \varphi_2^2 \end{pmatrix},$$

where λ is nonzero (positive) eigenvalue of the matrix C_1. Hence

$$\lambda = \frac{\varphi_1^2 \, \bar{\alpha}_1}{1 - \bar{\alpha}_1}, \quad \varphi_2 = 0. \tag{8.26}$$

Let η be the eigenvector of the matrix C_1 corresponding to the eigennumber λ. Then

$$\omega_1 = \tilde{\omega} \pm \sqrt{\lambda \frac{1 - \bar{\alpha}_1}{\bar{\alpha}_1}} \, \eta, \quad \omega_2 = \tilde{\omega} \mp \sqrt{\lambda \frac{\bar{\alpha}_1}{1 - \bar{\alpha}_1}} \, \eta. \tag{8.27}$$

We assume that $\tilde{\omega}$, η, λ are known. To solve uniquely (8.27), we apply the additional relation (8.20). The condition $0 \le \bar{\alpha}_1 \le 1$ in this case represents the condition of the choice of signs (8.27). The detailed analysis of the atmosphere circulation regimes with the use of above technique is given in [25,27,32].

It is easily seen that the results of dynamical theory of two-regime barotropic atmosphere circulation are closely similar to those of the statistical theory.

This similarity of the results is the reflection of more general relations between the eigenfunctions of dynamical operator and the empirical orthogonal components [24,26,34]. We shall assume that the low-frequency variability of the barotropic atmosphere circulation may be described by the linear equation (8.12):

$$L\left(\bar{\Psi}, \bar{\omega}\right)\omega' = f, \tag{8.28}$$

where the right side f is to be a random vector. Here it is natural to assume that L is the nondegenerate nonsimmetric matrix with the spectre lying in the right half-plane.

Let $C \equiv\, < \omega' \omega'^T >$ be the autocorrelation matrix of the random process ω', while $F \equiv\, < f\, f'^T >$ the autocorrelation matrix of the right side f. Then the next proposition is valid.

Proposition 8.4. If $F = \sigma E$, $\sigma > 0$, E is the identity matrix (f the white noise), then the eigenvectors of the matrix C coincide with the right singular eigenvectors of the matrix L (eigenvectors of the matrix $L^* L$), hence the vector corresponding to the maximal eigennumber of the matrix C coincides with the eigenvector of the matrix $L^* L$ corresponding to the minimal eigennumber of the matrix.

The proof of the proposition is evident. We formulate yet another proposition.

Proposition 8.5. The system of eigenvectors of autocorrelation matrix C will be the right system of singular eigenvectors of the matrix L, if the autocorrelation matrix of the right side F is function of the matrix $(L\, L^*)^{-1}$.

Indeed, the function of matrix is presented as the polynomial of the matrix, therefore the relations hold

$$C = L^{-1} \sum_{i=0}^{\infty} a_i \, G^i \, L^{*-1},$$

where $G = (L\,L^*)^{-1} = L^{*^{-1}} L^{-1}$. Hence

$$C = \sum_{i=0}^{\infty} a_i \, M^{i+1}, \quad \text{where} \quad M = (L^* \, L)^{-1},$$

i.e. C is the function of matrix $L^* L$, hence it possesses the same system of vectors.

Remark. If $\sum_{i=0}^{\infty} a_i \, G^i$ is positively defined, then C is also positively defined.

It follows from the above propositions that if the matrix L will have the close to zero eigennumber of multiplicity 1, which is separated from the others eigennumbers, then the relation $\lambda \gg \lambda_i$, for $i = \overline{2, N}$ will hold for the eigenvaluess of the autocorrelation matrix C, when the corresponding conditions posed on the right side are fulfilled, i.e. total "energy " will be concentrated on the subspace spanned by the eigenvector, which is associated to the eigennumber λ_1.

It should be noted that the models of the general atmosphere circulation having such properties exist (see, e.g., [78,79]) and their main feature is the localization of the response on the disturbances of the right side in the above-mentioned subspace without regard to the nature of disturbances (be it the aerosol anomalies, surface temperature of the ocean or the variations of the parametrizations on the subgrid scales). One can assume that the "ideal" model of climate may also behave itself in the similar way on some portions of its attractor.

8.4   S-Regimes of Atmosphere Circulation

In the present section we shall consider the pattern of the classification of the atmosphere circulation regimes according to Definition 8.2. As the main characteristic of the regime we take the informational entropy of the state, which by assumption must be defined by the barotropic circulation. This assumption seems to be sufficient to describe the low-frequency variability of the atmosphere circulation, which, as known, is equivalent to the barotropic circulation.

Since the barotropic circulation is wholly defined by the relative vorticity, the distribution function, through which the informational entropy of the state must be expressed, may be defined via the distribution of the relative vorticity on the surface which characterizes the barotropic atmosphere circulation (e.g., at 500 mb surface).

In the simplest case one can suppose that there is a statistical and informational equivalency of all elements of the surface, obtaining in this way the one-dimensional distribution function. Then the function of distribution density may be defined, for instance, as follows

$$\frac{d\,D}{D} = \rho\,(\omega,\,t)\,d\omega, \tag{8.29}$$

where ω is the relative vorticity, $\rho\,(\omega\,t)$ is the function of distribution density depending on time and relative vorticity, $d\,D$ is the area of surface, on which the relative vorticity belongs to the interval $(\omega, \omega + d\omega)$, and D is the whole area of the surface. From (8.29) the condition of normalization follows:

$$\int_{-\infty}^{\infty} \rho\,(\omega,\,t)\,d\omega = 1. \tag{8.30}$$

Using the definition of the distribution function (8.29) one can determine naturally the informational entropy of the state in the phase space

$$S\,(t) = -\int_{-\infty}^{\infty} \rho\,(\omega,\,t)\ln\,\rho\,(\omega,\,t)\,d\omega. \tag{8.31}$$

"Marking" in this way the state of the system in the phase space, we proceed from the infinite dimensional space (or finite dimensional space in the case if one considers the motion on the attractor) to the one-dimensional space by identification all the states of the system which possess one and the same informational entropy.

Let E be the set of values of the function S (of finite or infinite measure) and let $\varphi_A\,(S)$ be a characteristic function of the set $A \subset E$. Then, in the usual way, the residence time of the system on the set A in the time interval $[0, T]$ one can calculate as follows

$$\tau\,(S,\,T,\,A) = \int_0^T \varphi_A\,(S)\,d\,t. \tag{8.32}$$

It will be assumed that there exists a limit

$$\lim_{T \to \infty} \frac{\tau}{T} = b_a,$$

which may be called the probability of the fact that the system lies on the set A. Since one can define the set A as the set of values of entropy, which belong to the interval $(S, S + \Delta S)$, the function b_A will characterize in this case the probability of the fact that the system under consideration lies on the set of states, whose informational entropy belongs to the interval $(S, S + \Delta S)$.

This value may be used as the metric invariant of the system for the identification of the climate model. Let $\rho(\omega, t)$ be presented as the Gauss distribution:

$$\rho = \frac{1}{\sqrt{2\pi\sigma^2}} \exp\left[-\frac{(\omega - \bar{\omega})^2}{2\sigma^2}\right], \qquad (8.33)$$

where $\bar{\omega}$ is the mean value of the relative vorticity, σ^2 is the dispersion. By the definition of the enstrophy for the case of the normal low of the distribution we have

$$\eta = \frac{1}{D}\int_D \omega^2\, dt = \int_{-\infty}^{\infty} \omega^2 \rho(\omega)\, d\omega = \sigma^2 + \bar{\omega}^2.$$

In the other hand, the expression for the entropy has the form

$$S = -\int_{-\infty}^{\infty} \rho \ln \rho\, d\omega = \int_{-\infty}^{\infty} \rho(\ln\sqrt{2\pi\sigma^2} + \frac{(\omega - \bar{\omega})^2}{2\sigma^2})d\omega$$

$$= \ln\sqrt{2\pi\sigma^2} + \frac{1}{2\sigma^2}\int(\omega - \bar{\omega})^2 \rho\, d\omega = \frac{1}{2}\ln(\eta - \bar{\omega}^2) + \text{const.}$$

It is seen from the above relations that $\bar{\omega} = 0$ (that is very close to the reality for the barotropic circulation on sphere or semisphere), then in the case of the normal law of distribution $\rho(\omega)$ the entropy of state is proportional to the logarithm of the enstrophy.

In Chapter 5 it was shown that on the some assumption the enstrophy of the barotropic circulation strongly correlates with the instantaneous Lyapunov exponents. This means that within the framework of the established assumptions we can consider the classification of the regimes of circulation with respect to the values of its informational entropy as the classification with respect to the values of the instantaneous Lyapunov exponents [37].

Chapter 9

Solvability of Ocean and Atmosphere Models

9.1 Introduction

The considered in the previous chapters systems of equations governing the dynamics of barotropic and two-layer baroclinic fluid belong to the class of systems which are called by the meteorologists the quasigeostrophic approximations for the description of the atmosphere dynamics. It is well known that the modern models of the general circulation of atmosphere and ocean are based on so-called primitive equations [99], which are also only some approximate form of the complete, not simplified hydrotermodynamic equation. The main simplification in them is the use instead of the total third dynamical equation (projection onto vertical axis) of its hydrostatic approximation. With the use of hydrostatic approximation for the fulfillment of the integral laws of conservation in diabatic approximation one needs to make a number of simplifications which are less essential than the hydrostatic approximation [99].

The application of the hydrostatic approximation for the description of the general circulation of atmosphere and ocean set off the primary equations from the class of the nonlinear hydrodynamic system of the type of the three-dimensional Navier–Stockes equations. One can say that these equations are 2,5-dimensional so that the problems which arise in the process of the investigation of the unique solvability of the three-dimensional Navier–Stockes equations for the primitive equations of the atmosphere and ocean dynamics may be will not arise at all. Historically, the investigation of the problems on correctness of mathematical models in hydrodynamics based on the primitive equations began in the middle 60s from the works of G.I Marchuk and G.V. Demidov [16,17].

It should be noted that these investigations were connected with another fundamental problem of the meteorology – the problem of short-range weather forecast in p-system of coordinates on the infinite (x, y) plane

$$\frac{\partial u}{\partial t} + u \frac{\partial u}{\partial x} + v \frac{\partial u}{\partial y} - l v + \frac{\partial \Phi}{\partial x} = 0,$$

$$\frac{\partial v}{\partial t} + u \frac{\partial v}{\partial x} + v \frac{\partial v}{\partial y} + l u + \frac{\partial \Phi}{\partial y} = 0, \quad \frac{\partial \Phi}{\partial p} = - \frac{R T}{p},$$

$$\frac{\partial u}{\partial x} + \frac{\partial v}{\partial y} + \frac{\partial \tau}{\partial p} = 0,$$

$$\frac{\partial T}{\partial t} + u \frac{\partial T}{\partial x} + v \frac{\partial T}{\partial y} - \sigma \tau = 0, \quad \sigma > 0, \quad p \in (p_0, P),$$

with the boundary conditions

$$(\frac{\partial}{\partial t} + u \frac{\partial}{\partial x} + v \frac{\partial}{\partial y}) \frac{\partial \Phi}{\partial p} = 0 \text{ for } p = p_0,$$

$$p (\frac{\partial}{\partial t} + u \frac{\partial}{\partial x} + v \frac{\partial}{\partial y}) \frac{\partial \Phi}{\partial p} + \alpha \frac{\partial \Phi}{\partial t} = 0, \quad p = P < \infty,$$

and the next initial conditions: for $t = 0$, and $u = u_0 (x, y, p)$, $\Phi = \Phi_0(x, y, p)$. It is proved that this system of equations is solvable in the small on the assumption that the initial data u_0, v_0, $\partial \Phi_0 / \partial p$ have the generalized derivatives in the sense of Sobolev, to the third order inclusive, summed with the square.

Since the problems of unique solvability of the system of equations describing the climate model are necessary element in the construction of the mathematical theory of climate, we shall detailedly consider in the present section the results concerning with solvability of primitive equations. (We notice once again that these equations are the base for the construction of the global models of the general circulation of ocean and atmosphere which are the main components of the present models of climate).

9.2 Solvability of Ocean and Atmosphere Models in Bounded Domain

We shall present the results of works [120-123] on solvability as a whole of the models of the dynamics of atmosphere and ocean.

Atmosphere. We start with the three-dimensional model of atmosphere dynamics [120-123]. We denote by $\Omega = \Omega(x, y, p)$ a bounded cylinder with the base $p = p_1 > 0, p = p_2 > p_1$ and lateral area S.

The projection of the cylinder Ω onto the plane (x, y) will be denoted by G. Here (x, y) are Cartesian coordinates at the plane, p a "vertical" coordinate represents the pressure. We consider in the domain $Q = \Omega \times (0, T)$, $T > 0$ the following system of equations of the atmospheric dynamics

$$u_t + (\vec{u}\,\nabla)\,u - l\,v = -H_x + \nu\,\frac{\partial}{\partial p} p^2\,u_p + \mu\,\Delta\,u,$$

$$v_t + (\vec{u}\,\nabla), v + l\,u = -H_y + \nu\,\frac{\partial}{\partial p} p^2\,v_p + \mu\,\Delta\,v,$$

$$u_x + v_y + \tau_p = p\,H_p,$$

$$T_t + (\vec{u}\,\nabla)\,T - \kappa\,\frac{T\,\tau}{p} = f + \nu\,\frac{\partial}{\partial p} p^2\,T_p + \mu\,\Delta\,T. \tag{9.1}$$

Here $\vec{u} = (u, v, \tau)$, T is temperature, ρ density, $H = gz$, z height of the isobaric surface $(z = z(x, y, p, t))$,, while g acceleration of gravity, $f = f(x, y, p, t)$ is a given function, ν, μ, l, κ are positive constants, Δ is Laplacian with respect to the pair of variables x, y, while

$$(\vec{u}\,\nabla) = u\,\frac{\partial}{\partial x} + v\,\frac{\partial}{\partial y} + \tau\,\frac{\partial}{\partial p}.$$

We combine the system (9.1) with the following boundary and initial conditions:

$$u = v = T = 0 \ \text{ on } S,$$
$$u = v = T = 0, \ \tau = H_\tau \quad \text{for } p = p_2,$$
$$u_p = v_p = T_p = \tau = 0, \quad \text{for } p = p_1,$$
$$u = v_0, \ v = v_0, \ H = H \quad \text{for } t = 0. \tag{9.2}$$

We note that

$$T_0 = T\,|_{t=0} = p\,\frac{\partial}{\partial p}\,H_0,$$

while τ and H may be eliminated from (9.1), putting

$$\tau = -\int\limits_{p_1}^{p} (u_x + v_y)\,d\,p,$$

$$H = -\int\limits_{p_2}^{p_1} \frac{T}{p}\,dp - \int\limits_{0}^{t} \left(\int\limits_{p_1}^{p_2} (u_x + v_y)\,d\,p\right)\,d\,G + H_0\,(x,\,y,\,p_2).$$

There takes place the following theorem on solvability as a whole the problem (9.1)–(9.2) [120].

Theorem 9.1. *Let $f \in L_\infty(Q)$, $H_0(x, y, p_2) \in W_2'(G)$, $\partial G \in C^2$. Then if the initial functions*

$$u_0, \, v_0, \, t_0 \in W_2^1 \, (\Omega) \cap L_\infty \, (\Omega)$$

satisfy the boundary conditions (9.1)–(9.2), *the problem* (9.1)–(9.2) *has a unique solution u, v, $T \in W_2^{2,1} \, (Q)$.*

Let us consider now the case, when $p_1 = 0$ (the upper boundary layer of the atmosphere). The system (9.1) in this case is degenerated with respect to the variable p. Consider once again (9.1) for the next boundary and initial conditions:

$$
\begin{aligned}
&u_z = v_z = w = T_z = 0 \ \text{for} \ p = 0, \\
&u = v = T = 0, \ w = H_t \ \text{for} \ p = P, \\
&u = v = T_n = 0 \ \text{on} \ S, \\
&u = u_0, \ v = v_0, \ H = H_0, \ \text{for} \ t = t_0. \tag{9.3}
\end{aligned}
$$

The next problem on the global solvability of (9.1)-(9.3) takes place [120].

Theorem 9.2. *Let the initial functions be*

$$u_0, \, v_0, \, T_0/p^\kappa \in W_2^1 \, (\Omega) \cap L_\infty \, (\Omega)$$

and they satisfy the boundary conditions (9.3). *Let also*

$$H_0 \in W_2^1 \, (G), \quad F/p^\kappa \in L_\infty \, (\Omega).$$

Then the problem (9.1)–(9.3) *has a unique solution*

$$u, \, v, \, T/p^\kappa \in L_2 \, (0, \, T, \, B) \cap L_\infty \, (0, \, T, \, W_2^1 \, (\Omega),$$

where $B = \{\varphi : \|\Delta \, \varphi\|_\Omega + \|\nabla \, \varphi_p\|_\Omega + \|p \, \varphi_{pp}\|_\Omega < \infty\}$.

Ocean. We consider the equations of the ocean dynamics in the form [122]:

$$
\begin{aligned}
&u_t + (\vec{u} \cdot \nabla) \, u - l \, u = - P_x + \mu \, \Delta \, u + \nu \, \frac{\partial^2 \, u}{\partial \, z^2}, \\
&v_t + (\vec{u} \cdot \nabla) \, v + l \, v = - P_y + \mu \, \Delta \, \nu + \nu \, \frac{\partial^2 \, v}{\partial \, z^2}, \\
&u_x + u_y + w_t = 0, \\
&T_t + (\vec{u} \cdot \nabla) \, T + \gamma \, w = \mu \, \Delta \, T + \nu \, \frac{\partial^2 \, T}{\partial \, z^2} + F_1, \\
&S_t + (\vec{u} \cdot \nabla) \, S + \gamma \, w = \mu \, \Delta \, S + \nu \, \frac{\partial^2 \, S}{\partial \, z^2} + F_2, \\
&P_z = \beta_1 \, T + \beta_2 \, S. \tag{9.4}
\end{aligned}
$$

Here T, P, S are departures of temperature, pressure and salinity from the mean values, μ, ν, γ, β_1, β_2 are constants. The system (9.3) is considered in the domain $Q = \Omega \times (0,T)$, where $(0,T)$ is the arbitrary time interval, $\Omega = G(x,y) \times (0,H)$, $z = 0$ is the ocean surface, $z = H$ is its bottom; the lateral area of the cylinder Ω is denoted by Γ, by n the unit vector of the normal to Γ is denoted. We combine the system (9.3) with the boundary and initial conditions:

$$z = 0 : u_z = v_z = w = T_z = S_z = 0,$$
$$z = H : u = v = T = S = 0, \ w = \alpha \, P_t,$$
$$\Gamma : u = v = 0, \ T_n = S_n = 0,$$
$$t = 0 : u = u_0, \ v = v_0, \ T = T_0, \ S = S_0. \tag{9.5}$$

The following theorem takes place [122].

Theorem 9.3. *Let the initial functions be* u_0, v_0, T_0, $S_0 \in W_2^1(\Omega)$ $\cap L_\infty(\Omega)$ *and satisfy the boundary conditions* (9.3) *and* $F \in L_\infty(Q)$. *Then* (9.4)-(9.5) *has a unique solution* u, v, T, $S \in W_2^{2,1}(Q)$.

9.3 Solvability of Atmosphere Models on Sphere in p-System of Coordinates

We do not have at the present time the theorems on the global solvability (i.e. for the arbitrary time scales) of the primitive equations of the dynamics of atmosphere and ocean. We have at our disposal only the theorems on the local solvability (i.e. for the small time intervals) of the atmosphere and ocean models, which are based on the primitive equations. Below we give some of these theorems [112-114, 92-94].

On the sphere S_a of radius a embedded into Euclidian space R_3 we introduce the coordinates $\theta \, (0 \leq \theta \in \pi)$ and $\varphi (0 \leq \varphi \leq 2\pi)$. Let r is the distance from the center of the sphere to the point under consideration. Then the location of any point M of the atmosphere is defined by the use of coordinates θ, φ, r. The velocity of the point

$$\vec{V} = v_\theta \, e_\theta + v_\varphi \, e_\varphi + v_r \, e_r,$$

where e_θ is the unit vector tangent to the meridian and oriented southward, e_φ the unit vector tangent to the circle of latitude and oriented eastward, e_r the unit vector oriented from the center of Earth to a given point. Since the thickness of the atmosphere h is small in comparison with the radius of Earth a, one assumes in the atmospheric models

$$r = a + z = a(1 + z/a) \approx a, \ \partial/\partial r = \partial/\partial z,$$

where z is the distance from the surface of the sphere S_a counted off along the normal to it. Since in the atmosphere the relation fulfills sufficiently well

$$\partial p / \partial z = -\rho g,$$

where p is the pressure, ρ the density, g the acceleration of gravity, it follows from this relation that the pressure p is a monotonic function of z (since $g > 0$, $\rho > 0$). Therefore, one can take as a new coordinate z the pressure p. In the system (θ, φ, p) the equations of atmospheric dynamics have the form [99,53]:

$$\frac{d v_\theta}{d t} - \frac{v_\theta^2}{a} \operatorname{ctg} \theta - 2 \Omega \cos \theta \, v_\varphi = - \frac{1}{a} \frac{\partial \Phi}{\partial \theta} + D_\theta,$$

$$\frac{d v_\varphi}{d t} + \frac{v_\varphi v_\theta}{a} \operatorname{ctg} \theta + 2 \Omega \cos \theta \, v_\theta = - \frac{1}{a \sin \theta} \frac{\partial \Phi}{\partial \varphi} + D_\varphi,$$

$$\frac{\partial \Phi}{\partial p} + \frac{R}{p} T = 0,$$

$$\frac{\partial \omega}{\partial p} + \frac{1}{a \sin \theta} \left(\frac{\partial v_\theta \sin \theta}{\partial \theta} + \frac{\partial v_\varphi}{\partial \varphi} \right) = 0,$$

$$c_p \frac{d T}{d t} - \frac{R T}{p} \omega = \frac{d Q}{d t}, \qquad (9.6)$$

where

$$\omega = \frac{d p}{d t}, \quad \frac{d}{d t} = \frac{\partial}{\partial t} + \frac{v_\theta}{a} \frac{\partial}{\partial \theta} + \frac{v_\varphi}{a \sin \theta} \frac{\partial}{\partial \varphi} + \omega \frac{\partial}{\partial p},$$

while $\Phi = g z$ is the geopotential, T the temperature, c_p and R are the known constants, D_θ and D_φ the terms defining the dissipation and connected with the environment viscosity, $d \theta / d t$ is the rate of the energy flow per unit of the air mass, Ω is the angular velocity of the rotation of Earth. In the works [112-114] there was investigated a particular case of the system (9.6), when $D_\theta = D_\varphi = 0$, while the components of velocity u and v do not depend on p. In the above-mentioned works at the first time the theorems on the local solvability of the primitive equations on the sphere in p-system of coordinates were obtained and the necessary functional spaces on the sphere and in the spherical layer which are needed for the investigation of the system (9.6) were introduced. The definitions of these spaces will be given below.

We denote by p_0 the pressure at the upper boundary layer of atmosphere and by p_1 the pressure at the Earth surface. Let $\bar{T}(p)$, $p_0 \leq p \leq p_1$ is a standard distribution of temperature with respect to height satisfying the condition

$$R \left(R \bar{T} / c_p - p \, \partial \bar{T} / \partial p \right) = C^2 = \text{const}.$$

Further, let $\bar{\Phi}(p)$ be a standard distribution of geopotential with respect to the height satisfying the condition

$$\frac{\partial \bar{\Phi}}{\partial p} + \frac{R\bar{T}}{p} = 0.$$

Following [92], we shall assume that $|\, T - \bar{T}\,| \ll p/R\omega$. Therefore, one can suppose that in the first approximation $RT\omega/p \approx R\bar{T}\omega/p$. The equation of the thermal balance may be written in departures $T' = T - T'$, $\Phi' = \Phi - \Phi'$. Omitting primes, we get

$$\frac{R^2}{C^2}\frac{dT}{dt} - \frac{R}{p}\omega = \frac{R^2}{c_p C^2}\frac{dQ}{dt}.$$

For the terms of dissipation and heat fluxes we take the relations

$$D_\theta e_\theta + D_\varphi e_\varphi = \mu_1 \Delta v + \nu_1 \frac{\partial}{\partial p}\left[\beta(p)\frac{\partial v}{\partial p}\right],$$

$$\frac{R^2}{c_p C^2}\frac{d\theta}{dt} = \mu_2 \Delta T + \nu_2 \frac{\partial}{\partial p}\left[\beta(p)\frac{\partial T}{\partial p}\right] + \frac{R^2}{c_p C^2}\varepsilon,$$

where $\beta(p) = (gp/RT)^2$, $v = v_\theta e_\theta + v_\varphi e_\varphi$, ε are diabatic fluxes of heat (phase and radiation heat fluxes).

With respect to the above relations the equation of atmosphere dynamics takes the form

$$\frac{\partial v}{\partial t} + \nabla_v v + \omega\frac{\partial v}{\partial p} + 2\Omega\cos\theta k \times v + \mathrm{grad}\,\Phi = \mu_1\Delta v + \nu_1\frac{\partial}{\partial p}(\beta\frac{\partial v}{\partial p}),$$

$$\frac{\partial \Phi}{\partial p} + \frac{RT}{p} = 0, \quad \mathrm{div}\,v + \frac{\partial \omega}{\partial p} = 0,$$

$$\frac{R^2}{C^2}\left(\frac{\partial T}{\partial t} + \nabla_v T + \omega\frac{\partial T}{\partial p}\right) - \frac{R}{p}\omega$$

$$= \mu_2\Delta T + \nu 2\frac{\partial}{\partial p}(\beta\frac{\partial v}{\partial p}) + \varepsilon\frac{R^2}{c_p C^2}, \tag{9.7}$$

where

$$\nabla_v v = \left(\frac{v_\theta}{a}\frac{\partial v_\theta}{\partial \theta} + \frac{v_\varphi}{a\sin\varphi}\frac{\partial v_\theta}{\partial \varphi} - \frac{v_\varphi^2}{a}\mathrm{ctg}\,\theta\right)e_\theta$$

$$+ \left(\frac{v_\theta}{a}\frac{\partial v_\varphi}{\partial \theta} + \frac{v_\varphi}{a\sin\theta}\frac{\partial v_\varphi}{\partial \varphi} + \frac{v_\theta v_\varphi}{a}\mathrm{ctg}\,\theta\right)e_\varphi,$$

$$\nabla_v T = \frac{v_\theta}{a}\frac{\partial T}{\partial \theta} + \frac{v_\varphi}{a\sin\theta}\frac{\partial T}{\partial \varphi}.$$

Laplacians Δ of the scalar function T and of vector v are defined as follows

$$\Delta T = \frac{1}{a^2 \sin \theta} \left[\frac{\partial}{\partial \theta} (\sin \theta \, \frac{\partial T}{\partial \theta}) + \frac{1}{\sin \theta} \frac{\partial^2 T}{\partial \varphi^2} \right],$$

$$\Delta v = \left(\Delta v_\theta - \frac{2 \cos \theta}{a^2 \sin^2 \theta} \frac{\partial^2 v_\varphi}{\partial \varphi^2} - \frac{v_\theta}{a^2 \sin^2 \theta} \right) e_\theta$$

$$+ \left(\Delta v_\varphi + \frac{2 \cos \theta}{a^2 \sin^2 \theta} \frac{\partial v_\theta}{\partial \varphi} - \frac{v_\varphi}{a^2 \sin^2 \theta} \right) e_\varphi,$$

while k is a unit vector tangent to the sphere S_a. The system (9.7) is considered in domain $S_a \times (p_0, p_1)$. We apply the next boundary conditions at the upper and lower boundary layer of atmosphere:

$$p = p_0 : v_\theta = v_\varphi = \omega = \partial T / \partial p = 0,$$
$$p = p_1 : v_\theta = v_\varphi = \omega = 0, \, \partial T / \partial p = \alpha_s (T_s - T), \qquad (9.8)$$

where T_s is temperature of Earth surface, α_s is a constant. Following [92], we write (9.7) in the dimensionless form. Let

$$v = U v', \, \omega = \frac{p_1 - p_0}{a} U \omega', \, T = \bar{T}_0 T',$$

$$\Phi = U^2 \Phi', \, t = \frac{a}{U} t', \, p = (p_1 - p_0) \xi + p_0,$$

$$f' = 2 \cos \theta, \, R_0 = U/a\Omega, \, \alpha_1 = \mu_1/aU,$$

$$\alpha_2 = \frac{v_1 \, a \, g^2}{U \, R^2 \, \bar{T}_0^2} (\frac{p_1}{p_1 - p_0})^2, \, \beta_1 = \mu_2 \bar{T}_0^2 / a U^3,$$

$$\beta_2 = v_2 \, a \, g^2 U^3 \, R^2 (\frac{p_1}{p_1 - p_2})^2, \, a_1 = R^2 \bar{T}_0^2 / C^2 U^2, \, b = b_1 \, p_1,$$

$$b_1 = R \bar{T} (p_1 - p_0) / U^2 p_1, \, \bar{T}_s = T_s / \bar{T}_0, \, \bar{\alpha}_s = (p_1 - p_0) \alpha_s.$$

Proceeding to the dimensionless variables v', ω', T', Φ', t' and ξ and omitting primes, we write the system (9.7) in the dimensionless form:

$$\frac{\partial v}{\partial t} + \nabla_v v + \omega \frac{\partial v}{\partial \xi} + \frac{f}{R_0} k \times v + \nabla \Phi - \alpha_1 \Delta v - \alpha_2 \frac{\partial}{\partial \varepsilon} \left(\gamma \frac{\partial v}{\partial \xi} \right) = f_1,$$

$$\text{div } v + \frac{\partial \omega}{\partial \xi} = 0, \quad \frac{\partial \Phi}{\partial \xi} + \frac{b}{p} T = 0,$$

$$a_1 \left(\frac{\partial T}{\partial t} + \nabla_v T + \omega \frac{\partial T}{\partial \xi} \right) - \frac{b}{p} \omega - \beta_1 \Delta T - \beta_2 \frac{\partial}{\partial \xi} \left(\gamma \frac{\partial T}{\partial \xi} \right) = f_2, \quad (9.9)$$

where $\gamma = (p \bar{T}_0 / p_1 \bar{T})^2$. The boundary conditions (9.8) take the form:

$$\xi = 0 : v_\theta = v_\varphi = \omega = \partial T / \partial \xi = 0, \qquad (9.10)$$
$$\xi = 1 : v_\theta = v_\varphi = \omega = 0, \, \partial T / \partial \xi = \bar{\alpha}_s (\bar{T}_s - T).$$

The system (9.9)–(9.10) is considered in the domain $S \times (0, 1)$. We write the system (9.9) in some another way. From the second equation of the system (9.9) we get

$$\omega(t, \theta, \varphi, \xi) = \int_\xi^1 \operatorname{div} v \, d\xi \equiv W(v), \quad \int_0^1 \operatorname{div} v \, d\xi = 0.$$

From the third equation of (9.9) it follows that

$$\Phi(t, \theta, \varphi, \xi) = \Phi_s(t, \theta, \varphi) + \int_\xi^1 \frac{b}{p} T \, d\xi,$$

where $\Phi_s = \Phi \mid_{p=p_1}$. From the last equality we obtain

$$\nabla \Phi = \nabla \Phi_s + \int_\xi^1 \frac{b}{p} \nabla T \, d\xi.$$

With respect to the obtained relation the system (9.9) may be written as follows

$$\frac{\partial v}{\partial t} + \nabla_v v + W \frac{\partial v}{\partial \xi} + \frac{f}{R_0} k \times v + \int_\xi^1 \frac{b}{p} \nabla T \, d\xi + \nabla \Phi_s$$

$$-\alpha_1 \Delta v - \alpha_2 \frac{\partial}{\partial \xi} \left(\gamma \frac{\partial v}{\partial \xi} \right) = f_1,$$

$$a_1 \left(\frac{\partial T}{\partial t} + \nabla_v T + W \frac{\partial T}{\partial \xi} \right) - \frac{b}{p} W - \beta_1 \Delta T - \beta_2 \frac{\partial}{\partial \xi} \left(\gamma \frac{\partial T}{\partial \xi} \right) = f_2,$$

$$\int_0^1 \operatorname{div} v \, d\xi = 0. \tag{9.11}$$

The boundary conditions (9.10) take the form

$$\xi = 0 : v_\theta = v_\varphi = \partial T / \partial \xi = 0, \tag{9.12}$$
$$\xi = 0 : v_\theta = v_\varphi = 0, \partial T / \partial \xi = \alpha_s (\bar{T}_s - T).$$

We combine the system (9.11)–(9.12) with the initial conditions

$$t = 0 : v_\theta = v_{\theta 0}, v_\varphi = v_{\varphi 0}, T = T_0. \tag{9.13}$$

We proceed now to the description of the functional spaces in which the problem (9.11)–(9.13) will be considered. The procedure of definition of the functional spaces on the unit sphere S embedded into

R_3 may be presented by two ways [101,102]. The first consists in such a procedure. On the sphere S one takes the arbitrary point λ_0 and circles it by a small neighborhood $\sigma(\lambda_0) \subset S$. Then one draws in the point λ_0 the plane tangent to S and onto the plane one projects orthogonally the neighborhood $\sigma(\lambda_0)$. Let $G(\lambda_0)$ be so obtained projection. Evidently, $G(\lambda_0) \subset R_2$. We put to any point $\lambda \in \sigma(\lambda_0)$ in correspondence the point $y \in G(\lambda_0)$, which represents the projection of the point λ onto the mentioned tangent plane $\lambda = \Psi(y)$.

We put to any function $f(\lambda)$ given of the sphere S in correspondence the function $F(y) = f(\Psi(y))$, $\lambda \in \sigma(\lambda_0)$. The function $f(\lambda)$ is considered to be belonging to the class $C^{(k)}(S)$, if $F \subset C^{(k)}(G(\lambda_0))$ for any point $\lambda_0 \in S$. We give another definition of the class $C^{(k)}(S)$ of smooth functions on the sphere S. Consider the spherical layer $\Sigma = \{x : r_1 \leq |x| \leq r_2\}$, where $0 < r_1 < 1 < r_2$ are the arbitrary positive numbers $|x| = (x_1^2 + x_2^2 + x_3^2)^{1/2}$, $x = (x_1, x_2, x_3) \in R_3$. Continue the function $f(\lambda)$ into spherical layer, setting $f^*(x) = f(x/|x|)$. Notice that $f^*(x)$ is constant along any ray passing through the origin of coordinates and because of this does not depend on $r = |x|$. We shall say that $f(\lambda) \in C^{(k)}(S)$, if $f^*(x) \in C^{(k)}(\Sigma)$.

Now we can to define the space $W_2^{(k)}(S)$ as the space of functions which are given on S and are continued on Σ, as it was said above, and have the generalized derivatives to the order k with respect to the coordinates of the point $x \in \Sigma$, square summed over the spherical layer Σ. One can define the main differential operators on the sphere [101,102,113]:

$$\nabla f = \frac{\partial f}{\partial \theta} e_\theta + \frac{1}{\sin \theta} \frac{\partial f}{\partial \varphi},$$

$$\operatorname{div} v = \frac{1}{\sin \theta} \left(\frac{\partial v_\theta \sin \theta}{\partial \theta} + \frac{\partial v_\varphi}{\partial \varphi} \right), \quad v = v_\theta e_\theta + v_\varphi e_\varphi,$$

$$\operatorname{rot} v = \frac{1}{\sin \theta} \left(\frac{\partial v_\theta}{\partial \varphi} - \frac{\partial v_\varphi \sin \theta}{\partial \theta} \right) e_r,$$

$$\Delta v = (\operatorname{rot} \operatorname{rot} - \nabla \operatorname{div}) v$$

$$= \left(\Delta v_\theta - \frac{2 \cos \theta}{\sin^2 \theta} \frac{\partial v_\varphi}{\sin^2 \theta} \right) e_\theta + \left(\Delta v_\varphi + \frac{2 \cos \theta}{\sin^2 \theta} \frac{\partial v_\varphi}{\partial \varphi} - \frac{v_\varphi}{\sin^2 \theta} \right) e_\varphi,$$

$$\Delta f = \operatorname{div}(\nabla f) = \frac{1}{\sin \theta} \left[\frac{\partial}{\partial \theta} \left(\sin \theta \frac{\partial f}{\partial \theta} \right) + \frac{1}{\sin \theta} \frac{\partial^2 f}{\partial \varphi^2} \right].$$

Here the Laplace–Beltrami operator in the first case is applied to the vector field v tangent to S, in the second case it is applied to the scalar function f. We denote by $T S$ the vector field on S tangent to S.

Let $L_2(S)$ be the space of functions $f(\theta, \varphi)$ with a scalar product

$$(f\,g) = \frac{1}{4\,\pi} \int\limits_0^\pi \int\limits_0^{2\,\pi} f\,g\,\sin\,\theta\,d\,\theta\,d\,\varphi.$$

The space $L_2(T S)$ may be defined similarly. The Laplace–Beltrami operators as in the scalar so in the vector case are positively defined selfadjoint in $L_2(S)$ and $L_2(T S)$ operators. Therefore one can introduce their degrees $(-\Delta)^s$, $s \in R$ and define the spaces $H^s(S) = D((-\Delta)^{s/2})$, $H^s(T S) = D((-\Delta)^{s/2})$. For instance,

$$\|U\|_{H^2(T S)} = (-\Delta\,v,\,v) = \|\mathrm{rot}\,\mathrm{v}\|^2_{L_2(T S)} + \|\mathrm{div}\,v\|^2_{L_2(S)}.$$

In our case the domain in which the problem (9.11)–(9.13) is considered, has the form of the spherical layer $M = S \times (0,1)$. We denote by $T(\theta, \varphi, \xi)\,M = T_{(\lambda,\xi)}\,M$ the tangent space to M at the point (λ, ξ) and by $T_\lambda S$ and $T_\xi(0,1) = R$ the tangent spaces to S and $(0,1)$ at the points λ and ξ respectively. Then $T_{(\lambda,\xi)}\,M = T_\lambda S \times T_\xi(0,1)$. The scalar product in $T_{(\lambda,\xi)}\,M$ may be defined as follows [92]:

$$X \cdot Y = X'Y' + X^2 Y^2 + X^3 Y^3,\ |\,X\,| = (X \cdot X)^{1/2},$$

$$X = X'\,e_\theta + X^2\,e_\varphi + X^3\,e_\xi,\ Y = Y'\,e_\theta + Y^2\,e_\varphi + Y^3\,e_\xi.$$

Let $\bar{M} = S \times [0, 1]$. Let $C^\infty(\bar{M})$ denote the space of functions of the class C^∞ from $\bar{M}$ into R and let $C^\infty(T\,\bar{M})$ denote the space of vector fields on $\bar{M}$ of the class C^∞. The space $C^\infty(T\,\bar{M})$ allows the following expansion:

$$C^\infty(T\,\bar{M}) = C^\infty(T\,\bar{M}/T\,S) \times C^\infty(\bar{M})$$
$$= \{v : \bar{M} \to TS \in C^\infty, v(\theta, \varphi, \xi) \in T_{(\theta,\varphi)}S\} \times \{h : \bar{M} \to R \in C^\infty\}.$$

Further, we denote by $C_0^\infty(M)$ the space of functions of the class C^∞ from M into R with a compact support, while by $C_0^\infty(T\,M)$ the space of vector fields on M of the class C^∞ with a compact support. If $f \in C^\infty(M)$, then also

$$\nabla_M f = \nabla f + \frac{\partial f}{\partial \xi}\,e_\xi \in C^\infty(T\,M).$$

Similarly, if

$$X = v + v_\xi\,e_\xi = v_\theta\,e_\theta + v_\varphi\,e_\varphi + v_\xi\,e_\xi \in C^\infty(T\,M),$$

then

$$\mathrm{div}\,X = \mathrm{div}\,v + \frac{\partial v_\xi}{\partial \xi},\quad \Delta_M f = \mathrm{div}_M(\nabla_M f) = \Delta f + \frac{\partial^2 f}{\partial \xi^2}.$$

The spaces $H^s(M)$, $H^s(TM)$ and $H^s(TM/TS)$ are defined for integer positive s as closure of the spaces $C^\infty(\bar{M})$, $C^\infty(T\bar{M})$ and $C^\infty(T\bar{M}/TS)$ with respect to [92] relatively

$$\|f\|^2_{H^s} = \sum_{m+n\le s} \|(-\Delta)^{m/2}(\nabla_\xi)^n f\|^2_{L_2}.$$

If one has here $s = 0$, then there are spaces $L_2(M)$, $L_2(TM)$ and $L_2(TM/TS)$ of the scalar functions on M, vector fields on M and vector fields with the two first components (v_θ, v_φ) on M relatively. The scalar product within these spaces is given in the usual way

$$(f\,g) = \int_M f\,g\,dM,$$

where f and g from $L_2(M)$ (or from $L_2(TM)$ and $L_2(TM/TS)$).
   Let

$$v_\infty = \{v \in C_0^\infty(TM/TS), \int_0^1 \operatorname{div} v\,d\xi = 0\}.$$

Let v_1 denote the closure v_∞ with respect to the norm H^1 and H_1 denote the closure v_∞ with respect to the norm L_2. Let $v_2 = H^1(M)$, $H_2 = l_2(M)$, $v = v_1 \times v_2$, $H = H_1 \times H_2$. There takes place the next theorem [92].
   Theorem 9.4. *There exists a unique solution of* (9.11)– (9.13)

$$(v, T) \in C([0, \tau], H^3 \cap V) \cap L_2(0, \tau; H^4 \cap V)$$

for the sufficiently small τ.
   As it was noted above, for the case when $D_\theta = D_\varphi = 0$ and v_θ, v_φ do not depend on p, Theorem 9.4 is established in [113].
   Remark. The theorems on the local solvability of coupled atmosphere and ocean models are detailly presented in work [94].

Bibliography

[1] Abarbanel H.D.I., Brown R., Kennel M.B., Lyapunov exponents in chaotic system. Their importance and their evaluation using observed data, *Modern Physics Letters*, 1991.

[2] Abarbanel H.D.I., Brown R., Sidorovich J.J., Tsimring L.Sh., Analysis of observed chaotic data in physical systems, *Rev. Mod. Phys.*, **65**, No 4, 1331–1390, 1993.

[3] Babin A.V., Vishik V.A., Attractors of evolutionary partial differential equations and estimates of their attractors, *Uspekhi Mat. Nauk.*, **38**, No 4, 134-187, 1983.

[4] Babin A.V., Vishik V.A., *Attractors of evolution equations*, Amsterdam, London, New York, Tokyo, North-Holland, 1992.

[5] Bekriaev R.V., Alternate chaos and large-scale regimes of atmospheric circulation, *Izvestija RAN, Fizika Atmospheri i Okeana*, **31**, No 4, 1995.

[6] Bogolubov N.N., Mitropolskii Yu.A., *Asymptotical methods for theory of nonlinear oscillations*, Nauka, Moscow, 1974.

[7] Charney J.G., De Vore J.G., Multiple flow equilibria in the atmosphere and blocking, *J. Atm. Sci.*, **36**, 1205–1216, 1979.

[8] Chepyzhev V.V., Vishik M.I., Attractors of non-autonomous dynamical systems and their dimension, *J. Mat. Pures Appl.*, **73**, 299–333, 1994.

[9] Constantin P., Foias C., Nicolaenko B., Temam R., Spectral barriers and inertial manifolds for dissipative partial differential equations, *J. Dyn. Diff. Eqs*, No 1, 45–73, 1989.

[10] Constantin P., Foias C., Nicolaenko B., Temam R., Nouveaux résultats sur les variétés inertielles pour les équations différentielles dissipatives, *C.R. Acad. Sci.*, Paris. Série I, **302**, 375–378, 1986.

[11] Constantin P., Foias C., Nicolaenko B., Temam R., *Integral and inertial manifolds for dissipative partial differential equations*, Springer Verlag, New York, 1988.

[12] Constantin P., Foias C., Temam R., On the dimension of attractors in two-dimensional turbulence, *Physics D.*, **30**, No 3, 284–296, 1988.

[13] Debussche A., Marion M., On the construction of families of approximate inertial manifolds, *J. Diff. Eqs* (to appear).

[14] Debussche A., Temam R., Inertial manifolds and their dimension. Dynamical systems. Theory and applications, Ed. by S.I.Andersson, A.E. Andersson and O.Ottoson. – World Scientific Co., 1993.

[15] Debussche A., Temam R., Inertial manifolds and the slow manifolds in meteorology, *Diff. and Int. Eqs*, **4**, No 5, 897–931, 1991.

[16] Demidov G.V., Some theorems on solutions of one problem of the meteorology, *Doklady Acad. Nauk SSSR*, **166**, No 4, 771-774, 1966.

[17] Demidov G.V., Marchuk G.I., Existence theorem for the solution of the problem of the short-term weather prediction, *Doklady Acad. Nauk SSSR*, **170**, No 5, 1006-1008, 1966.

[18] Demengel F., Ghidaglia J.M., Inertial manifolds for partial differential evolution equations under time-discretization: existence, convergence, and applications, *J. Math. Anal. and Appl.*, **155**, 177–225, 1991.

[19] Demengel F., Ghidaglia J.M., Some remarks on the smoothness of inertial manifolds, *Nonlinear Anal., T.M.A.*, **16**, No 1, 79–87, 1991.

[20] Dole R.M., *Persistent anomalies of the extratropical northern hemisphere wintertime circulation. Large-scale dynamical processes in the atmosphere*, Academic Press, 95–109, 1983.

[21] Doering C.R., Gibbon J.D., Note of the Constantin–Foias–Temam attractor dimension estimate for two-dimensional turbulence, *Physics D.*, **48**, 471–480, 1991.

[22] Dutton J.A., Fundamental theorems of climate theory - some proved, some conjectured, *Review*, **24**, No 1, 1–33, SIAM, 1982.

[23] Dzedzeevsky B.N., Collection of the meteorological investigations, Moscow, 1956, 240.

[24] Dymnikov V.P., On the connection of the orthogonal components of the fields of meteoelements with the eigenfunctions of dynamical operators, *Izvestiya Acad. Nauk SSSR, Fizika Atmosphery i Okeana*, **24**, No 7, 675-683, 1988.

[25] Dymnikov V.P., Some problems of the theory of climate in advanced mathematics: Computations and Applications, *NCC Publisher*, 32-48, Novosibirsk, 1995, (in Russian).

[26] Dymnikov V.P., Dynamical-stochastical models of low-frequency variability of atmospheric circulation, *Rus. J. Numer. Anal. Math. Modelling*, **11**, No 5, 1996 (in Russian).

[27] Dymnikov V.P., Alexeev V.A., Volodin E.M., Galin V.J., Diansky N.A., Lykosov V.N., Esau I.N., Modelling of the general atmosphere circulation and the upper layer of the ocean, *Izvestiya RAN, Fizika Atmosphery i Okeana*, **31**, No 3, 324–346, 1995.

[28] Dymnikov V.P., Filatov A.N., *Stability of the large-scale atmospheric processes*, Gidrometeoizdat, Leningrad, 1990.

[29] Dymnikov V.P., Filatov A.N., *Introduction to the mathematical theory of climate*, IVM RAN, Moscow, 1993.

[30] Dymnikov V.P., Filatov A.N., Averaging the equations of the atmosphere dynamics and the influence function for empirical orthogonal components of low frequency atmospheric processes, *Rus. J. Numer. Anal. Math. Modelling*, **10**, No 3, 173-186, 1995.

[31] Dymnikov V.P., Filatov A.N., *Foundations of mathematical theory of climate*, Moscow, 1994 (in Russian).

[32] Dymnikov V.P., Filatov A.N., On the several problems of mathematical theory of climate, *Izvestia RAN, Fizika Atmosphery i Okeana*, **31**, No 3, 313–323, 1995 (in Russian).

[33] Dymnikov V.P., Gritsun A.S., Lyapunov's exponents and attractor dimension of two-layer baroclinic atmospheric circulation model, *Doklady RAN, Geofizika*, No 3, 1996 (in Russian).

[34] Dymnikov V.P., Gritsun A.S., Barotropic instability and structure of low-frequency circulation generated by two-layer baroclinic atmospheric model, *Izvestia RAN, Fizika Atmosphery i Okeana*, No 5, 1996 (in Russian).

[35] Dymnikov V.P., Kazantsev E.V., On the structure of the attractor generated by the system of barotropic atmospheric equations, *Izvestiya RAN, Fizika Atmosphery i Okeana*, **29**, No 5, 581-595, 1993.

[36] Dymnikov V.P., Kazantsev E.V., Kharin V.V., Characteristics of the stability and the life-time of the regimes of the atmospheric circulation, *Izvestiya Acad. Nauk SSSR, Fizika Atmosphery i Okeana*, **26**, No 4, 339-349, 1990.

[37] Dymnikov V.P., Kazantsev E.V., Kharin V.V., Informational entropy and local Lyapunov exponents of barotropic atmospheric circulation, *Izvestiya RAN, Fizika Atmosphery i Okeana*, **28**, No 6, 563-573, 1992.

[38] Dymnikov V.P., Kazantsev E.V., Simulation of the quasi stationary barotropic circulation regimes, *Rus. J. Comput. Mech.*, (in press).

[39] Egger J., Schilling H.D., On the theory of long-term variability of the atmosphere, *J. Atm. Sci.*, **40**, 1073–1085, 1983.

[40] Fabes E., Luskin M., Sell G.R., Construction of inertial manifolds by elliptic regularization, *J. Diff. Eqs*, **89**, 355–387, 1991.

[41] Filatov A.N., *Averaging methods for differential and integrodifferential equations*, FAN, Tashkent, 1971 (in Russian).

[42] Filatov A.N., On the closeness of the solutions of original and averaged nonlinear dissipative systems with high-frequency forcing, *Rus. J. Numer. Anal. Math. Modelling*, **11**, No.5, 1996.

[43] Filatov A.N., On the closeness of the attractors of initial and averaged nonlinear dissipative systems, *Doklady RAN*, **347**, No 5, 601-603, 1996.

[44] Filatov A.N., Estimation of number unstable stationary solutions of spectral atmospheric models, *Trudy Gidrometzentra*, **323**, 134-142, Sanct-Peterburg, 1992.

[45] Filatov A.N. Gorelov A.S., Inertial manifolds of the barotropic atmospheric equations, *Doklady Acad. Nauk SSSR*, **318**, No 6, 1991.

[46] Filatov A.N., Gorelov A.S., Inertial manifolds of the equations of a barotropic atmosphere on a rotating sphere, *Trudy Gidrometzentra*, **232**, 189–195, Gidromateoizdat, Leningrad, 1992.

[47] Filatov A.N., Ilyin A.A., Stability of nonstationary solutions of barotropic atmosphere equations, *Trudy Gidrometzentra*, **323**, 142–149, Sanct-Peterburg, 1992.

[48] Filatov A.N., Ilyin A.A., Navier–Stockes equations of the sphere and their unique solvability. Stability of the stationary solutions, *Math. Phys.*, 128-146, Leningrad, 1987.

[49] Filatov A.N., Ilyin A.A., On the unique solvability of Navier–Stockes equations of two-dimensional sphere, *Doklady Acad. Nauk SSSR*, **301**, No 1, 1988.

[50] Filatov A.N., Ilyin A.A., Stability of the stationary solutions of the barotropic atmospheric equations using linear approximation, *Doklady Acad. Nauk SSSR*, **306**, No 6, 1989.

[51] Filatov A.N., Ipatova V.M., On globally stable difference schemes for the barotropic vorticity equation on a sphere, *Rus. J. Numer. Anal. Math. Modelling*, **11**, No 1, 1–26, 1996.

[52] Filatov A.N., Ipatova V.M., On globally stable difference schemes for the barotropic vorticity equation on a sphere with almost periodic forcing, *Rus. J. Numer. Anal. Math. Modelling*, **11**, No 4, 1996.

[53] Filatov A.N., Shershkov V.V., *Asymptotic methods in atmospheric models*, Gidrometeoizdat, Leningrad, 1988 (in Russian).

[54] Filatov A.N, Sharova L.V., *Integral inequalities and the theory of nonlinear oscillations*, Nauka, Moscow, 1976 (in Russian).

[55] Foias C., Manley O., Temam R., Modelling of the interaction of small and large eddies in two dimensional turbulent flow, *Math. Modelling and Numer. Anal.*, **22**, No 1, 93–118, 1988.

[56] Foias C., Nicolaenko B, Sell G.R., Temam R., Inertial manifolds of the Kuramoto–Sivashinsky equation and an estimate of their lowest dimension, *J. Math. Pures Appl.*, No 67, 197–222, 1988.

[57] Foias C., Sell G.R., Temam R., Inertial manifolds for dissipative differential equations, *C.R. Acad. Sci.*, **301**, Ser. 1, No 5, 139–141, Paris, 1985.

[58] Foias C., Sell G.R., Temam R., Inertial manifolds for nonlinear evolutionary equations, *J. Diff. Eqs*, No 73, 309–353, 1988.

[59] Foias C., Temam R., The algebraic approximation of attractors: the finite dimensional case, *Physica D.*, **32**, 163–182, 1988.

[60] Foias C., Temam R., Approximation of attractors by algebraic or analytic sets, Preprint N 9004 (The Institute for Applied Mathematics and Scientific Computing).

[61] Fraedrich K., Estimating the weather and climate predictability on attractors, *J. Atm. Sci.*, **44**, 722–728, 1987.

[62] Freadrich K., Ziehmann C., Sielmann, Estimates of spatial degrees of freedom, *J. Climate*, **8**, No 2, 361-369.

[63] Frankignoul C., Hasselman K., Stochastic climate models. Part II: Application to sea-surface temperature anomalies and termocline variability, *Tellus*, **29**, 289–305, 1977.

[64] Fursikov A.V., On the problem of the closeness of the chain of momentum equations for the case of large Reynolds numbers, *Nonclassic Equations and Mixed Type Equations*, 231–250, Int. math. SO AN SSSR, Novosibirsk, 1990.

[65] Fursikov A.V., The problem of closeness of the chains of momentum equations corresponding to the three-dimensional Navier–Stockes system for the case of large Reynolds numbers, *Doklady Acad. Nauk SSSR*, **319**, No 1, 1991.

[66] Ghidaglia J.M., Weekly damped forced Kortweg–de Vries equations behave as a finite dimensional system in the long time, *J. Diff. Eqs*, **74**, 369–390, 1988.

[67] Girs A.A., *Macrocirculation method for long-term meteorological forecasts*, Gidrometeoizdat, Leningrad, 1974, 343.

[68] Gorelov A.S., Attractor dimension of the baroclinic model, *Doklady RAN*, 1996.

[69] Gorelov A.S., Filatov A.N., Inertial manifolds for equations of barotropic atmosphere, *Rus. J. Numer. Anal. Math. Modelling*, **7**, No 1, 25–43, 1992.

[70] Grassberger P., Procaccia I., Measuring the strangeness of strange attractors, *Phisica D.*, **90**, No 1, 189–208, 1983.

[71] Haraux A., Attractors of asymptotically compact processes and applications to nonlinear partial differential equations. *Comm. in Partial Diff. Eqs*, **13**, 1383–1414, 1988.

[72] Haraux A., *Systemés dynamiques dissipatifs et applications*, Paris – Milan – Barcelona – Roma – Masson, 1991.

[73] Henri D., *Geometrical theory of semilinear parabolic equations*, Mir, Moscow, 1985.

[74] Ilyin A.A., Partially dissipative semigroups generated by Navier–Stockes system on two-dimensional manifolds, and their attractors, *Math. Sbornik*, **184**, No 1, 55–88, 1993.

[75] Ilyin A.A., Navier–Stokes equations on the rotating sphere. A simple proof of the attractor dimension estimate, *Nonlinearity*, **7**, 31–39, 1994.

[76] Kapitanskii L.V., Kostin I.N., Attractors of nonlinear evolutionary equations and their approximations, *Algebra Anal.*, **2**, No 1, 114–140, 1990.

[77] Kaplan J.L, Yorke J.A., Chaotic behaviour in multidimensional difference equations, *Lecture notes in mathematics*, **730**, 228, 1979.

[78] Kharin V., The relationship between sea surface temperature anomalies and atmospheric circulation in general circulation model experiments, *Report* No 136, Max-Plank-Institut für Meteorologie, 1994.

[79] Kirchner I., Graf H.F., Volcanos and El-Niño-signal separation in winter, *Report* No 121, Max-Plank-Institut für Meteorologie, 1993.

[80] Kloeden P.E., Lorenz J., Stable attracting sets in dynamical systems and their one-step discretizations, *J. Numer. Anal.*, **23**, No 5, SIAM, 1986.

[81] Kolmogorov A.N., A new metric invariant of transitive dynamical systems and automorphisms in the Lebesgue space, *Doklady Acad. Nauk SSSR*, **119**, No 4, 861–864, 1958.

[82] Kolmogorov A.N., Fomin S.V., *Elements of the theory of functions and the functional analysis*, Nauka, Moscow, 1989.

[83] Kostin I.N., On one method of approximating the attractors, *Zap. Nauch. Semin.*, **17**, 87–103, LOMI (Leningrad Branch of V.A. Steklov Institute of Mathematics, the USSR Academy of Sciences), 1991,

[84] Kostin I.N., Approximation of the attractors of nonlinear evolutionary equations by the attractors of finite systems, *Zap. Nauch. Semin.*, **197**, 71–86, LOMI, 1992.

[85] Krylov N.M., Bogolyubov N.N., The general theory of measure in nonlinear mechanics, (Bogolyubov N.N. *Selected works. In 3 volumes vol. 1*, Naukova Dumka, Kiev, 1969.)

[86] Kwak M., Finite dimensional inertial forms for the 2D Navier–Stokes equations, *Indiana Univ. Math. J.*, 1993.

[87] Ladyzhenskaya O.A., On dynamical system generated by the Navier–Stockes equations, *Zap. Nauch. Semin.*, **27**, 91–114, LOMI, 1972.

[88] Ladyzhenskaya O.A., On determination of minimal global attractors for the Navier–Stockes equations and other partial differential equations, *Uspekhi Math. Nauk*, **42**, No 6, 25–60, 1987.

[89] Ladyzhenskaya O.A., Globally stable difference schemes and their attractors, (Preprint LOMI: 5–91), Leningrad, 1991.

[90] Legras B., Ghil M., Persistent anomalies, blocking and variations in atmospheric predictability, *J. Atm. Sci.*, **42**, 433–471, 1985.

[91] Levitan B.M., Zhikov V.V., *Almost periodic functions and differential equations*, Izd. MGU, Moscow, 1978.

[92] Lions J.L., Temam R., Wang S., New formalities of the primitive equations of atmosphere and applications, *Nonlinearity*, **5**, 237–288, 1992.

[93] Lions J.L., Temam R., Wang S., On the equations of the large-scale ocean, *Nonlinearity*, **5**, No 5, 1007–1053, 1992.

[94] Lions J.L., Temam R., Wang S., Models of the coupled atmosphere and ocean, No. 9206, No 9302, No 9305, Institute for Applied Mathematics and Scientific Computing, Indiana university-Bloomington.

[95] Lorenz E.N., Deterministic nonperiodic flow, *J. Atm. Sci.*, **20**, No 2, 130–141, 1963.

[96] Malinetskii G.G., Potapov A.B., On calculation of dimension of the strange attractor, (Preprint IPM Keldysh, No 101), Moscow, 1987.

[97] Malinetskii G.G., Potapov A.B., On calculation of the dimension of the strange attractors, *JVM and MPH*, **28**, No 7, 1021–1037, 1988.

[98] Marchuk G.I., Numerical forecast on the sphere, *Doklady Acad. Nauk SSSR*, **156**, No 4, 810–813, 1966.

[99] Marchuk G.I., Dymnikov V.P. Zalesny V.B., *Mathematic models in geophysical hydrodynamics and numerical methods of their realization*, Gidrometeoizdat, Leningrad, 1987.

[100] Marshall J., Molteni F., Toward a dynamical understanding of planetary-scale flow regimes, *J. Atm. Sci.*, **50**, No 12, 1792–1818, 1993.

[101] Mikhlin S.G., *Multidimensional singular integrals and integral equations*, Moscow, 1962.

[102] Mikhlin S.G., *Linear equations in partial derivatives*, Moscow, 1977.

[103] (96. Mitropolskii Yu.A. Lykova O.B., *Integral manifolds in nonlinear mechanics*, Nauka, Moscow, 1973.

[104] Mitropolskii Yu.A., Khoma G.P., *Mathematical backgrounds of the asymptotical methods for nonlinear mechanics*, Naukova Dumka, Kiev, 1983.

[105] Mo K., Ghil M., Statistic and Dynamics of persistent anomalies, *J. Atm. Sci.*, **44**, No 5, 877–901, 1987.

[106] Mo K., Ghil M., Cluster analysis of multiple planetary flow regimes, *J. Geoph. Res.*, **93**, No D9, 10927–10952, 1988.

[107] Molteni F., Tibaldi S., Palmer T.W., Regimes in the wintertime circulation over Northern extratropics. I. Observational evidence, *Q.Y.R. Met. Soc.*, **116**, 31–67, 1990.

[108] Nemytskii V.V., Stepanov V.V., *Qualitative theory of differential equations*, Moscow, 1949, (Prinston University Press, Prinston, No 3.)

[109] Oseledets V.I., Multiplicative ergodic theorem. Characteristic Lyapunov exponents for dynamical systems, *Trudy Moskovskogo matematicheskogo obschestva*, **19**, 179–210, 1969.

[110] Pesin B.J., Characteristic Lyapunov exponents and smooth ergodic theory, *Uspekhi Mathem. Nauk*, **32**, No.1, 4-55, 1977.

[111] Piterbarg L.I., *Dynamics and forecast of the large-scale temperature anomalies of ocean surface*, Gidrometeoizdat, Leningrad, 1989.

[112] Potapov A.B., Programms for calculation of the correlation exponents and the estimate of the generalized entropy over the time series, (Preprint IPM Keldysh: No 27), Moscow, 1991.

[113] Raputa V.F., On solvability of some problems of the weather forecast on the sphere, *Comput. Math. and Programming*, 64–73, 1974.

[114] Raputa V.F., The existence theorem for the short-range weather forecast problems on the sphere, (Preprint VC SO AN SSSR: No 131), Novosibirsk, 1978 (in Russian).

[115] Romanov A.V., Sharp estimation of dimension of inertial manifolds for nonlinear parabolic equations, *Izvestiya RAN, Matematika*, **57**, No 4, 36-54, 1993.

[116] Ruelle R., Deterministic chaos: the science and the fiction, *Proc. R. Soc.*, **A427**, 241–24, London.

[117] Scherbakov B.A., *Topological dynamics and stability according to Poisson of the solutions of differential equations*, Kishinev, 1972.

[118] Sell G.R., You Y., Inertial manifolds: the non-self-adjoint case. *J. Diff. Eqs*, **96**, 203–255, 1992.

[119] Sibirskii K.C., *Introduction into topological dynamics*, Kishinev, 1970.

[120] Sukhonosov V.I., On the solvability of three-dimensional problems of atmospheric dynamics, *Numer. Methods of Continuum Mech.*, **II**, No 4, 1980.

[121] Sukhonosov V.I., On correctness in general boundary problems for the models of atmosphere and ocean dynamics, *Doklady Acad. Nauk SSSR*, No 3, 556–560, 1983.

[122] Sukhonosov V.I., On the correctness in general three-dimensional problem of the ocean dynamics, *Inhomogeneuos Fluid Dynamics*, 119–126, 1981.

[123] Sukhonosov V.I., On existence and uniqueness theorem in general for any finite time interval for the atmospheric dynamics, *Inhomogeneous Fluid Dynamics*, 90–112, 1982.

[124] Takens F., Detecting strange attractors in turbulence, *Dynamical Systems and Turbulence*, Springer-Verlag, 366–381, 1981.

[125] Temam R., *Infinite dimensional dynamical systems in mechanics and physics*, Springer-Verlag, New-York, 1988.

[126] Temam R., Wang S., Inertial forms of Navier–Stokes equations on the sphere, *J. Func. Anal.*, 1993.

[127] Ter-Mkrtchian M.G., On the definition of the number of independent stations which are equivalent to a given system of correlated stations, *Meteorologia i Gidrologia*, No 2, 24-36, 1969 (in Russian).

[128] Thung K.K., Rosental A.J., Theories of multiple equilibria – a critical reexamination. Part 1: Barotropic models, *J. Atm. Sci.*, **42**, No 24, 2804–2819, 1985.

[129] Vilenkin N.J., *Especial functions and representation group theory,* Nauka, Moscow, 1965.

[130] Vishik V.B., Fursikov A.V., *Mathematical problems of the statistical hydromechanics,* Nauka, Moscow, 1980, 271.

[131] Voevodin V.V., Kuznetsov Yu. A., *Matrices and calculations,* Nauka, Moscow, 1984, 143.

[132] Wallace J.M., Blackmon M.L., *Observation of low-frequency atmospheric variability. Large-scale dynamical processes in the Atmosphere,* (Ed. by B.J.Hoskins., R.P.Pearce), Academic Press, 1983.

[133] Wallace J.M., Cheng X., Sun D., Does low-frequency atmospheric variability exhibit regime-like behavior? *Tellus,* **43**, AB, 16–26, 1991.

[134] Wang S., Approximate inertial manifolds for the 2D model of atmosphere, *Numer. Funct. Anal. and Optimiz.,* No 11, 1043–1070, 1990.

[135] Wang S., On the 2nd of large-scale atmospheric motion: well-posedness and attractors, *Nonl. Anal.,* **18**, No 1, 17–60, 1992.

[136] Yoshizawa T., *Stability theory by Lyapunov's second method,* The Mathematical Society of Japan, 1966.

Index